STATA MULTIVARIATE STATISTICS
REFERENCE MANUAL
RELEASE 9

A Stata Press Publication
StataCorp LP
College Station, Texas

Stata Press, 4905 Lakeway Drive, College Station, Texas 77845

The suggested citation for this software is

StataCorp. 2005. *Stata Statistical Software: Release 9*. College Station, TX: StataCorp LP.

Table of Contents

Cross-Referencing the Documentation

When reading this manual, you will find references to other Stata manuals. For example,

[U] **26 Overview of Stata estimation commands**
[R] **regress**
[D] **reshape**

The first is a reference to chapter 26, *Overview of Stata estimation commands* in the *Stata User's Guide*, the second is a reference to the `regress` entry in the *Base Reference Manual*, and the third is a reference to the `reshape` entry in the *Data Management Reference Manual*.

All the manuals in the Stata Documentation have a shorthand notation, such as [U] for the *User's Guide* and [R] for the *Base Reference Manual*.

The complete list of shorthand notations and manuals is as follows:

[GSM]	*Getting Started with Stata for Macintosh*
[GSU]	*Getting Started with Stata for Unix*
[GSW]	*Getting Started with Stata for Windows*
[U]	*Stata User's Guide*
[R]	*Stata Base Reference Manual*
[D]	*Stata Data Management Reference Manual*
[G]	*Stata Graphics Reference Manual*
[P]	*Stata Programming Reference Manual*
[XT]	*Stata Longitudinal/Panel Data Reference Manual*
[MV]	*Stata Multivariate Statistics Reference Manual*
[SVY]	*Stata Survey Data Reference Manual*
[ST]	*Stata Survival Analysis and Epidemiological Tables Reference Manual*
[TS]	*Stata Time-Series Reference Manual*
[I]	*Stata Quick Reference and Index*
[M]	*Mata Reference Manual*

Detailed information about each of these manuals may be found online at

http://www.stata-press.com/manuals/

Title

intro — Introduction to multivariate statistics manual

Description

This entry describes this manual and what has changed since Stata 8.

Remarks

This manual documents Stata's multivariate analysis features and is referred to as the [MV] manual in cross-references.

Following this entry, [MV] **multivariate** provides an overview of the multivariate analysis features in Stata and Stata's multivariate analysis commands. The other parts of this manual are arranged alphabetically.

Stata is continually being updated, and Stata users are always writing new commands. To find out about the latest multivariate analysis features, type `search multivariate analysis` after installing the latest official updates; see [R] **update**.

What's new

This section is intended for previous Stata users. If you are new to Stata, you may as well skip it.

Multivariate statistics

Stata has four all-new methods for analyzing multivariate data and many more extensions to existing methods. In addition, most methods now support direct analysis of matrices as well as raw data.

Be sure to check the postestimation documentation for the multivariate estimators you use; many important new features are documented there. In particular, all the multivariate commands make extensive use of new command `estat` for providing additional statistics and results after estimation.

1. New commands `mds`, `mdslong`, and `mdsmat` perform classic metric multidimensional scaling: `mds` performs the scaling with respect to the distances (dissimilarities) between observations, `mdslong` performs the scaling on a long dataset where each observation represents the distance between two points or objects, and `mdsmat` performs the scaling on a matrix of distances. See [MV] **mds**, [MV] **mdslong**, and [MV] **mdsmat**.

   `mds` supports all 33 similarity/dissimilarity measures available in Stata; see [MV] *measure_option*.

   The following new `estat` commands work after `mds`, `mdslong`, and `mdsmat` and provide additional statistics and results:

   a. `estat config` reports the coordinates of the approximating configuration.

   b. `estat correlations` reports the Pearson and Spearman correlations between the dissimilarities and the approximating distances for each object.

   c. `estat pairwise` reports a set of statistics for each pairwise comparison; it reports the dissimilarities, the approximating distances, and the raw residuals.

   d. `estat quantiles` reports the quantiles of the residuals for each observation (after `mds`) or object (after `mdslong` or `mdsmat`).

1

    e. `estat stress` reports the Kruskal stress (loss) measure between the transformed dissimilarities and fitted distances per object.

See [MV] **mds postestimation** for more information.

In addition, there are two new commands for graphing results from a multidimensional scaling:

    a. `mdsconfig` plots the approximating Euclidean configuration of the first two dimensions; see [MV] **mds postestimation**.

    b. `mdsshepard` produces a Shepard diagram of the dissimilarities against the approximating Euclidean distances; see [MV] **mds postestimation**.

    `predict` after any multidimensional-scaling command will produce

        a. variables containing the approximating configuration (`predict` *newvarlist*, `config`);

        b. variables containing the dissimilarity, distance, and raw residuals (`predict` *newvarlist*, `pairwise`)

See [MV] **mds postestimation** for more information.

2. New commands `ca` and `camat` perform two-way correspondence analysis using any of several available forms of normalization. `ca` performs the analysis on the cross-tabulation of two categorical variables; `camat` performs the analysis on a matrix of counts; see [MV] **ca** for more information on both.

The following new `estat` commands work after `ca` and `camat` and provide additional statistics and results:

    a. `estat coordinates` reports the coordinates in both the row space and the column space.

    b. `estat distances` reports the chi-squared distances between the row profiles and between the column profiles, including the distances to the marginal distributions (commonly called centers). Both observed or fitted profiles are available.

    c. `estat inertia` reports the inertia contributions of the individual cells.

    d. `estat profiles` reports the row profiles and column profiles—the conditional distributions, given the other dimension.

    e. `estat summarize` reports summary information of the row and column variables over the estimation sample.

    f. `estat table` reports the fitted correspondence table, the observed "correspondence" table, or the expected table under the assumption of independence.

See [MV] **ca postestimation** for more information.

In addition, there are two new commands for graphing results from a correspondence analysis:

    a. `cabiplot` produces a biplot of each row category and each column category; see [MV] **ca postestimation**.

    b. `caprojection` produces a graph that shows the ordering of row categories and column categories on each principal dimension of the analysis. Each principal dimension is represented by a vertical line; markers are plotted on the lines where the row categories and column categories project onto the dimensions; see [MV] **ca postestimation**.

`predict` after `ca` and `camat` computes fitted values and row or column scores for any dimension; see [MV] **ca postestimation**.

3. The new command `procrustes` performs Procrustean analysis for comparing and measuring the similarity between two sets of variables: source and target. Two datasets can also be compared if the datasets are first merged by record.

   The following new `estat` commands work after `procrustes` and provide additional statistics and results:

   a. `estat compare` reports fit statistics of the three transformations available in Procrustean analysis: orthogonal, oblique, and unrestricted.

   b. `estat mvreg` reports the multivariate regression that is related to the current Procrustean analysis.

   c. `estat summarize` reports summary information of the two sets of variables over the estimation sample.

   See [MV] **procrustes postestimation** for more information.

   New command `procoverlay` after `procrustes` creates an overlay graph comparing the target variables with the fitted values derived from the source variables; see [MV] **procrustes postestimation**.

   `predict` after `procrustes` produces fitted values for all variables, residuals for all variables, or residual sums of squares for a specified target variable; see [MV] **procrustes postestimation**.

4. New command `biplot` performs a biplot analysis of a dataset and produces a two-dimensional biplot of the results. A biplot simultaneously displays the observations (rows) and the relative positions of the variables (columns). Observations are projected to two dimensions such that the distance between the observations is approximately preserved. The variables are plotted as arrows, with the cosine of the angle between arrows approximating the correlation between the variables. See [MV] **biplot**.

5. New command `tetrachoric` computes a tetrachoric correlation matrix for a set of binary variables. `tetrachoric` is documented in [R] but will often be used in multivariate analyses; see [R] **tetrachoric**.

   `tetrachoric` results can be used in subsequent factor analyses or principal component analyses using the new `factormat` and `pcamat` commands; see [MV] **factor** and [MV] **pca**.

6. Existing command `canon` now allows analysis and presentation of more than one linear combination and has new options for reporting the raw or standardized coefficients and for reporting significance tests of the canonical correlations; see [MV] **canon**.

   The following new `estat` commands work after `canon` and provide additional statistics and results:

   a. `estat correlations` reports the correlations among all variables.

   b. `estat loadings` reports the matrices of canonical loadings.

   See [MV] **canon postestimation** for more information.

7. Existing command `cluster dendrogram` has many new features, including horizontal dendrograms and the ability to label branch counts. The look of the graph can now be changed (titles, axes, colors, etc.); see [MV] **cluster dendrogram**.

8. The existing hierarchical cluster commands have new option `measure()` that specifies the proximity measure to use in computing dissimilarities between observations. Any of 33 measures may be specified; see [MV] *measure_option*. Previously most of the measures were available under other option names; those options continue to work but are undocumented. See [MV] **cluster**.

9. Existing command `cluster stop` has new option `varlist()` that specifies alternative variables to use when computing the stopping rules; see [MV] **cluster stop**.

What's new: Analysis of proximity matrices

All of Stata's multivariate analysis facilities that rely on pairwise comparisons of distance, similarity, dissimilarity, covariance, correlation, or other proximity measures can now work directly with proximity matrices that you compute or obtain from other sources.

Previously, all these facilities worked only with raw datasets. The new commands implement analyses on matrices. They share the common ability to accept either full matrices or vectors representing the lower or upper triangle of a symmetric proximity matrix.

10. New command clustermat extends all of Stata's hierarchical clustering facilities to the analysis of matrices of a dissimilarity measure (sometimes called a distance or proximity measure). This includes all seven linkage methods and the ability to create dendrograms of the results; see [MV] **clustermat**.

11. New command factormat performs factor analysis on a matrix of correlations, extending all the new and previously available capabilities of the existing command factor to precomputed matrices of correlations; see [MV] **factor**.

12. New command pcamat performs principal component analysis on an existing correlation or covariance matrix; see [MV] **pca**.

13. New matrix subcommand dissimilarity computes similarity, dissimilarity, or distance matrices using any of 19 proximity measures for continuous data and 14 measures for binary data; see [MV] *measure_option* and see [MV] **matrix dissimilarity**.

What's new: Factor and principal component analysis additions

In addition to allowing direct analysis of correlation and covariance matrices using factormat and pcamat, Stata's factor analysis and principal component analysis (PCA) methods have been expanded, particularly through the addition of postestimation commands for reporting and graphing results.

14. Command factor has new reporting option altdivisor that specifies the trace of the correlation matrix be used as the divisor for proportions, rather than the default (the sum of all eigenvalues).

15. New estat commands for use after factor and factormat provide additional statistics and results:

   a. estat common reports the correlation matrix of the common factors and is more of interest after oblique rotations.

   b. estat factors reports model-selection criteria (AIC and BIC) over all the factors retained in an analysis.

   c. estat rotatecompare reports the unrotated factor loadings next to the most-recent rotated loadings.

   d. estat structure reports the factor structure—the correlations between the variables and the common factors.

   See [MV] **factor postestimation** for more information.

16. Existing command pca allows several new options:

   a. Option vce(normal) computes the VCE of the eigenvalues and eigenvectors, assuming multivariate normality.

   This gives you access to many of Stata's postestimation facilities for analyzing estimation results, including tests of eigenvalue and eigenvector significance, tests of linear and nonlinear combinations (test and testnl), linear and nonlinear combinations with confidence intervals (lincom and nlcom), and nonlinear predictions with confidence intervals (predictnl).

vce(normal) also produces the ingredients for adding confidence intervals to screeplots; see [MV] **screeplot**.

b. Options level(), blanks(), novce, and norotated allow more flexible control of the displayed results.

c. Option components(#) specifies the number of components to retain and is a synonym for old option factor().

d. Options tol() and ignore provide advanced control for computationally difficult problems.

See [MV] **pca** for more information.

17. New estat commands for use after pca and pcamat provide additional statistics and results:

a. estat loadings reports the component loading matrix in any of several available normalizations of the columns (eigenvectors).

b. estat rotatecompare reports the unrotated (principal) components next to the most recent rotated components.

See [MV] **pca postestimation** for more information.

18. New estat commands for use after any factor analysis or any principal component analysis (that is, after factor or factormat or after pca or pcamat) provide additional statistics and results:

a. estat anti reports the anti-image correlation and anti-image covariance matrices.

b. estat kmo reports the Kaiser–Meyer–Olkin measure of sampling adequacy.

c. estat residuals reports the difference between the observed correlation or covariance matrix and the fitted (reproduced) matrix using the retained factors.

d. estat smc reports the squared multiple correlations (SMC) between each variable and all other variables. SMC is a theoretical lower bound for communality, so it is an upper bound for the unexplained variance.

See [MV] **factor postestimation** and [MV] **pca postestimation** for more information.

19. Three new graphs are available after any factor analysis (factor and factormat) or after any principal component analysis (pca and pcamat):

a. scoreplot graphs scatterplots comparing each pair of factors or components; see [MV] **scoreplot**.

b. loadingplot graphs scatterplots comparing loadings for each pair of factors or components; see [MV] **scoreplot**.

c. screeplot plots the eigenvalues of a covariance or correlation matrix; see [MV] **screeplot**. (screeplot replaces greigen and has more features; greigen continues to work but is undocumented.)

20. New command rotate performs orthogonal and oblique rotations after factor, factormat, pca, and pcamat. Available rotations include varimax, quartimax, equamax, parsimax, minimum entropy, Comrey's tandem 1 and 2, promax power, biquartimax, biquartimin, covarimin, oblimin, factor parsimony, Crawford–Ferguson family, Bentler's invariant pattern, oblimax, quartimin, and target and partial-target matrices; see [MV] **rotate**.

New command rotatemat performs these same linear transformations (rotations) on any Stata matrix.

For a complete list of all the new features in Stata 9, see [U] **1.3 What's new**.

Also See

Background: [R] **intro**

Title

multivariate — Introduction to multivariate commands

Description

The *Multivariate Reference Manual* organizes the commands alphabetically, which makes it easy to find individual command entries if you know the name of the command. This overview organizes and presents the commands conceptually, that is, according to the similarities in the functions that they perform.

The commands listed under the heading **Cluster analysis** perform cluster analysis on variables or the similarity or dissimilarity values within a matrix. An introduction to cluster analysis and a description of the `cluster` and `clustermat` subcommands is provided in [MV] **cluster** and [MV] **clustermat**.

The commands listed under the heading **Factor analysis and principal component analysis** provide factor analysis of a correlation matrix and principal component analysis (PCA) of a correlation or covariance matrix. The correlation or covariance matrix can be provided directly or computed from variables.

The commands listed under the heading **Rotation** provide methods for rotating a factor or PCA solution or for rotating a matrix. Also provided is Procrustean rotation analysis for rotating a set of variables to best match another set of variables.

The commands listed under **Multivariate analysis of variance and related techniques** provide canonical correlation analysis, multivariate regression, multivariate analysis of variance (MANOVA), and comparison of multivariate means.

The commands listed under **Multidimensional scaling and biplots** provide classic MDS and two-dimensional biplots. MDS can be performed on the variables or on proximity data in a matrix or as proximity data in long format.

The commands listed under **Correspondence analysis** provide simple correspondence analysis (CA) on the cross-tabulation of two categorical variables or on a matrix.

Cluster analysis

cluster	Introduction to cluster-analysis commands
cluster ⋯	(See [MV] **cluster** for details)
clustermat	Introduction to clustermat commands
matrix dissimilarity	Compute similarity or dissimilarity measures; may be used by clustermat

Factor analysis and principal component analysis

factor	Factor analysis
factor postestimation	Postestimation tools for factor and factormat
pca	Principal component analysis
pca postestimation	Postestimation tools for pca and pcamat
rotate	Orthogonal and oblique rotations after factor and pca
screeplot	Scree plot of eigenvalues
scoreplot	Score and loading plots after factor and pca

7

Rotation

rotate	Orthogonal and oblique rotations after factor and pca
rotatemat	Orthogonal and oblique rotation of a Stata matrix
procrustes	Procrustes transformation
procrustes postestimation	Postestimation tools for procrustes

Multivariate analysis of variance and related techniques

canon	Canonical correlations
canon postestimation	Postestimation tools for canon
mvreg	See [R] **mvreg** — Multivariate regression
mvreg postestimation	See [R] **mvreg postestimation** — Postestimation tools for mvreg
manova	Multivariate analysis of variance and covariance
manova postestimation	Postestimation tools for manova
hotelling	Hotelling's T-squared generalized means test

Multidimensional scaling and biplots

mds	Multidimensional scaling for two-way data
mds postestimation	Postestimation tools for mds, mdsmat, and mdslong
mdslong	Multidimensional scaling of proximity data in long format
mdsmat	Multidimensional scaling of proximity data in a matrix
biplot	Biplots

Correspondence analysis

ca	Simple correspondence analysis
ca postestimation	Postestimation tools for ca and camat

Remarks

Remarks are presented under the headings

Cluster analysis
Factor analysis and principal component analysis
Rotation
Multivariate analysis of variance and related techniques
Multidimensional scaling and biplots
Correspondence analysis

Cluster analysis

Cluster analysis is concerned with finding natural groupings or clusters. Stata's cluster-analysis commands provide several hierarchical and partition clustering methods, postclustering summarization methods, and cluster-management tools. The hierarchical clustering methods may be applied to the data using the cluster command or to a user-supplied dissimilarity matrix using the clustermat command. See [MV] **cluster** for an introduction to cluster analysis and the cluster and clustermat suite of commands.

A wide variety of similarity and dissimilarity measures are available for comparing observations; see [MV] *measure_option*. Dissimilarity matrices, for use with clustermat, are easily obtained using the matrix dissimilarity command; see [MV] **matrix dissimilarity**. This provides the building blocks necessary for clustering variables instead of observations, or for clustering using a dissimilarity not automatically provided by Stata; [MV] **clustermat** provides examples.

Factor analysis and principal component analysis

Factor analysis and principal component analysis (PCA) have dual uses. They may be used as a dimension-reduction technique, and they may be used in describing the underlying data.

In PCA, the leading eigenvectors from the eigen decomposition of the correlation or covariance matrix of the variables describe a series of uncorrelated linear combinations of the variables that contain most of the variance. For data reduction, a small number of these leading components are retained. For describing the underlying structure of the data, the magnitudes and signs of the eigenvector elements are interpreted in relation to the original variables (rows of the eigenvector).

pca uses the correlation or covariance matrix computed from the dataset. pcamat allows the correlation or covariance matrix to be directly provided. The vce(normal) option provides standard errors for the eigenvalues and eigenvectors, which aids in their interpretation. See [MV] **pca** for details.

Factor analysis finds a small number of common factors that linearly reconstruct the original variables. Reconstruction is defined in terms of prediction of the correlation matrix of the original variables, unlike PCA where reconstruction means minimum residual variance summed across all variables. Factor loadings are examined for interpretation of the structure of the data.

factor computes the correlation from the dataset, whereas factormat is supplied the matrix directly. They both display the eigenvalues of the correlation matrix, the factor loadings, and the "uniqueness" of the variables. See [MV] **factor** for details.

To perform factor analysis or PCA on binary data, compute the tetrachoric correlations and use these with factormat or pcamat. Tetrachoric correlations are available with the tetrachoric command; see [R] **tetrachoric**.

After factor analysis and PCA a suite of commands are available that provide for rotation of the loadings; generation of score variables; graphing of scree plots, loading plots, and score plots; display of matrices and scalars of interest such as anti-image matrices, residual matrices, Kaiser–Meyer–Olkin measures of sampling adequacy, squared multiple correlations, and more. See [MV] **factor postestimation**, [MV] **pca postestimation**, [MV] **rotate**, [MV] **screeplot**, and [MV] **scoreplot** for details.

Rotation

Rotation provides a modified solution that is rotated from an original multivariate solution in such a way that interpretation is enhanced. Rotation is provided through three commands: rotate, rotatemat, and procrustes.

rotate works directly after pca, pcamat, factor, and factormat. It knows where to obtain the component- or factor-loading matrix for rotation, and after rotating the loading matrix, it places the rotated results in e() so that all the postestimation tools available after pca and factor may be applied to the rotated results. See [MV] **rotate** for details.

Perhaps you have the component or factor loadings from a published source and want to investigate various rotations, or perhaps you wish to rotate a loading matrix from some other multivariate command. rotatemat provides rotations for a specified matrix. See [MV] **rotatemat** for details.

A large selection of orthogonal and oblique rotations are provided for `rotate` and `rotatemat`. These include varimax, quartimax, equamax, parsimax, minimum entropy, Comrey's tandem 1 and 2, promax power, biquartimax, biquartimin, covarimin, oblimin, factor parsimony, Crawford–Ferguson family, Bentler's invariant pattern simplicity, oblimax, quartimin, target, and weighted target rotations. Horst normalization is also available.

The `procrustes` command provides Procrustean analysis. The goal is to transform a set of source variables to be as close as possible to a set of target variables. The permitted transformations are any combination of dilation (uniform scaling), rotation and reflection (orthogonal and oblique transformations), and translation. Closeness is measured by the residual sum of squares. See [MV] **procrustes** for details.

A set of postestimation commands are available after `procrustes` for generating fitted values and residuals; for providing fit statistics for orthogonal, oblique, and unrestricted transformations; and for providing a Procrustes overlay graph. See [MV] **procrustes postestimation** for details.

Multivariate analysis of variance and related techniques

The first canonical correlation is the maximum correlation that can be obtained between a linear combination of one set of variables and a linear combination of another set of variables. The second canonical correlation is the maximum correlation that can be obtained between linear combinations of the two sets of variables subject to the constraint that these second linear combinations are orthogonal to the first linear combinations, and so on.

`canon` estimates these canonical correlations and provides the loadings that describe the linear combinations of the two sets of variables that produce the correlations. Standard errors of the loadings are provided, and tests of the significance of the canonical correlations are available. See [MV] **canon** for details.

Postestimation tools are available after `canon` for generating the variables corresponding to the linear combinations underlying the canonical correlations. Various matrices and correlations may also be displayed. See [MV] **canon postestimation** for details.

In canonical correlation, there is no real distinction between the two sets of original variables. In multivariate regression, however, the two sets of variables take on the roles of dependent and independent variables. multivariate regression is an extension of regression that allows for multiple dependent variables. See [R] **mvreg** for multivariate regression, and see [R] **mvreg postestimation** for the postestimation tools available after multivariate regression.

Just as analysis of variance (ANOVA) can be formulated in terms of regression where the categorical independent variables are represented by indicator (sometimes called dummy) variables, multivariate analysis of variance (MANOVA), a generalization of ANOVA that allows for multiple dependent variables, can be formulated in terms of multivariate regression where the categorical independent variables are represented by indicator variables. Multivariate analysis of covariance (MANCOVA) allows for both continuous and categorical independent variables.

The `manova` command fits MANOVA and MANCOVA models for balanced and unbalanced designs, including designs with missing cells, and for factorial, nested, or mixed designs, or designs involving repeated measures. Four multivariate test statistics—Wilks' lambda, Pillai's trace, Lawley–Hotelling trace, and Roy's largest root—are computed for each term in the model. See [MV] **manova** for details.

Postestimation tools are available after `manova` that provide for univariate Wald tests of expressions involving the coefficients of the underlying regression model and that provide for multivariate tests involving terms or linear combinations of the underlying design matrix. Linear combinations of the dependent variables are also supported. See [MV] **manova postestimation** for details.

Related to MANOVA is Hotelling's T-squared test of whether a set of means is zero, or two sets of means are equal. It is a multivariate test that reduces to a standard t test if only one variable is involved. The `hotelling` command provides Hotelling's T-squared test; see [MV] **hotelling**.

Multidimensional scaling and biplots

Multidimensional scaling (MDS) is a dimension-reduction and visualization technique. Dissimilarities (for instance, Euclidean distances) between observations in a high-dimensional space are represented in a lower-dimensional space (typically two dimensions) so that the Euclidean distance in the lower-dimensional space approximates the dissimilarities in the higher-dimensional space.

The `mds` command provides classical metric MDS for dissimilarities between observations with respect to the variables; see [MV] **mds**. A wide variety of similarity and dissimilarity measures are allowed (the same ones available for the `cluster` command); see [MV] *measure_option*.

`mdslong` and `mdsmat` provide classical metric MDS directly on the dissimilarities recorded either as data in long format (`mdslong`) or as a dissimilarity matrix (`mdsmat`); see [MV] **mdslong** and [MV] **mdsmat**.

Postestimation tools available after `mds`, `mdslong`, and `mdsmat` provide MDS configuration plots and Shepard diagrams; generation of the approximating configuration or the dissimilarities, distances, and raw residuals; and various matrices and scalars, such as Kruskal stress (loss), quantiles of the residuals per object, and correlations between dissimilarities and approximating distances. See [MV] **mds postestimation** for details.

Biplots are two-dimensional representations of data. Both the observations and the variables are represented. The observations are represented by marker symbols, and the variables are represented by arrows from the origin. Observations are projected to two dimensions so that the distance between the observations is approximately preserved. The cosine of the angle between arrows approximates the correlation between the variables. A biplot aids in understanding the relationship between the variables, the observations, and the observations and variables jointly. The `biplot` command produces biplots; see [MV] **biplot**.

Correspondence analysis

Simple correspondence analysis (CA) is a technique for jointly exploring the relationship between rows and columns in a cross-tabulation. It is known by many names, including dual scaling, reciprocal averaging, and canonical correlation analysis of contingency tables.

`ca` performs CA on the cross-tabulation of two integer-valued variables. `camat` performs CA on a matrix with non-negative entries—perhaps from a published table. See [MV] **ca** for details.

A suite of commands are available following `ca` and `camat`. These include commands for producing CA biplots and dimensional projection plots; for generating fitted values, row coordinates, and column coordinates; and for displaying distances between row and column profiles, individual cell inertia contributions, χ^2 distances between row and column profiles, and the fitted correspondence table. See [MV] **ca postestimation** for details.

Also See

Complementary:	[U] **1.3 What's new**
Background:	[R] **intro**

Title

> **biplot** — Biplots

Syntax

biplot *varlist* [*if*] [*in*] [, *options*]

options	description
Main	
std	use standardized instead of centered variables
alpha(#)	row weight = #; column weight = 1 − #; default is 0.5
mahalanobis	approximate Mahalanobis distance; implies alpha(0)
stretch(#)	stretch the column (variable) arrows
xnegate	negate the data relative to the x-axis
ynegate	negate the data relative to the y-axis
autoaspect	adjust aspect ratio based on the data; default aspect ratio is 1
separate	produce separate plots for rows and columns
nograph	suppress graph
table	display tables with biplot coordinates
Rows	
rowopts(*row_options*)	affect rendition of rows (observations)
Columns	
colopts(*col_options*)	affect rendition of columns (variables)
Y-Axis, X-Axis, Title, Caption, Legend, Overall	
twoway_options	any options other than by() documented in [G] ***twoway_options***

row_options	description
nolabel	remove the default row (observation) label from the graph
name(*name*)	override the default name given to rows (observations)
marker_options	change look of markers (color, size, etc.)
marker_label_options	change look or position of marker labels

col_options	description
nolabel	remove the default column (variable) label from the graph
name(*name*)	override the default name given to columns (variables)
pcarrow_options	affect the rendition of paired-coordinate arrows

Description

biplot displays a two-dimensional biplot of a dataset. A biplot simultaneously displays the observations (rows) and the relative positions of the variables (columns). Marker symbols (points) are displayed for observations, and arrows are displayed for variables. Observations are projected to two dimensions such that the distance between the observations is approximately preserved. The cosine of the angle between arrows approximates the correlation between the variables.

Options

⌐ Main ⌐

std produces a biplot of the standardized variables instead of the centered variables.

alpha(#) specifies that the variables be scaled by $\lambda^{\#}$ and the observations by $\lambda^{(1-\#)}$, where λ are the singular values. It is required that $0 \le \# \le 1$. The most common values are 0, 0.5, and 1. The default is alpha(0.5) and is known as the symmetrically scaled biplot or symmetric factorization biplot. The result with alpha(1) is the principal component biplot, also called the row preserving metric (RPM) biplot. The biplot with alpha(0) is referred to as the column-preserving metric (CPM) biplot.

mahalanobis implies alpha(0) and additionally scales the positioning of points (observations) by $\sqrt{n-1}$ and positioning of arrows (variables) by $1/\sqrt{n-1}$. This additional scaling causes the distances between observations to change from being approximately proportional to the Mahalanobis distance to instead being approximately equal to the Mahalanobis distance. Also, the inner products between variables approximate their covariance.

stretch(#) causes the length of the arrows to be multiplied by #. For example, stretch(1) would leave the arrows the same length, stretch(2) would double their length, and stretch(0.5) would halve their length.

xnegate specifies that dimension-one (x-axis) values be negated (multiplied by negative one).

ynegate specifies that dimension-two (y-axis) values be negated (multiplied by negative one).

autoaspect specifies that the aspect ratio be automatically adjusted based on the range of the data to be plotted. This option can make some biplots more readable. By default, biplot uses an aspect ratio of one, producing a square plot. Some biplots will have little variation in the y-axis direction, and using the autoaspect option will better fill the available graph space while preserving the equivalence of distance in the x- and y-axes.

As an alternative to autoaspect, the *twoway_option* aspectratio() can be used to override the default aspect ratio. biplot accepts the aspectratio() option as a suggestion only and will override it when necessary to produce plots with balanced axes; i.e., distance on the x-axis equals distance on the y-axis.

twoway_options, such as xlabel(), xscale(), ylabel(), and yscale(), should be used with caution. These options are accepted but may have unintended side effects on the aspect ratio.

separate produces separate plots for the row and column categories. The default is to overlay the plots.

nograph suppresses displaying the graph.

table displays tables with the biplot coordinates.

⌐ Rows ⌐

rowopts(*row_options*) affects the rendition of rows (observations).

    nolabel removes the default row label from the graph.

    name(*name*) overrides the default name given to rows.

    marker_options affect the rendition of markers drawn at the plotted points, including their shape, size, color, and outline; see [G] ***marker_options***.

    marker_label_options specify if and how the markers are to be labeled; see [G] ***marker_label_options***.

⌐ Columns ⌐

colopts(*col_options*) affects the rendition of columns (variables).

    nolabel removes the default column label from the graph.

    name(*name*) overrides the default name given to columns.

    pcarrow_options affect the rendition of paired-coordinate arrows; see [G] **graph twoway pcarrow**.

⌐ Y-Axis, X-Axis, Title, Caption, Legend, Overall ⌐

twoway_options are any of the options documented in [G] ***twoway_options***, excluding by(). These include options for titling the graph (see [G] ***title_options***) and options for saving the graph to disk (see [G] ***saving_option***).

Remarks

The biplot command produces what Cox and Cox (2001) refer to as "the classic biplot". Biplots were introduced by Gabriel (1971); also see Gabriel (1981). Gower and Hand (1996) discuss extensions and generalizations to biplots and place many of the well-known multivariate techniques into a generalized biplot framework extending beyond the classic biplot implemented by Stata's biplot command. Cox and Cox (2001), Jolliffe (2002), Gordon (1999), Jacoby (1998), Rencher (2002), and Seber (1984) discuss the classic biplot. Kohler (2004) provided a Stata implementation of biplots.

Let $\mathbf{X}$ be the centered (or standardized if the std option is specified) data. A biplot splits the information in $\mathbf{X}$ into a portion related to the observations (rows of $\mathbf{X}$) and a portion related to the variables (columns of $\mathbf{X}$)

$$\mathbf{X} \approx (\mathbf{U}_2\, \boldsymbol{\Lambda}_2^{\alpha})(\mathbf{V}_2\, \boldsymbol{\Lambda}_2^{1-\alpha})'$$

where $0 \leq \alpha \leq 1$; see *Methods and Formulas* for details. $\mathbf{U}_2\, \boldsymbol{\Lambda}_2^{\alpha}$ contains the plotting coordinates corresponding to observations (rows), and $\mathbf{V}_2\, \boldsymbol{\Lambda}_2^{1-\alpha}$ contains the plotting coordinates corresponding to variables (columns). In a biplot, the row coordinates are plotted as symbols, and the column coordinates are plotted as arrows from the origin.

The commonly used values for α are 0, 0.5, and 1. The default is 0.5. The alpha() option allows you to set α.

Biplots with an α of 1 are also called "principal component biplots" since $\mathbf{U}_2\, \boldsymbol{\Lambda}_2$ contains the principal component scores and $\mathbf{V}_2$ contains the principal component coefficients. Euclidean distance between points in this kind of biplot approximates the Euclidean distance between points in the original higher-dimensional space.

Using an α of 0, Euclidean distances in the biplot are approximately proportional to Mahalanobis distances in the original higher-dimensional space. Also, the inner product of the arrows is approximately proportional to the covariances between the variables.

When you set α to 0 and also specify the `mahalanobis` option, the Euclidean distances are not just approximately proportional, but are approximately equal to Mahalanobis distances in the original space. Likewise, the inner products of the arrows are approximately equal (not just proportional) to the covariances between the variables. This means that the length of an arrow is approximately equal to the standard deviation of the variable it represents. Additionally, the cosine of the angle between two arrows is approximately equal to the correlation between the two variables.

A biplot with an α of 0.5 is called a symmetric factorization biplot or symmetrically scaled biplot. It often produces reasonable looking biplots where the points corresponding to observations and the arrows corresponding to variables are given equal weight. Using an α of 0 (or 1) causes the points (or the arrows) to be bunched tightly around the origin while the arrows (or the points) are predominant in the graph. In this case, many authors recommend picking a scaling factor for the arrows to bring them back into balance. The `stretch()` option allows you to do this.

Regardless of your choice of α, the position of a point in relation to an arrow indicates whether that observation is relatively large, medium, or small for that variable. Additionally, while the special conditions mentioned earlier may not strictly hold for all α, the biplot still aids in understanding the relationship between the variables, the observations, and the observations and variables jointly.

▷ Example 1

Gordon (1999, 176) provides a simple example of a biplot based on data having five rows and three columns.

```
. input v1 v2 v3

            v1       v2       v3
  1.   60    80 -240
  2. -213    66  180
  3.  123 -186  180
  4.   -9    38  -60
  5.   39     2  -60
  6. end
```

(Continued on next page)

```
. biplot v1 v2 v3
```

Biplot of 5 observations and 3 variables

```
        Explained variance by component 1  0.6283
        Explained variance by component 2  0.3717
                Total explained variance   1.0000
```

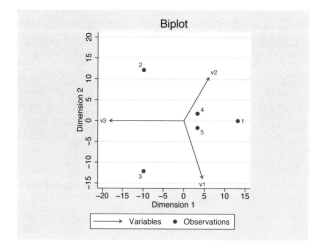

The first component accounts for 63% of the variance, and the second component accounts for the remaining 37%. All the variance is accounted for because, in this case, the 5-by-3 data matrix is only of rank 2.

Gordon actually used an α of 0 and performed the scaling to better match Mahalanobis distance. We do the same using the options `alpha(0)` and `mahalanobis`. (We could just use `mahalanobis` since it implies `alpha(0)`.) With an α of 0, Gordon decided to scale the arrows by a factor of .01. We accomplish this with the `stretch()` option and also add options to provide a title and subtitle in place of the default title obtained previously.

```
. biplot v1 v2 v3, alpha(0) mahalanobis stretch(.01) title(Simple biplot)
> subtitle(See figure 6.10 of Gordon (1999))
Biplot of 5 observations and 3 variables
```

```
    Explained variance by component 1   0.6283
    Explained variance by component 2   0.3717
            Total explained variance   1.0000
```

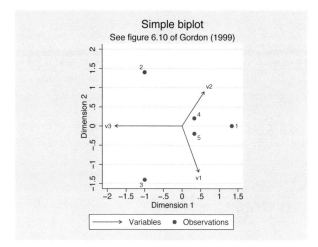

Little changed between the first and second biplot except the additional titles and the scale of the x- and y-axes.

◁

▷ Example 2

Table 7.1 of Cox and Cox (2001) provides the scores of ten Renaissance painters on four attributes using a scale from 0 to 20, as judged by Roger de Piles in the seventeenth century.

```
. use http://www.stata-press.com/data/r9/renpainters
(Scores by Roger de Piles for Renaissance Painters)
. list
```

	painter	compos~n	drawing	colour	expres~n
1.	Del Sarto	12	16	9	8
2.	Del Piombo	8	13	16	7
3.	Da Udine	10	8	16	3
4.	Giulio Romano	15	16	4	14
5.	Da Vinci	15	16	4	14
6.	Michelangelo	8	17	4	8
7.	Fr. Penni	0	15	8	0
8.	Perino del Vaga	15	16	7	6
9.	Perugino	4	12	10	4
10.	Raphael	17	18	12	18

```
. biplot composition-expression, alpha(1) stretch(10) table rowopts(name(painters)
> mlabel(painter)) colopts(name(attributes)) title(Renaissance painters)

Biplot of 10 painters and 4 attributes
```

```
    Explained variance by component 1   0.6700
    Explained variance by component 2   0.2375
            Total explained variance    0.9075
```

Biplot coordinates

Painters	dim1	dim2
Del Sarto	1.2120	0.0739
Del Piombo	-4.5003	5.7309
Da Udine	-7.2024	7.5745
Giulio Rom~o	8.4631	-2.5503
Da Vinci	8.4631	-2.5503
Michelangelo	0.1284	-5.9578
Fr Penni	-11.9449	-5.4510
Perino del~a	2.2564	-0.9193
Perugino	-7.8886	-0.8757
Raphael	11.0131	4.9251

Attributes	dim1	dim2
composition	6.4025	3.3319
drawing	2.4952	-3.3422
colour	-2.4557	8.7294
expression	6.8375	1.2348

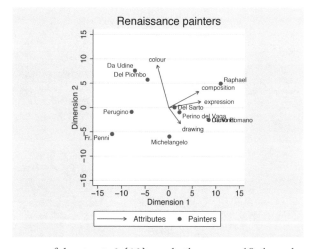

alpha(1) gave us an α of 1. stretch(10) made the arrows 10 times longer. table requested that the biplot coordinate tables be displayed. rowopts() and colopts() affected the rendition of the rows (observations) and columns (variables). The name() suboption provided a name to use instead of the default names "Observations" and "Variables" in the graph legend and in the biplot coordinate tables. The mlabel(painter) suboption provided to rowopts() requested that the variable painter be used to label the row points (observations) in both the graph and table. The title() option was used to override the default title.

The default is to produce a square graphing region. Since the x-axis containing the first component has more variability than the y-axis containing the second component, there are often no observations or arrows appearing in the upper and lower regions of the graph. The autoaspect option sets the aspect ratio and the x-axis and y-axis scales so that more of the graph region is used while maintaining the equivalent interpretation of distance for the x- and y-axes.

Here is the previous biplot with the omission of the `table` option and the addition of the `autoaspect` option. We also add the `ynegate` option to invert the orientation of the data in the y-axis direction to match the orientation shown in figure 7.1 of Cox and Cox (2001).

```
. biplot composition-expression, autoaspect alpha(1) stretch(10) ynegate
> rowopts(name(painters) mlabel(painter)) colopts(name(attributes))
> title(Renaissance painters)
Biplot of 10 painters and 4 attributes
      Explained variance by component 1  0.6700
      Explained variance by component 2  0.2375
               Total explained variance  0.9075
```

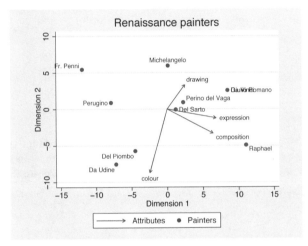

◁

Saved Results

`biplot` saves in `r()`:

Scalars
`r(rho1)`	explained variance by component 1
`r(rho2)`	explained variance by component 2
`r(rho)`	total explained variance
`r(alpha)`	value of `alpha()` option

Matrices
`r(U)`	biplot coordinates for the observations
`r(V)`	biplot coordinates for the variables
`r(Vstretch)`	biplot coordinates for the variables times `stretch()` factor

Methods and Formulas

biplot is implemented as an ado-file.

Let $\mathbf{X}$ be the centered (standardized if std is specified) data with N rows (observations) and p columns (variables). A biplot splits the information in $\mathbf{X}$ into a portion related to the observations (rows of $\mathbf{X}$) and a portion related to the variables (columns of $\mathbf{X}$). This is done using the singular value decomposition (SVD).

$$\mathbf{X} = \mathbf{U}\mathbf{\Lambda}\mathbf{V}'$$

The biplot formula is derived from this SVD by first splitting $\mathbf{\Lambda}$, a diagonal matrix, into

$$\mathbf{\Lambda} = \mathbf{\Lambda}^{\alpha}\,\mathbf{\Lambda}^{1-\alpha}$$

and then retaining the first 2 columns of $\mathbf{U}$, the first 2 columns of $\mathbf{V}$, and the first 2 rows and columns of $\mathbf{\Lambda}$. Using the subscript 2 to denote this, the biplot formula is

$$\mathbf{X} \approx \mathbf{U}_2\,\mathbf{\Lambda}_2^{\alpha}\,\mathbf{\Lambda}_2^{1-\alpha}\,\mathbf{V}_2'$$

where $0 \leq \alpha \leq 1$. This is then written as

$$\mathbf{X} \approx (\mathbf{U}_2\,\mathbf{\Lambda}_2^{\alpha})(\mathbf{V}_2\,\mathbf{\Lambda}_2^{1-\alpha})'$$

$\mathbf{U}_2\,\mathbf{\Lambda}_2^{\alpha}$ contains the plotting coordinates corresponding to observations (rows) and $\mathbf{V}_2\,\mathbf{\Lambda}_2^{1-\alpha}$ contains the plotting coordinates corresponding to variables (columns). In a biplot, the row coordinates are plotted as symbols and the column coordinates are plotted as arrows from the origin.

Let λ_i be the ith diagonal of $\mathbf{\Lambda}$. The explained variance for component 1 is

$$\rho_1 = \left\{ \sum_{i=1}^{p} \lambda_i^2 \right\}^{-1} \lambda_1^2$$

and for component 2 is

$$\rho_2 = \left\{ \sum_{i=1}^{p} \lambda_i^2 \right\}^{-1} \lambda_2^2$$

The total explained variance is

$$\rho = \rho_1 + \rho_2$$

Acknowledgment

Several biplot options were based on the work of Ulrich Kohler (2004) from WZB, Berlin.

References

Cox, T. F. and M. A. A. Cox. 2001. *Multidimensional Scaling*. 2nd ed. Boca Raton, FL: Chapman & Hall.

Gabriel, K. R. 1971. The biplot graphical display of matrices with application to principal component analysis. *Biometrika* 58: 453–467.

———. 1981. Biplot display of multivariate matrices for inspection of data and diagnosis. In *Interpreting Multivariate Data*, ed. V. Barnett, 147–174. Chichester, UK: Wiley.

Gordon, A. D. 1999. *Classification.* 2nd ed. Boca Raton, FL: Chapman & Hall.

Gower, J. C. and D. J. Hand. 1996. *Biplots.* London: Chapman & Hall.

Jacoby, W. G. 1998. *Statistical Graphics for Visualizing Multivariate Data.* Thousand Oaks, CA: Sage.

Jolliffe, I. T. 2002. *Principal Component Analysis.* 2nd ed. New York: Springer.

Kohler, U. 2004. Biplots, revisited. *April 2004 Stata Users Group Meeting in Berlin.*

Rencher, A. C. 2002. *Methods of Multivariate Analysis.* 2nd ed. New York: Wiley.

Seber, G. A. F. 1984. *Multivariate Observations.* New York: Wiley.

Also See

Related: [MV] **ca**, [MV] **mds**, [MV] **pca**

Title

ca — Simple correspondence analysis

Syntax

Simple correspondence analysis of data

ca *rowvar colvar* $\left[\,if\,\right]$ $\left[\,in\,\right]$ $\left[\,weight\,\right]$ $\left[\,,\,options\,\right]$

Simple correspondence analysis of a $n_r \times n_c$ *matrix*

camat *matname* $\left[\,,\,options\,\right]$

options	description
Model 2	
<u>dimensions</u>(#)	number of dimensions (factors, axes); default is dim(2)
<u>normalize</u>(*nopts*)	normalization of row and column coordinates
<u>rowsupp</u>(*matname$_r$*)	matrix of supplementary rows
<u>colsupp</u>(*matname$_c$*)	matrix of supplementary columns
<u>missing</u>	treat missing values as ordinary values (ca only)
<u>rowname</u>(*string*)	label for rows; default is rowname(rows) (camat only)
<u>colname</u>(*string*)	label for columns; default is colname(columns) (camat only)
Reporting	
<u>norowpoints</u>	suppress table for row points
<u>nocolpoints</u>	suppress table for column points
<u>compact</u>	display tables in a compact format
<u>plot</u>	plot the row and column coordinates
<u>maxlength</u>(#)	maximum number of characters for labels; default is maxlength(12)

nopts	description
<u>sy</u>mmetric	symmetric coordinates; the default
<u>row</u>	row principal coordinates
<u>col</u>umn	column principal coordinates
<u>principal</u>	principal coordinates
#	power $0 \leq \# \leq 1$ for row coordinates; seldom used

bootstrap, by, jackknife, rolling, and statsby may be used with ca; see [U] **11.1.10 Prefix commands**.
fweights, aweights, and iweights are allowed with ca; see [U] **11.1.6 weight**.
See [MV] **ca postestimation** for features available after estimation.

Description

ca performs a simple correspondence analysis (CA) of the cross-tabulation of the integer-valued variables *rowvar* and *colvar* with n_r and n_c categories with n_r, $n_c \geq 2$. CA is formally equivalent to various other geometric approaches, including "dual scaling", "reciprocal averaging", and "canonical correlation analysis of contingency tables" (Greenacre 1984, chapter 4).

camat performs a simple correspondence analysis of a $n_r \times n_c$ matrix *matname* having non-negative entries and strictly positive margins. The correspondence table need not contain frequencies. The labels for the row and column categories are obtained from the matrix row and column names.

Optionally, a CA biplot may be produced. The biplot displays the row and column coordinates within the same two-dimensional graph.

Results may be replayed using ca or camat; there is no difference.

Options

────────┐ Model 2 ┌──

dimensions(#) specifies the number of dimensions (= factors = axes) to be extracted. The default is dimensions(2). If you may specify dimensions(1), the row and column categories are placed "on a single dimension". # should be strictly smaller than the number of rows and the number of columns, excluding supplementary rows and columns (see options rowsupp() and colsupp()).

CA is a hierarchical method so that extracting additional dimensions does not affect the coordinates and decomposition of inertia of dimensions already included. The percentages of inertia accounting for the dimensions are in decreasing order as indicated by "singular values". The first dimension accounts for the most inertia, followed by the second dimension, and then the third dimension, etc.

normalize(*nopt*) specifies the normalization method, i.e., how the row and column coordinates are obtained from the singular vectors and singular values of the matrix of standardized residuals. See *Normalization and the interpretation of correspondence analysis* in *Remarks* for a discussion of these different normalization methods.

symmetric, the default, distributes the inertia equally over rows and columns, treating the rows and columns symmetrically. The symmetric normalization is also known as the standard or canonical normalization. This is the most common normalization when making a biplot. normalize(symmetric) is equivalent to normalize(0.5); see below. <u>canonical</u> is a synonym for symmetric.

row should be chosen if you want to compare row categories. Similarity of column categories should not be interpreted. The biplot interpretation of the relationship between row and column categories is appropriate. normalize(row) is equivalent to normalize(1).

column should be chosen if you want to compare column categories. Similarity of row categories should not be interpreted. The biplot interpretation of the relationship between row and column categories is appropriate. normalize(column) is equivalent to normalize(0).

principal is the normalization to choose if you want to make comparisons among the row categories and among the column categories. In this normalization, it is not appropriate to compare row and column points. Thus a biplot in this normalization is best avoided. In the principal normalization, the row and column coordinates are obtained from the left and right singular vectors, multiplied by the singular values.

#, $0 \leq \# \leq 1$, is seldom used; it specifies that the row coordinates are obtained as the left singular vectors multiplied by the singular values to the power *#*, while the column coordinates equal the right singular vectors multiplied by the singular values to the power $1 - \#$.

rowsupp(*matname$_r$*) specifies a matrix of supplementary rows. *matname$_r$* should have n_c columns. The row names of *matname$_r$* are used for labeling. Supplementary rows do not affect the computation of the dimensions and the decomposition of inertia. They are, however, included in the plots and in the table with statistics of the row points. Since supplementary points do not contribute to the dimensions, their entries under the column labeled contrib are left blank.

colsupp(*matname$_c$*) specifies a matrix of supplementary columns. *matname$_c$* should have n_r rows. The column names of *matname$_c$* are used for labeling. Supplementary columns do not affect the computation of the dimensions and the decomposition of inertia. They are, however, included in the plots and in the table with statistics of the column points. Since supplementary points do not contribute to the dimensions, their entries under the column labeled contrib are left blank.

missing, allowed only with ca, treats missing values of *rowvar* and *colvar* as ordinary categories to be included in the analysis. Observations with missing values are omitted from the analysis by default.

rowname(*string*), allowed only with camat, specifies a label to refer to the rows of the matrix. The default is rowname(rows).

colname(*string*), allowed only with camat, specifies a label to refer to the columns of the matrix. The default is colname(columns).

> **Reporting**

norowpoints suppresses the table with row category statistics.

nocolpoints suppresses the table with column category statistics.

compact specifies that the table with category statistics be displayed multiplied by 1000 as proposed by Greenacre (1993), enabling the display of more columns without wrapping output. The compact tables can be displayed without wrapping for models with two dimensions at linesize 77 and with three dimensions at linesize 97.

plot displays a plot of the row and column coordinates in two dimensions. With row principal normalization, only the row points are plotted. With column principal normalization, only the column points are plotted. In the other normalizations, both row and column points are plotted. You can use cabiplot directly if you need another selection of points to be plotted or if you wish to otherwise refine the plot; see [MV] **ca postestimation**.

maxlength(*#*) specifies the maximum number of characters for row and column labels. The default is maxlength(12).

Remarks

Remarks are presented under the headings

> *Introduction*
> *A first example*
> *How many dimensions?*
> *Statistics on the points*
> *Normalization and the interpretation of correspondence analysis*
> *Plotting the points*
> *Supplementary points*
> *Matrix input*

Introduction

Correspondence analysis, or CA for short, offers a geometric representation of the rows and columns of a two-way frequency table that is helpful in understanding the similarities between the categories of variables and the association between the variables. For an informal introduction to CA and related metric approaches, see Weller and Romney (1990). Greenacre (1993) provides a much more thorough introduction with little mathematical prerequisites. More advanced treatments are given by Greenacre (1984) and Gower and Hand (1996).

In some respects, CA can be thought of as an analogue to principal components for the case of nominal variables. It is also possible to interpret CA in the context of reciprocal averaging (Greenacre 1984, 96–102; Cox and Cox 2001, 193–200), in the context of optimal scaling (Greenacre 1984, 102–108) and in the context of canonical correlations (Greenacre 1984, 108–116; Gower and Hand 1996, 183–185). Scaling refers to the assignment of scores to the categories of the row and column variables. Different criteria for the assignment of scores have been proposed, generally with different solutions. If the aim is to maximize the correlation between the scored row and column, the problem can be formulated in terms of correspondence analysis. The optimal scores are the coordinates on the first dimension. The coordinates on the second and subsequent dimensions maximize the correlation between row and column scores subject to orthogonality constraints. See also [MV] **ca postestimation**.

A first example

▷ Example 1

We illustrate CA with an example of smoking behavior by different ranks of personnel. This example is often used in the CA literature (e.g., Greenacre 1984, 55; Greenacre 1993, 64), so likely you have encountered these (artificial) data before. By using these familiar data, we make it easier to relate the literature on CA to the output of the ca command.

```
. use http://www.stata-press.com/data/r9/ca_smoking
. tabulate rank smoking
```

rank	smoking intensity				Total
	none	light	medium	heavy	
senior_mngr	4	2	3	2	11
junior_mngr	4	3	7	4	18
senior_empl	25	10	12	4	51
junior_empl	18	24	33	13	88
secretary	10	6	7	2	25
Total	61	45	62	25	193

ca displays the results of a CA on two categorical variables in a multipanel format.

(*Continued on next page*)

```
. ca rank smoking
```

Correspondence analysis

				Number of obs	=	193
				Pearson chi2(12)	=	16.44
				Prob > chi2	=	0.1718
				Total inertia	=	0.0852
5 active rows				Number of dim.	=	2
4 active columns				Expl. inertia (%)	=	99.51

Dimensions	singular values	principal inertia	chi2	percent	cumul percent
dim 1	.2734211	.0747591	14.43	87.76	87.76
dim 2	.1000859	.0100172	1.93	11.76	99.51
dim 3	.0203365	.0004136	0.08	0.49	100.00
total		.0851899	16.44	100	

Statistics for row and column categories in symmetric normalization

Categories	mass	overall quality	inertia	dimension_1 coord	sqcorr	contrib
rank						
senior mngr	0.057	0.893	0.003	0.126	0.092	0.003
junior mngr	0.093	0.991	0.012	-0.495	0.526	0.084
senior empl	0.264	1.000	0.038	0.728	0.999	0.512
junior empl	0.456	1.000	0.026	-0.446	0.942	0.331
secretary	0.130	0.999	0.006	0.385	0.865	0.070
smoking						
none	0.316	1.000	0.049	0.752	0.994	0.654
light	0.233	0.984	0.007	-0.190	0.327	0.031
medium	0.321	0.983	0.013	-0.375	0.982	0.166
heavy	0.130	0.995	0.016	-0.562	0.684	0.150

Categories	dimension_2 coord	sqcorr	contrib
rank			
senior mngr	0.612	0.800	0.214
junior mngr	0.769	0.465	0.551
senior empl	0.034	0.001	0.003
junior empl	-0.183	0.058	0.152
secretary	-0.249	0.133	0.081
smoking			
none	0.096	0.006	0.029
light	-0.446	0.657	0.463
medium	-0.023	0.001	0.002
heavy	0.625	0.310	0.506

The order in which we specify the variables is mostly immaterial. The first variable (`rank`) is also called the row variable, and the second (`smoking`) is the column variable. This is only important as far as the interpretation of some options and some labeling of output are concerned. For instance, the option `norowpoints` suppresses the table with row points, i.e., the categories of rank. `ca` requires two integer-valued variables. The rankings of the categories and the actual values used to code categories are not important. Thus rank may be coded 1, 2, 3, 4, 5, or 0, 1, 4, 9, 16, or −2, −1, 0, 1, 2, it does not matter. We do suggest assigning value labels to the variables to improve the interpretability of tables and plots.

Correspondence analysis seeks to offer a low-dimensional representation describing how the row and column categories contribute to the inertia in a table. ca reports Pearson's test of independence, just like tabulate with the chi2 option. Inertia is Pearson's chi-squared statistic divided by the sample size, $16.44/193 = 0.0852$. Note that Pearson's chi-squared test has significance level $p = 0.1718$, casting doubt on any association between rows and columns. Still, given the prominence of this example in the CA literature, we will continue.

The first panel produced by ca displays the decomposition of total inertia in orthogonal dimensions—analogous to the decomposition of the total variance in principal component analysis (see [MV] **pca**). The first dimension accounts for 87.76% of the inertia, the second dimension accounts for 11.76% of the inertia. Since the dimensions are orthogonal, we may add the contributions of the two dimensions and say that the two leading dimensions account for $87.76\% + 11.76\% = 99.51\%$ of the total inertia. A two-dimensional representation seems in order. The remaining output is discussed later.

◁

How many dimensions?

▷ Example 2

In the example above, we displayed coordinates and statistics for a two-dimensional approximation of the rows and columns. This is the default. We can specify more or fewer dimensions with the option dimensions(). The maximum number is $\min(n_r - 1, n_c - 1)$. At this maximum, the chi-squared distances between the rows and columns are exactly represented by CA; 100% of the inertia is accounted for. This is called the saturated model; the fitted values of the CA model equal the observed correspondence table.

The minimum number of dimensions is 1; the model with 0 dimensions would be a model of independence of the rows and columns. With 1 dimension, the rows and columns of the table are identified by points on a line, with distance on the line approximating the chi-squared distance in the table, and a biplot is no longer feasible.

```
. ca rank smoking, dim(1)
Correspondence analysis              Number of obs      =        193
                                     Pearson chi2(12)   =      16.44
                                     Prob > chi2        =     0.1718
                                     Total inertia      =     0.0852
    5 active rows                    Number of dim.     =          1
    4 active columns                 Expl. inertia (%)  =      87.76
```

Dimensions	singular values	principal inertia	chi2	percent	cumul percent
dim 1	.2734211	.0747591	14.43	87.76	87.76
dim 2	.1000859	.0100172	1.93	11.76	99.51
dim 3	.0203365	.0004136	0.08	0.49	100.00
total		.0851899	16.44	100	

(Continued on next page)

Statistics for row and column categories in symmetric normalization

Categories	mass	overall quality	inertia	dimension_1 coord	sqcorr	contrib
rank						
senior mngr	0.057	0.092	0.003	0.126	0.092	0.003
junior mngr	0.093	0.526	0.012	-0.495	0.526	0.084
senior empl	0.264	0.999	0.038	0.728	0.999	0.512
junior empl	0.456	0.942	0.026	-0.446	0.942	0.331
secretary	0.130	0.865	0.006	0.385	0.865	0.070
smoking						
none	0.316	0.994	0.049	0.752	0.994	0.654
light	0.233	0.327	0.007	-0.190	0.327	0.031
medium	0.321	0.982	0.013	-0.375	0.982	0.166
heavy	0.130	0.684	0.016	-0.562	0.684	0.150

The first panel produced by ca does not depend on the number of dimensions extracted; thus, we will always see all singular values and the percentage of inertia explained by the associated dimensions. In the second panel, the only thing that depends on the number of dimensions is the overall quality of the approximation. The overall quality is the sum of the quality scores on the extracted dimensions, and so increases with the number of extracted dimensions. The higher the quality, the better the chi-squared distances with other rows (columns) are represented by the extracted number of dimensions. In a saturated model, the overall quality is 1 for each row and column category.

So, how many dimensions should we retain? It is not uncommon for researchers to extract the minimum number of dimensions in a CA to explain at least 90% of the inertia, analogous to similar heuristic rules on the number of components in principal component analysis. Likely, we could also search for a scree, the number of dimensions where the singular values flatten out (see [MV] **screeplot**). A screeplot of the singular values can be obtained by typing

```
. screeplot e(Sv)
(output omitted )
```

Note that e(Sv) is where ca has stored the singular values.

◁

Statistics on the points

▷ Example 3

We now turn our attention to the second panel. The overall section of the panel lists the following statistics:

- The mass of the category, i.e., the proportion in the marginal distribution. The masses of all categories of a variable add up to 1.

- The quality of the approximation for a category, expressed as a number between 0 (very bad) and 1 (perfect). In a saturated model, quality is 1.

- The inertia contained in the category. The inertias of the categories of a variable add up to the total inertia.

For each of the dimensions, the panel lists

- The coordinate of the category.

- The squared residuals between the profile and the categories. The sum of the squared residuals over the dimensions add up to the quality of the approximation for the category.

- The contribution made by the categories to the dimensions. These add up to 1 over all categories of a variable.

The table with point statistics becomes pretty large, especially with more than two dimensions. ca can also list the second panel in a more compact form, saving space by multiplying all entries by 1000; see Greenacre (1993).

```
. ca rank smoking, dim(2) compact
Correspondence analysis                      Number of obs     =       193
                                             Pearson chi2(12)  =     16.44
                                             Prob > chi2       =    0.1718
                                             Total inertia     =    0.0852
        5 active rows                        Number of dim.    =         2
        4 active columns                     Expl. inertia (%) =     99.51
```

Dimensions	singular values	principal inertia	chi2	percent	cumul percent
dim 1	.2734211	.0747591	14.43	87.76	87.76
dim 2	.1000859	.0100172	1.93	11.76	99.51
dim 3	.0203365	.0004136	0.08	0.49	100.00
total		.0851899	16.44	100	

Statistics for row and column categories in symmetric norm. (x 1000)

Categories	overall			dimension 1			dimension 2		
	mass	qualt	inert	coord	sqcor	contr	coord	sqcor	contr
rank									
senior mngr	57	893	3	126	92	3	612	800	214
junior mngr	93	991	12	-495	526	84	769	465	551
senior empl	264	1000	38	728	999	512	34	1	3
junior empl	456	1000	26	-446	942	331	-183	58	152
secretary	130	999	6	385	865	70	-249	133	81
smoking									
none	316	1000	49	752	994	654	96	6	29
light	233	984	7	-190	327	31	-446	657	463
medium	321	983	13	-375	982	166	-23	1	2
heavy	130	995	16	-562	684	150	625	310	506

◁

Normalization and the interpretation of correspondence analysis

The normalization method used in CA determines whether and how the similarity of the row categories, the similarity of the column categories, and the relationship (association) between the row and column variables can be interpreted in terms of the row and column coordinates and the origin of the plot.

How does one "compare row points"—provided that the normalization method allows such a comparison? Formally, the Euclidean distance between the row points approximates the chi-squared distances between the corresponding row profiles. Thus in the biplot, row categories mapped close together have similar row profiles; i.e., the distributions on the column variable are similar. Row categories mapped widely apart have dissimilar row profiles. Moreover, the Euclidean distance between a row point and the origin approximates the chi-squared distance from the row profile and the row centroid, so it indicates how different a category is from the population.

An analogous interpretation applies to column points.

For the association between the row and column variables: In the CA biplot, you should not interpret the distance between a row point r and a column point c as the relationship of r and c. Instead, think in terms of the vectors origin-to-r (OR) and origin-to-c (OC). Remember that CA decomposes scaled deviations $d(r, c)$ from independence and $d(r, c)$ is approximated by the inner product of OR and OC. The larger the absolute value of $d(r, c)$, the stronger the association between r and c. In geometric terms $d(r, c)$ can be written as the product of the length of OR, the length of OC, and the cosine of the angle between OR and OC.

What does this mean? First, consider the effects of the angle. The association in (r, c) is strongly positive if OR and OC point in roughly the same direction; the frequency of (r, c) is much higher than expected under independence, so r tends to flock together with c. Note that this is the case if the points r and c are close together. Similarly, the association is strongly negative if OR and OC point in opposite directions. In this case, the frequency of (r, c) is much lower than expected under independence, so r and c are unlikely to occur simultaneously. Finally, if OR and OC are roughly orthogonal (angle = ±90), the deviation from independence is small.

Second, the association of r and c increases with the lengths of OR and OC. Points far away from the origin tend to have large associations. If a category is mapped close to the origin, all its associations with categories of the other variable are small; in other words, its distribution resembles the marginal distribution.

Here are the interpretations enabled by the main normalization methods as specified in the `normalize()` option.

normalization method	similarity row cat.	similarity column cat.	association row vs. column
`symmetric`	no	no	yes
`principal`	yes	yes	no
`row`	yes	no	yes
`column`	no	yes	yes

If we say that a comparison between row categories or between column categories is not possible, we really mean to say that the chi-squared distance between row profiles or column profiles is actually approximated by a weighted Euclidean distance between the respective plots in which the weights depend on the inertia of the dimensions, rather than the standard Euclidean distance.

You may want to do a CA in principal normalization to study the relationship between the categories of a variable, and a CA in symmetric normalization to study the inter-relation of the row and column categories.

Plotting the points

▷ Example 4

In our discussion of normalizations, we stated that CA offers simple geometric interpretations to the similarity of categories and the association of the variables. We may specify the option `plot` with `ca` during estimation or during replay.

```
. ca, norowpoint nocolpoint plot
Correspondence analysis                          Number of obs      =       193
                                                 Pearson chi2(12)   =     16.44
                                                 Prob > chi2        =    0.1718
                                                 Total inertia      =    0.0852
        5 active rows                            Number of dim.     =         2
        4 active columns                         Expl. inertia (%)  =     99.51
```

Dimensions	singular values	principal inertia	chi2	percent	cumul percent
dim 1	.2734211	.0747591	14.43	87.76	87.76
dim 2	.1000859	.0100172	1.93	11.76	99.51
dim 3	.0203365	.0004136	0.08	0.49	100.00
total		.0851899	16.44	100	

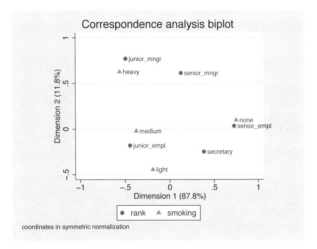

The options `norowpoint` and `nocolpoint` suppress the large tables of statistics for the rows and columns. If we did not request the plot during estimation, we can still obtain it with the `cabiplot` postestimation command. Unlike requesting the plot at estimation time, `cabiplot` allows us to fine tune the plot; see [MV] **ca postestimation**.

The horizontal dimension seems to distinguish smokers from nonsmokers, while the vertical dimensions can be interpreted as intensity of smoking. Since the orientations from the origin to `none` and from the origin to `senior_empl` are so close, we conclude that senior employees tend not to smoke. Similarly, junior managers tend to be heavy smokers, and junior employees tend to be medium smokers.

◁

Supplementary points

A useful feature of CA is the ability to locate supplementary rows and columns in the space generated by the "active" rows and columns (see Greenacre 1984, 70–74; Greenacre 1993, chapter 12, for an extensive discussion). Think of supplementary rows and columns as having mass 0; as a consequence, supplementary points do not influence the approximating space—their contribution values are zero.

▷ Example 5

In our example, we want to include the national distribution of smoking intensity as a supplementary row.

ca requires that we define the supplementary row distributions as rows of a matrix. In this example, we only have one supplementary row, with the percentages of the smoking categories in a national sample. The matrix should have one row per supplementary row category, and as many columns as there are active columns. We define the row name to obtain appropriately labeled output.

```
. matrix S_row = ( 42, 29, 20, 9 )
. matrix rowname S_row = national
```

Before we show the CA analysis with the supplementary row, we also include two supplementary columns for the rank distribution of alcoholic beverage drinkers and nondrinkers. It will be interesting to see where smoking is located relative to drinking and nondrinking.

```
. matrix S_col = (  0, 11 \
                    1, 19 \
                    5, 44 \
                   10, 78 \
                    7, 18 )
. matrix colnames S_col = nondrink drink
```

We now invoke ca, specifying the names of the matrices with supplementary rows and columns with the options rowsupp() and colsupp().

```
. ca rank smoking, rowsupp(S_row) colsupp(S_col) plot
```

Correspondence analysis

				Number of obs	=	193
				Pearson chi2(12)	=	16.44
				Prob > chi2	=	0.1718
				Total inertia	=	0.0852
5 active + 1 supplementary rows				Number of dim.	=	2
4 active + 2 supplementary columns				Expl. inertia (%)	=	99.51

Dimensions	singular values	principal inertia	chi2	percent	cumul percent
dim 1	.2734211	.0747591	14.43	87.76	87.76
dim 2	.1000859	.0100172	1.93	11.76	99.51
dim 3	.0203365	.0004136	0.08	0.49	100.00
total		.0851899	16.44	100	

Statistics for row and column categories in symmetric normalization

Categories	mass	overall quality	inertia	dimension_1 coord	sqcorr	contrib
rank						
senior mngr	0.057	0.893	0.003	0.126	0.092	0.003
junior mngr	0.093	0.991	0.012	-0.495	0.526	0.084
senior empl	0.264	1.000	0.038	0.728	0.999	0.512
junior empl	0.456	1.000	0.026	-0.446	0.942	0.331
secretary	0.130	0.999	0.006	0.385	0.865	0.070
suppl_rows						
national	0.518	0.761	0.055	0.494	0.631	
smoking						
none	0.316	1.000	0.049	0.752	0.994	0.654
light	0.233	0.984	0.007	-0.190	0.327	0.031
medium	0.321	0.983	0.013	-0.375	0.982	0.166
heavy	0.130	0.995	0.016	-0.562	0.684	0.150
suppl_cols						
nondrink	0.119	0.439	0.039	0.220	0.040	
drink	0.881	0.838	0.008	-0.082	0.202	

Categories	dimension_2 coord	sqcorr	contrib
rank			
senior mngr	0.612	0.800	0.214
junior mngr	0.769	0.465	0.551
senior empl	0.034	0.001	0.003
junior empl	-0.183	0.058	0.152
secretary	-0.249	0.133	0.081
suppl_rows			
national	-0.372	0.131	
smoking			
none	0.096	0.006	0.029
light	-0.446	0.657	0.463
medium	-0.023	0.001	0.002
heavy	0.625	0.310	0.506
suppl_cols			
nondrink	-1.144	0.398	
drink	0.241	0.636	

(Continued on next page)

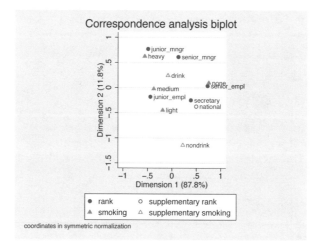

The first panel and the information about the five active rows and the four active columns have not changed—the approximating space is fully determined by the active rows and columns and is independent of the location of the supplementary rows and columns.

The table with statistics for the row and column categories now also contains entries for the supplementary rows and columns. The `contrib` entries for the supplementary points are blank. Supplementary points do not "contribute to" the location of the dimensions—or in other words, their contribution is 0.000, but displaying blanks makes the point more clearly. All other columns for the supplementary points are informative. The inertia of supplementary points is the chi-squared distance to the respective centroid. The coordinates of supplementary points are obtained by applying the translation equations of the CA. Correlations of the supplementary profiles with the dimensions are also well defined. Finally, we may consider the quality of the two-dimensional approximation for the supplementary points. These are lower than for the active points, which will be the case in most applications—the active points exercise influence on the dimensions to improve their quality, while the supplementary points simply have to accept the dimensions as determined by the active points.

If we look at the biplot, the supplementary points are shown along with the active points. We may interpret the supplementary points just like the active points. Secretaries are close to the national sample in terms of smoking. Drinking alcohol is closer to the smoking categories than to nonsmoking, indicating that alcohol consumption and smoking are similar behaviors—but, it is of course not possible to conclude that the *same* people smoke and drink since we do not have three-way data.

◁

Matrix input

▷ Example 6

If we want to do a correspondence analysis of a published two-way frequency table, we typically do not have immediate access to the data in the form of a dataset. We could enter the data with frequency weights.

```
. input rank smoking freq
  1.      1        1      4
  2.      1        2      2
  3.      1        3      3
 (output omitted )
 19.      5        3      7
 20.      5        4      2
 21.  end
. label define vl_rank  1  "senior_mngr" ...
. label value rank vl_rank
. label define vl_smoke 1  "none" ...
. label value smoke vl_smoke
. ca rank smoking [fw=freq]
 (output omitted )
```

Or we may enter the data as a matrix and use `camat`. First, we enter the frequency matrix with proper column and row names and then list the matrix for verification.

```
. matrix F = ( 4,2,3,2 \ 4,3,7,4 \ 25,10,12,4 \ 18,24,33,13 \ 10,6,7,2 )
. matrix colnames F = none light medium heavy
. matrix rownames F = senior_mngr junior_mngr senior_empl junior_empl secretary
. matlist F, border
```

	none	light	medium	heavy
senior_mngr	4	2	3	2
junior_mngr	4	3	7	4
senior_empl	25	10	12	4
junior_empl	18	24	33	13
secretary	10	6	7	2

We can use `camat` on F to obtain the same results as from the raw data. We use the `compact` option for a more compact table.

```
. camat F, compact
Correspondence analysis                   Number of obs     =      193
                                          Pearson chi2(12)  =    16.44
                                          Prob > chi2       =   0.1718
                                          Total inertia     =   0.0852
        5 active rows                     Number of dim.    =        2
        4 active columns                  Expl. inertia (%) =    99.51

                   singular    principal                          cumul
    Dimensions       values      inertia      chi2    percent    percent

         dim 1    .2734211    .0747591      14.43      87.76      87.76
         dim 2    .1000859    .0100172       1.93      11.76      99.51
         dim 3    .0203365    .0004136       0.08       0.49     100.00

         total                .0851899      16.44        100
```

Statistics for row and column categories in symmetric norm. (x 1000)

Categories	overall			dimension 1			dimension 2		
	mass	qualt	inert	coord	sqcor	contr	coord	sqcor	contr
rows									
senior mngr	57	893	3	126	92	3	612	800	214
junior mngr	93	991	12	-495	526	84	769	465	551
senior empl	264	1000	38	728	999	512	34	1	3
junior empl	456	1000	26	-446	942	331	-183	58	152
secretary	130	999	6	385	865	70	-249	133	81
columns									
none	316	1000	49	752	994	654	96	6	29
light	233	984	7	-190	327	31	-446	657	463
medium	321	983	13	-375	982	166	-23	1	2
heavy	130	995	16	-562	684	150	625	310	506

◁

▷ Example 7

The command `camat` may also be used for a correspondence analysis of nonfrequency data. The data should be non-negative, with strictly positive margins. An example are the compositional data on the distribution of government R&D funds over 11 areas in five European countries in 1989; the data are listed in Greenacre (1993, 82). The expenditures are scaled to 1000 within country, to focus the analysis on the intranational distribution policies. Moreover, with absolute expenditures, small countries, such as the Netherlands, would have been negligible in the analysis.

We enter the data as a Stata matrix. The command `matrix input` allows us to input row entries separated by blanks, rather than by commas; rows are separated by the backward slash (\).

```
. matrix input RandD = (
    18   19  14  14    6 \
    12   34   4  15   31 \
    44   33  36  58   25 \
    37   88  67 101   40 \
    42   20  36  28   43 \
    90  156 107 224  176 \
    28   50  59  88   28 \
   165  299 120 303  407 \
    48  128 147  62  103 \
   484  127 342  70   28 \
    32   46  68  37  113 )
. matrix colnames RandD = Britain West_Germany France Italy Netherlands
. matrix rownames RandD = earth_exploration pollution human_health
                          energy agriculture industry space university
                          nonoriented defense other
```

We perform a CA, suppressing the voluminous row- and column-point statistics. We want to show a biplot, and therefore we select symmetric normalization.

```
. camat RandD, dim(2) norm(symm) rowname(source) colname(country) norowpoints
> nocolpoints plot
```

```
Correspondence analysis                         Number of obs    =      5000
                                                Pearson chi2(40) =   1321.55
                                                Prob > chi2      =    0.0000
                                                Total inertia    =    0.2643
        11 active rows                          Number of dim.   =         2
        5 active columns                        Expl. inertia (%) =    89.08
```

Dimensions	singular values	principal inertia	chi2	percent	cumul percent
dim 1	.448735	.2013631	1006.82	76.18	76.18
dim 2	.1846219	.0340852	170.43	12.90	89.08
dim 3	.1448003	.0209671	104.84	7.93	97.01
dim 4	.0888532	.0078949	39.47	2.99	100.00
total		.2643103	1321.55	100	

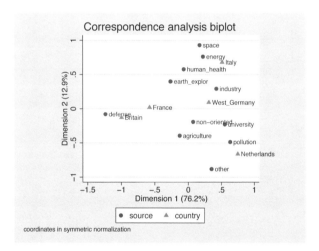

The two dimensions account for 89% of the inertia in this example, justifying an interpretation of the biplot. Let us focus on the position of the Netherlands. The orientation of the Netherlands from the origin is in the same direction as the orientation of pollution and university from the origin, indicating that the Netherlands spends more on academic research and on research to reduce environmental pollution than the average country. Earth exploration and human health are in the opposite direction, indicating investments much lower than average in these areas. Industry and agriculture are approximately orthogonal to the orientation of the Netherlands, indicating average investments by the Netherlands in these areas. Notice that Britain and France have big military investments, while Germany and Italy have more of an industrial orientation.

◁

❑ Technical Note

You may have noticed that the interpretation of the biplot is not fully in line with what we easily see in the row and column profiles—surprisingly, Greenacre does not seem to feel the need to comment on this. Why is this the case? The clue is in the statistics we did not show. While the two dimensions account for 90% of the total inertia, this does not mean that all rows and columns are approximated to this extent. There are some row and column categories that are not well described in two dimensions. For instance, the quality of the source categories nonoriented, agriculture, and earth_exploration are only 0.063, 0.545, and 0.584, respectively, indicating that these rows are poorly represented in a two-dimensional space. The quality of West_Germany is also rather low at 0.577. Adding a third dimension improves the quality of the category nonoriented but hardly affects the other two problematic categories. This can only be seen from the squared correlations between the third dimension and the profiles of the row and column categories—these correlations are small for all categories but nonoriented. Thus nonoriented does not seem to really belong with the other categories and should probably be omitted from the analysis.

❏

Saved Results

Let r be the number of rows, c be the number of columns, and f be the number of retained dimensions. ca and camat save in e():

Scalars

e(N)	number of observations
e(f)	number of dimensions (factors, axes); maximum of $\min(r-1, c-1)$
e(inertia)	total inertia = e(X2)/e(N)
e(pinertia)	inertia explained by e(f) dimensions
e(X2)	chi-squared statistic
e(X2_df)	degrees of freedom $(r-1)(c-1)$
e(X2_p)	p-value for e(X2)

Macros

e(cmd)	ca (even for camat)
e(title)	correspondence analysis
e(varlist)	the row and column variable names (ca only)
e(Cname)	name for columns
e(Rname)	name for rows
e(norm)	normalization method
e(sv_unique)	1 if the singular values are unique, 0 otherwise
e(estat_cmd)	ca_estat
e(predict)	ca_p
e(properties)	nob noV eigen
e(wtype)	weight type (ca only)
e(wexp)	weight expression (ca only)

Matrices

e(Ccoding)	column categories $(1 \times c)$ (ca only)
e(Rcoding)	row categories $(1 \times r)$ (ca only)
e(GSC)	column statistics $(c \times 3(1+f))$
e(GSR)	row statistics $(r \times 3(1+f))$
e(TC)	normalized column coordinates $(c \times f)$
e(TR)	normalized row coordinates $(r \times f)$
e(SV)	singular values $(1 \times f)$
e(C)	column coordinates $(c \times f)$
e(R)	row coordinates $(r \times f)$
e(c)	column mass (margin) $(c \times 1)$
e(r)	row mass (margin) $(r \times 1)$
e(P)	analyzed matrix $(r \times c)$
e(GSC_supp)	supplementary column statistics
e(GSR_supp)	supplementary row statistics
e(PC_supp)	principal coordinates supplementary column points
e(PR_supp)	principal coordinates supplementary row points
e(TC_supp)	normalized coordinates supplementary column points
e(TR_supp)	normalized coordinates supplementary row points

Functions

e(sample)	marks estimation sample (ca only)

Methods and Formulas

ca and camat are implemented as ado-files.

Our presentation of simple correspondence analysis follows Greenacre (1984, 83–125); see also Blasius and Greenacre (1994) and Rencher (2002, 514–530). See Greenacre and Blasius (1994) for a concise presentation of CA from a computational perspective. Simple CA seeks a geometric representation of the rows and column of a (two-mode) matrix with non-negative entries in a common low-dimensional space so that chi-squared distances between the rows and between the columns are well approximated by the Euclidean distances in the common space.

Let N be an $I \times J$ matrix with non-negative entries and strictly positive margins. N may be frequencies of a two-way cross-tabulation, but this is not assumed in most of CA. Let $n = N_{++}$ be the overall sum of N_{ij} ("number of observations"). Define the *correspondence table* as the matrix $\mathbf{P}$ where $P_{ij} = N_{ij}/n$, so the overall sum of P_{ij} is $P_{++} = 1$. Let $\mathbf{r} = \mathbf{P} \mathbf{1}$ be the row margins, also known as the *row masses*, with elements $r_i > 0$. Similarly, $\mathbf{c} = \mathbf{P}'\mathbf{1}$ contains the column margins or *column masses*, with elements $c_j > 0$.

CA is defined in terms of the generalized singular value decomposition (GSVD) of $\mathbf{P} - \mathbf{rc}'$ with respect to the inner products normed by $\mathbf{D}_r^{-1}$ and $\mathbf{D}_c^{-1}$, where $\mathbf{D}_r = \mathrm{diag}(\mathbf{r})$ and $\mathbf{D}_c = \mathrm{diag}(\mathbf{c})$. The GSVD can be expressed in terms of the orthonormal (or standard) SVD of the standardized residuals

$$\mathbf{Z} = \mathbf{D}_r^{-\frac{1}{2}}(\mathbf{P} - \mathbf{rc}')\mathbf{D}_c^{-\frac{1}{2}} \quad \text{with elements} \quad Z_{ij} = \frac{P_{ij} - r_i c_j}{\sqrt{r_i c_j}}$$

Denote by $\mathbf{Z} = \mathbf{R}\mathbf{\Lambda}\mathbf{C}'$ the SVD of $\mathbf{Z}$ with $\mathbf{R}'\mathbf{R} = \mathbf{C}'\mathbf{C} = \mathbf{I}$ and $\mathbf{\Lambda}$ a diagonal matrix with singular values in decreasing order. ca displays a warning message if $\mathbf{Z}$ has common singular values.

The *total principal inertia* of the correspondence table $\mathbf{P}$ is defined as $X^2/n = \sum_{i,j} Z_{ij}^2$, where X^2 is Pearson's chi-squared statistic. We can express the inertia of $\mathbf{P}$ in terms of the singular values of $\mathbf{Z}$:

$$\text{inertia} = \frac{1}{n} X^2 = \sum_{k=1}^{\min(I,J)} \lambda_k^2$$

The inertia accounted for by d dimensions is $\sum_{k=1}^{d} \lambda_k^2$. The fraction of inertia accounted for (explained) by the d dimensions is defined as

$$\text{explained inertia} = \frac{\sum_{k=1}^{d} \lambda_k^2}{\sum_{k=1}^{\min(I-1,J-1)} \lambda_k^2}$$

Principal row ($\widetilde{R}_{ik}$) and principal column coordinates ($\widetilde{C}_{jk}$) are defined as

$$\widetilde{R}_{ik} = \frac{R_{ik}\lambda_k}{\sqrt{r_i}} = (\mathbf{D}_r^{-\frac{1}{2}} \mathbf{R}\mathbf{\Lambda})_{ik} \qquad \widetilde{C}_{jk} = \frac{C_{jk}\lambda_k}{\sqrt{c_j}} = (\mathbf{D}_c^{-\frac{1}{2}} \mathbf{C}\mathbf{\Lambda})_{jk}$$

The α-normalized row and column coordinates are defined as

$$R_{ik}^{(\alpha)} = \frac{R_{ik}\lambda_k^{\alpha}}{\sqrt{r_i}} \qquad C_{jk}^{(\alpha)} = \frac{C_{jk}\lambda_k^{1-\alpha}}{\sqrt{c_j}}$$

The row principal coordinates are obtained with $\alpha = 1$. The column principal coordinates are obtained with $\alpha = 0$. The symmetric coordinates are obtained with $\alpha = 1/2$.

Decomposition of inertia by rows ($\text{In}^{(r)}$) and by columns ($\text{In}^{(c)}$) is defined as

$$\text{In}_i^{(r)} = \sum_{j=1}^{J} Z_{ij}^2 \qquad \text{In}_j^{(c)} = \sum_{i=1}^{I} Z_{ij}^2$$

Quality of subspace approximations for the row and column categories are defined as

$$Q_i^{(r)} = \frac{r_i}{\text{In}_i^{(r)}} \sum_{k=1}^{d} \widetilde{R}_{ik}^2 \qquad Q_j^{(c)} = \frac{c_j}{\text{In}_j^{(c)}} \sum_{k=1}^{d} \widetilde{C}_{jk}^2$$

If $d = \min(I-1, J-1)$, the quality index satisfies $Q_i^{(r)} = Q_j^{(c)} = 1$.

CA provides a number of diagnostics for a more detailed analysis of inertia: What do the categories contribute to the inertia explained by the dimensions, and what do the dimensions contribute to the inertia explained for the categories?

The relative contributions of row i ($G_{ik}^{(r)}$) and of column j ($G_{jk}^{(c)}$) to the inertia of principal dimension k are defined as

$$G_{ik}^{(r)} = \frac{r_i \widetilde{R}_{ik}}{\lambda_k^2} \qquad G_{jk}^{(c)} = \frac{c_j \widetilde{C}_{jk}}{\lambda_k^2}$$

Note that $G_{+k}^{(r)} = G_{+k}^{(c)} = 1$.

The correlations $H_{ik}^{(r)}$ of the ith row profile and kth principal row dimension and, analogously, $H_{jk}^{(c)}$ for columns are

$$H_{ik}^{(r)} = \frac{r_i}{\text{In}_i^{(r)}} \widetilde{R}_{ik}^2 \qquad H_{jk}^{(c)} = \frac{c_j}{\text{In}_j^{(c)}} \widetilde{C}_{jk}^2$$

We now define the quantities returned by the `estat` subcommands after `ca`. The row profiles are $\mathbf{U} = \mathbf{D}_r^{-1}\mathbf{P}$. The chi-squared distance between rows i_1 and i_2 of $\mathbf{P}$ is defined as the Mahalanobis distance between the respective row profiles $\mathbf{U}_{i_1}$ and $\mathbf{U}_{i_2}$ with respect to $\mathbf{D}_c$,

$$(\mathbf{U}_{i_1} - \mathbf{U}_{i_2})\mathbf{D}_c^{-1}(\mathbf{U}_{i_1} - \mathbf{U}_{i_2})'$$

The column profiles and the chi-squared distances between columns are defined analogously. The chi-squared distances for the approximated correspondence table are defined analogously in terms of $\widehat{\mathbf{P}}$ (see below).

The fitted or reconstructed values $\widehat{P}_{ij}$ are

$$\widehat{P}_{ij} = r_i c_j \left(1 + \lambda_k^{-1} \sum_{k=1}^{d} \widetilde{R}_{ik} \widetilde{C}_{jk} \right)$$

References

Blasius J. and M. J. Greenacre. 1994. Computation of correspondence analysis. In *Correspondence Analysis in the Social Sciences—Recent Developments and Applications*, ed. M. Greenacre and J. Blasius. London: Academic Press.

Cox, T. F. and M. A. A. Cox. 2001. *Multidimensional Scaling*. 2nd ed. Boca Raton, FL: Chapman & Hall.

Gower, J. C. and D. J. Hand. 1996. *Biplots*. London: Chapman & Hall.

Greenacre, M. J. 1984. *Theory and Applications of Correspondence Analysis*. London: Academic Press.

——. 1993. *Correspondence Analysis in Practice*. London: Academic Press.

Greenacre, M. J. and J. Blasius (eds.). 1994. *Correspondence Analysis in the Social Sciences—Recent Developments and Applications*. London: Academic Press.

Kerm, P. van. 1998. sg78: Simple and multiple correspondence analysis in Stata. *Stata Technical Bulletin* 42: 32–37. Reprinted in *Stata Technical Bulletin Reprints*, vol. 7, pp. 210–217.

Rencher, A. C. 2002. *Methods of Multivariate Analysis*. 2nd ed. New York: Wiley.

Weller, S.C. and A. K. Romney. 1990. *Metric Scaling: Correspondence Analysis*. Newbury Park, CA: Sage.

Also See

Complementary:	[MV] **ca postestimation**
Related:	[R] **tabulate twoway**
Background:	[U] **11.1.10 Prefix commands**,
	[U] **20 Estimation and postestimation commands**

Title

> **ca postestimation** — Postestimation tools for ca and camat

Description

The following postestimation commands are of special interest after `ca` and `camat`:

command	description
cabiplot	biplot of row and column points
caprojection	CA dimension projection plot
estat coordinates	display row and column coordinates
estat distances	display χ^2 distances between row and column profiles
estat inertia	display inertia contributions of the individual cells
estat profiles	display row and column profiles
† estat summarize	estimation sample summary
estat table	display fitted correspondence table

† `estat summarize` is not available after `camat`.

For information about these commands, see below.

In addition, the following standard postestimation commands are available:

command	description
* estimates	cataloging estimation results
† predict	fitted values, row coordinates, or column coordinates

* All `estimates` subcommands except `table` and `stats` are available.

† `predict` is not available after `camat`.

See the corresponding entries in the *Stata Base Reference Manual* for details.

Special-interest postestimation commands

`cabiplot` produces a plot of the row points or column points, or a biplot of the row and column points. In this plot, the (Euclidean) distances between row (column) points approximates the χ^2 distances between the associated row (column) profiles if the CA is properly normalized. Similarly, the association between a row and column point is approximated by the inner product of vectors from the origin to the respective points (see [MV] **ca**).

`caprojection` produces a line plot of the row and column coordinates. The goal of this graph is to show the ordering of row and column categories on each principal dimension of the analysis. Each principal dimension is represented by a vertical line; markers are plotted on the lines where the row and column categories project onto the dimensions.

`estat coordinates` displays the row and column coordinates.

estat distances displays the χ^2 distances between the row profiles and between the column profiles. In addition, the χ^2 distances between the row and column profiles to the respective centers (marginal distributions) are displayed. Optionally, the fitted profiles rather than the observed profiles are used.

estat inertia displays the inertia (χ^2/N) contributions of the individual cells.

estat profiles displays the row and column profiles; the row (column) profile is the conditional distribution of the row (column) given the column (row). This is equivalent to specifying the row and column options with the tabulate command.

estat summarize displays summary information about the row and column variables over the estimation sample.

estat table displays the fitted correspondence table. Optionally, the observed "correspondence table" and the expected table under independence are displayed.

Syntax for predict

predict $[$ *type* $]$ *newvar* $[$ *if* $]$ $[$ *in* $]$ $[$, *statistic* $]$

statistic	description
Main	
fit	fitted values; the default
rowscore(#)	row score for dimension #
colscore(#)	column score for dimension #

predict is not available after camat.

Options for predict

⌐ Main ⌐

fit specifies that fitted values for the correspondence analysis model be computed. fit displays the fitted values p_{ij} according to the correspondence analysis model. fit is the default.

rowscore(#) generates the row score for dimension #, i.e., the appropriate elements from the normalized row coordinates.

colscore(#) generates the column score for dimension #, i.e., the appropriate elements from the normalized column coordinates.

Syntax for estat

Display row and column coordinates

estat coordinates $[$, norow nocolumn format(%*fmt*) $]$

Display chi-squared distances between row and column profiles

estat distances $[$, norow nocolumn approx format(%*fmt*) $]$

Display inertia contributions of cells

    estat <u>in</u>ertia [, <u>tot</u>al <u>nosca</u>le <u>form</u>at(%*fmt*)]

Display row and column profiles

    estat <u>pro</u>files [, norow <u>nocol</u>umn <u>form</u>at(%*fmt*)]

Display summary information

    estat <u>su</u>mmarize [, <u>lab</u>els <u>noh</u>eader <u>nowe</u>ights]

Display fitted correspondence table

    estat <u>t</u>able [, fit obs <u>ind</u>ependence <u>nosca</u>le <u>form</u>at(%*fmt*)]

options	description
Main	
norow	suppress display of row results
<u>noco</u>lumn	suppress display of column results
<u>form</u>at(%*fmt*)	display format; default is format(%9.4f)
<u>approx</u>	display distances between fitted (approximated) profiles
<u>total</u>	add row and column margins
<u>nosca</u>le	display χ^2 contributions; default is inertias $= \chi^2/N$ (with estat inertia)
<u>lab</u>els	display variable labels
<u>noh</u>eader	suppress the header
<u>nowe</u>ights	ignore weights
fit	display fitted values from correspondence analysis model
obs	display correspondence table ("observed table")
<u>ind</u>ependence	display expected values under independence
<u>nosca</u>le	suppress scaling of entries to 1 (with estat table)

Options for estat

    ⌐ Main ⌐

norow, an option used with estat coordinates, estat distances, and estat profiles, suppresses the display of row results.

nocolumn, an option used with estat coordinates, estat distances, and estat profiles, suppresses the display of column results.

format(%*fmt*), an option used with many of the subcommands of estat, specifies the display format for the matrix, e.g., format(%8.3f). The default is format(%9.4f).

approx, an option used with estat distances, computes distances between the fitted profiles. The default is to compute distances between the observed profiles.

total, an option used with estat inertia, adds row and column margins to the table of inertia or χ^2 (χ^2/N) contributions.

noscale, as an option used with estat inertia, displays χ^2 contributions rather than inertia ($= \chi^2/N$) contributions. (See below for the description of noscale with estat table.)

labels, an option used with estat summarize, displays variable labels.

noheader, an option used with estat summarize, suppresses the header.

noweights, an option used with estat summarize, ignores the weights, if any. The default when weights are present is to perform a weighted summarize on all variables except the weight variable itself. An unweighted summarize is performed on the weight variable.

fit, an option used with estat table, displays the fitted values for the correspondence analysis model; see the description of fit under *Options for predict* for further details. fit is implied if obs and independence are not specified.

obs, an option used with estat table, displays the observed table with non-negative entries (the "correspondence table").

independence, an option used with estat table, displays the expected values p_{ij} assuming independence of the rows and columns, $p_{ij} = r_i c_j$, where r_i is the mass of row i and c_j is the mass of column j.

noscale, as an option used with estat table, normalizes the displayed tables to the sum of the original table entries. The default is to scale the tables to overall sum 1. (See above for the description of noscale with estat inertia.)

Syntax for cabiplot

cabiplot [, *options*]

options	description
Main	
dim(# #)	the two dimensions to be displayed; default is dim(2 1)
norow	suppress row coordinates
nocolumn	suppress column coordinates
xnegate	negate the data relative to the x-axis
ynegate	negate the data relative to the y-axis
maxlength(#)	maximum number of characters for labels; default is maxlength(12)
Rows	
rowopts(*row_opts*)	affect rendition of rows
Columns	
colopts(*col_opts*)	affect rendition of columns
Y-Axis, X-Axis, Title, Caption, Overall	
twoway_options	any options other than by() documented in [G] *twoway_options*

(Continued on next page)

row_opts and *col_opts*	description
plot_options	change look of markers (color, size, etc.) and look or position of marker labels
suppopts(*plot_options*)	change look of supplementary markers and look or position of supplementary marker labels

plot_options	description
marker_options	change look of markers (color, size, etc.)
marker_label_options	add marker labels; change look or position

Options for cabiplot

dim(*# #*) identifies the dimensions to be displayed. For instance, dim(3 2) plots the third dimension (vertically) versus the second dimension (horizontally). The dimension number cannot exceed the number of extracted dimensions. The default is dim(2 1).

norow suppresses plotting of row points.

nocolumn suppresses plotting of column points.

xnegate specifies that dimension-one (x-axis) values are to be negated (multiplied by negative one).

ynegate specifies that dimension-two (y-axis) values are to be negated (multiplied by negative one).

maxlength(*#*) specifies the maximum number of characters for row and column labels; the default is maxlength(12).

rowopts(*row_opts*) affects the rendition of rows. The following *row_opts* are allowed:

plot_options affect the rendition of row markers, including their shape, size, color, and outline (see [G] ***marker_options***) and specify if and how the row markers are to be labeled (see [G] ***marker_label_options***).

suppopts(*plot_options*) affects supplementary markers and supplementary marker labels; see above for description of *plot_options*.

colopts(*col_opts*) affects the rendition of columns. The following *col_opts* are allowed:

plot_options affect the rendition of column markers, including their shape, size, color, and outline (see [G] ***marker_options***) and specify if and how the column markers are to be labeled (see [G] ***marker_label_options***).

suppopts(*plot_options*) affects supplementary markers and supplementary marker labels; see above for description of *plot_options*.

Y-Axis, X-Axis, Title, Caption, Overall

twoway_options are any of the options documented in [G] ***twoway_options***, excluding by(). These include options for titling the graph (see [G] ***title_options***) and options for saving the graph to disk (see [G] ***saving_option***).

cabiplot automatically adjusts the aspect ratio based on the range of the data and ensures that the axes are balanced. As an alternative, the *twoway_option* aspectratio() can be used to override the default aspect ratio. cabiplot accepts the aspectratio() option as a suggestion only and will override it when necessary to produce plots with balanced axes, i.e., distance on the x-axis equals distance on the y-axis.

twoway_options, such as xlabel(), xscale(), ylabel(), and yscale() should be used with caution. These options are accepted but may have unintended side effects on the aspect ratio.

Syntax for caprojection

caprojection [, *options*]

options	description
Main	
dim(*numlist*)	dimensions to be displayed; default is all
norow	suppress row coordinates
nocolumn	suppress column coordinates
alternate	alternate labels
maxlength(#)	number of characters displayed for labels; default is maxlength(12)
combine_options	affect the rendition of the combined column and row graphs
Rows	
rowopts(*row_opts*)	affect rendition of rows
Columns	
colopts(*col_opts*)	affect rendition of columns
Y-Axis, X-Axis, Title, Caption, Overall	
twoway_options	any options other than by() documented in [G] ***twoway_options***

row_opts and *col_opts*	description
plot_options	change look of markers (color, size, etc.) and look or position of marker labels
suppopts(*plot_options*)	change look of supplementary markers and look or position of supplementary marker labels

plot_options	description
marker_options	change look of markers (color, size, etc.)
marker_label_options	add marker labels; change look or position

Options for caprojection

⌐ Main ┌───

dim(*numlist*) identifies the dimensions to be displayed. By default, all dimensions are displayed.

norow suppresses plotting of rows.

nocolumn suppresses plotting of columns.

alternate causes adjacent labels to alternate sides.

maxlength(#) specifies the maximum number of characters for row and column labels; the default is maxlength(12).

combine_options affect the rendition of the combined plot; see [G] **graph combine**. *combine_options* may not be specified with either norow or nocolumn.

⌐ Rows ┌───

rowopts(*row_opts*) affects the rendition of rows. The following *row_opts* are allowed:

> *plot_options* affect the rendition of row markers, including their shape, size, color, and outline (see [G] *marker_options*) and specify if and how the row markers are to be labeled (see [G] *marker_label_options*).

> suppopts(*plot_options*) affects supplementary markers and supplementary marker labels; see above for description of *plot_options*.

⌐ Columns ┌──

colopts(*col_opts*) affects the rendition of columns. The following *col_opts* are allowed:

> *plot_options* affect the rendition of column markers, including their shape, size, color, and outline (see [G] *marker_options*) and specify if and how the column markers are to be labeled (see [G] *marker_label_options*).

> suppopts(*plot_options*) affects supplementary markers and supplementary marker labels; see above for description of *plot_options*.

⌐ Y-Axis, X-Axis, Title, Caption, Overall ┌───────────────────────────────

twoway_options are any of the options documented in [G] *twoway_options*, excluding by(). These include options for titling the graph (see [G] *title_options*) and options for saving the graph to disk (see [G] *saving_option*).

Remarks

Remarks are presented under the headings

> *Postestimation statistics*
> *Postestimation graphs*
> *Predicting new variables*

Postestimation statistics

After you conduct a correspondence analysis, there are a number of additional tables to help you understand and interpret your results. Some of these tables resemble tables produced by other Stata commands but are provided as part of the `ca` postestimation suite of commands for a unified presentation style.

▷ Example 1

We continue with the classic example of correspondence analysis, namely the data on smoking in organizations. We extract only one dimension.

```
. use http://www.stata-press.com/data/r9/ca_smoking
. ca rank smoking, dim(1)
```

Correspondence analysis

			Number of obs	=	193
			Pearson chi2(12)	=	16.44
			Prob > chi2	=	0.1718
			Total inertia	=	0.0852
5 active rows			Number of dim.	=	1
4 active columns			Expl. inertia (%)	=	87.76

Dimensions	singular values	principal inertia	chi2	percent	cumul percent
dim 1	.2734211	.0747591	14.43	87.76	87.76
dim 2	.1000859	.0100172	1.93	11.76	99.51
dim 3	.0203365	.0004136	0.08	0.49	100.00
total		.0851899	16.44	100	

Statistics for row and column categories in symmetric normalization

Categories	mass	overall quality	inertia	dimension_1 coord	sqcorr	contrib
rank						
senior mngr	0.057	0.092	0.003	0.126	0.092	0.003
junior mngr	0.093	0.526	0.012	-0.495	0.526	0.084
senior empl	0.264	0.999	0.038	0.728	0.999	0.512
junior empl	0.456	0.942	0.026	-0.446	0.942	0.331
secretary	0.130	0.865	0.006	0.385	0.865	0.070
smoking						
none	0.316	0.994	0.049	0.752	0.994	0.654
light	0.233	0.327	0.007	-0.190	0.327	0.031
medium	0.321	0.982	0.013	-0.375	0.982	0.166
heavy	0.130	0.684	0.016	-0.562	0.684	0.150

CA analyzes the similarity of row and of column categories by comparing the row profiles and the column profiles—some may prefer to talk about conditional distributions in the case of a two-way frequency distribution, but CA is not restricted to this type of data.

(Continued on next page)

```
. estat profiles
```
Row profiles (rows normalized to 1)

	none	light	medium	heavy	mass
senior mngr	0.3636	0.1818	0.2727	0.1818	0.0570
junior mngr	0.2222	0.1667	0.3889	0.2222	0.0933
senior empl	0.4902	0.1961	0.2353	0.0784	0.2642
junior empl	0.2045	0.2727	0.3750	0.1477	0.4560
secretary	0.4000	0.2400	0.2800	0.0800	0.1295
mass	0.3161	0.2332	0.3212	0.1295	

Column profiles (columns normalized to 1)

	none	light	medium	heavy	mass
senior mngr	0.0656	0.0444	0.0484	0.0800	0.0570
junior mngr	0.0656	0.0667	0.1129	0.1600	0.0933
senior empl	0.4098	0.2222	0.1935	0.1600	0.2642
junior empl	0.2951	0.5333	0.5323	0.5200	0.4560
secretary	0.1639	0.1333	0.1129	0.0800	0.1295
mass	0.3161	0.2332	0.3212	0.1295	

The tables also include the row and column masses—marginal probabilities. Two row categories are similar to the extent that their row profiles (i.e., their distribution over the columns) are the same. Similar categories could be collapsed without distorting the information in the table. In CA, similarity or dissimilarity of the row categories is expressed in terms of the χ^2 distances between the rows. These are sums of squares, weighted with the inverse of the column masses. Thus a difference is counted "heavier" (inertia!), the smaller the respective column mass. In the table, we also add the χ^2 distances of the rows to the row centroid, i.e., to the marginal distribution. This allows us to easily see which row categories are similar to each other as well as which row categories are similar to the population.

```
. estat distances, nocolumn
```
Chi2 distances between the row profiles

rank	junior_~r	senior_~l	junior_~l	secretary	center
senior_mngr	0.3448	0.3721	0.3963	0.3145	0.2166
junior_mngr		0.6812	0.3044	0.5622	0.3569
senior_empl			0.6174	0.2006	0.3808
junior_empl				0.4347	0.2400
secretary					0.2162

We see that senior employees are especially dissimilar from junior managers in terms of their smoking behavior but are rather similar to secretaries. Also the senior employees are least similar to the average staff member among all staff categories.

One of the goals of CA is to come up with a low-dimensional representation of the rows and columns in a common space. One way to see the adequacy of this representation is to inspect the implied approximation for the χ^2 distances—are the similarities between the row categories and between the column categories adequately represented in lower dimensions?

```
. estat distances, nocolumn approx
```

Chi2 distances between the dim=1 approximations of the row profiles

rank	junior_~r	senior_~l	junior_~l	secretary	center
senior_mngr	0.3247	0.3148	0.2987	0.1353	0.0658
junior_mngr		0.6396	0.0260	0.4600	0.2590
senior_empl			0.6135	0.1795	0.3806
junior_empl				0.4340	0.2330
secretary					0.2011

Some of the row distances are obviously poorly approximated, while the quality of other approximations is hardly affected. The dissimilarity in smoking behavior between junior managers and junior employees is particularly poorly represented in one dimension. From the CA with two dimensions, it is clear that the second dimension is crucial to adequately represent the senior managers and the junior managers. By itself, this does not explain where the one-dimensional approximation fails; for this, we would have to take a closer look at the representation of the smoking categories, as well.

A correspondence analysis can also be seen as equivalent to fitting the model

$$P_{ij} = r_i c_j (1 + R_{i1} C_{j1} + R_{i2} C_{j2} + \cdots)$$

to the correspondence table $\mathbf{P}$ by some sort of least squares, with parameters r_i, c_j, R_{ij}, and C_{jk}. We may compare the (observed) table $\mathbf{P}$ with the fitted table $\widehat{\mathbf{P}}$ to assess goodness-of-fit informally. In this case we extract only one dimension, and so the fitted table is

$$\widehat{P}_{ij} = r_i c_j (1 + \widehat{R}_{i1} \widehat{C}_{j1})$$

with $\mathbf{R}$ and $\mathbf{C}$ the coordinates in symmetric (or row principal or column principal) normalization. We display the observed and fitted tables.

```
. estat table, fit obs
```

Correspondence table (normalized to overall sum = 1)

	none	light	medium	heavy
senior_mngr	0.0207	0.0104	0.0155	0.0104
junior_mngr	0.0207	0.0155	0.0363	0.0207
senior_empl	0.1295	0.0518	0.0622	0.0207
junior_empl	0.0933	0.1244	0.1710	0.0674
secretary	0.0518	0.0311	0.0363	0.0104

Approximation for dim = 1 (normalized to overall sum = 1)

	none	light	medium	heavy
senior_mngr	0.0197	0.0130	0.0174	0.0069
junior_mngr	0.0185	0.0238	0.0355	0.0154
senior_empl	0.1292	0.0531	0.0617	0.0202
junior_empl	0.0958	0.1153	0.1710	0.0738
secretary	0.0528	0.0280	0.0356	0.0132

Interestingly, some categories (e.g., the junior employees, the nonsmokers, and the medium smokers) are very well represented in one dimension, whereas the quality of the fit of other categories is rather poor. This can, of course, also be inferred from the quality column in the ca output. We would consider the fit unsatisfactory and would refit the model with a second dimension.

◁

❑ Technical Note

If the data are two-way cross-classified frequencies, as with ca, it may make sense to assume that the data are multinomial distributed, and the parameters can be estimated by maximum likelihood. The estimator has well-established properties in contrast to the estimation method commonly used in CA. One advantage is that sampling variability, e.g., in terms of standard errors of the parameters, can be easily assessed. Also, the likelihood-ratio test against the saturated model may be used to select the number of dimensions to be extracted. See Van der Heijden and De Leeuw (1985).

❑

Postestimation graphs

In [MV] **ca**, we showed that plots can be obtained simply by specifying the plot option during estimation (or replay). If the default plot is not exactly what you want, the cabiplot postestimation command provides control over the appearance of the plot.

▷ Example 2

For instance, if we constructed a CA in row principal normalization, we would only want to look at the (points for the) row categories, omitting the column categories. Recall that in this normalization, the Euclidean distances between the row points approximate the χ^2 distances between the corresponding row profiles, but the Euclidean distances between the column categories are a distortion of the χ^2 distances of the column profiles. We can use cabiplot with the nocolumn option to suppress the graphing of the column points.

```
. quietly ca rank smoking, norm(principal)
. cabiplot, nocolumn legend(on label(1 rank))
```

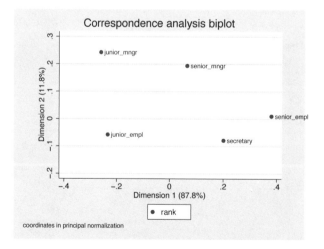

The default graph would not have provided a legend, so we included legend(on label(1 rank)) to produce one. We see that secretaries have smoking behavior that is rather similar to senior employees, but rather dissimilar to the junior managers, with the other two ranks taking intermediate positions. Since we actually specified the principal normalization, we may also interpret the distances between the smoking categories as approximations to χ^2 distances.

```
. cabiplot, norow legend(on label(1 smoking))
```

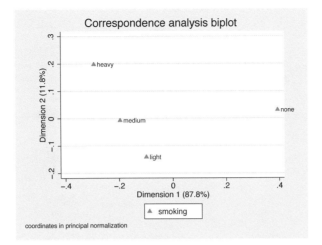

You may not like the orientation of the dimensions. For instance, in this plot, the smokers are on the left and the nonsmokers are on the right. It is more natural to locate the nonsmokers on the left and the smokers on the right so that smoking increases from left to right. This is accomplished with the `xnegate` option.

```
. cabiplot, xnegate norow legend(on label(1 smoking))
```

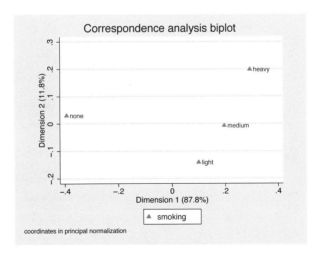

◁

❏ Technical Note

To see that negating is possible think in terms of the fitted values

$$\widehat{P}_{ij} = r_i c_j (1 + \widehat{R}_{i1}\widehat{C}_{j1} + \widehat{R}_{i2}\widehat{C}_{j2} + \cdots)$$

If the sign of the first column of $\mathbf{R}$ and $\mathbf{C}$ is changed at the same time, the fitted values are not affected. This is true for all CA statistics, and it holds true for other columns of $\mathbf{R}$ and $\mathbf{C}$ as well.

❏

▷ Example 3

Using the symmetric normalization allows us to display a biplot where row categories may be compared with column categories. We execute ca again, with the normalize(symmetric) option, but suppress the output. This normalization somewhat distorts the interpretation of the distances between row points (or column points) as approximations to χ^2 distances. Thus the similarity of the staff categories (or smoking categories) cannot be adequately assessed. However, this plot allows us to study the association between smoking and rank.

```
. quietly ca rank smoking, normalize(symmetric) dim(2)
. cabiplot, xline(0) yline(0)
```

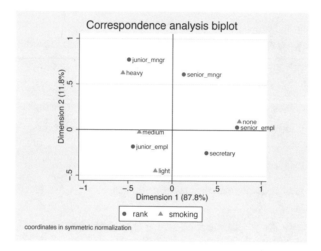

With this symmetric normalization, we do not interpret the distances between categories of smoking and rank. Rather, we have to think in terms of vectors from the origin. The inner product of vectors approximates the residuals from a model of independence of the rows and columns. The inner product depends on the lengths of the vectors and the (cosine of the) angle between the vectors. If the vectors point in the same direction, the residuals are positive—these row and column categories tend to occur together. In our example, we see that senior employees tend do be nonsmokers. If the vectors point in opposite directions, the residuals are negative—these row and column categories tend to be exclusive. In our example, senior managers tend not to be light smokers. Finally, if the vectors are orthogonal (± 90 degrees), the residuals tend to be small; that is, the observed frequencies correspond to what we expect under independence. For instance, junior managers have an average rate of light smoking.

Using various graph options, we can enhance the look of the plot.

```
. cabiplot, xline(0) yline(0) subtitle("Fictitious data, N = 193")
  legend(pos(2) ring(0) col(1) lab(1 Employee rank) lab(2 Smoking status))
```

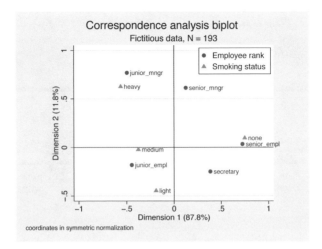

◁

▷ Example 4

`caprojection` produces a projection plot of the row and column coordinates after `ca` or `camat` and is especially useful if we think of CA as optimal scaling of the categories of the variables in order to maximize the correlations between the row and column variables. We continue where we left off with our previous example.

```
. caprojection
```

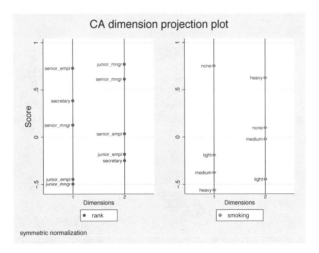

This example has relatively few categories, so we could visualize the orderings of the rows and columns from the previous biplots. However, CA is often used with larger problems, and in those cases, a projection plot is a useful presentation device.

◁

Predicting new variables

If you use `ca` to obtain the optimal scaling positions for the rows and columns, you may use `predict` to obtain the corresponding scores in the normalization used.

▷ Example 5

First, we obtain scores for the first dimension.

```
. quietly ca rank smoking, normalize(symmetric) dim(2)
. predict r1, row(1)
. predict c1, col(1)
. describe r1 c1
```

variable name	storage type	display format	value label	variable label
r1	float	%9.0g		rank score(1) in symmetric norm.
c1	float	%9.0g		smoking score(1) in symmetric norm.

```
. correlate r1 c1
(obs=193)
```

	r1	c1
r1	1.0000	
c1	0.2734	1.0000

The correlation of `r1` and `c1` is 0.2734, which equals the first singular value reported in the first panel by `ca`. In the same way, we may obtain scores for the second dimension.

```
. predict r2, row(2)
. predict c2, col(2)
. correlate r1 r2 c1 c2
(obs=193)
```

	r1	r2	c1	c2
r1	1.0000			
r2	−0.0000	1.0000		
c1	0.2734	0.0000	1.0000	
c2	0.0000	0.1001	0.0000	1.0000

The correlation between the row and column scores `r2` and `c2` for the second dimension is 0.1001, which is the same as the second singular value. Moreover, the row scores for dimensions 1 and 2 are not correlated, nor are the column scores.

◁

It is also possible to obtain the fitted values of the CA model

$$\pi_{ij} = r_i c_j (1 + R_{i1} C_{i1} + R_{i2} C_{i2})$$

where $\mathbf{R}$ and $\mathbf{C}$ are the row and column scales in symmetric normalization. These may be used, say, to compute fit measures, for instance, from the Cressie–Read power family to analyze the fit of the CA model (Weesie 1997).

Saved Results

estat distances saves in r():

Matrices

    r(Dcolumns) χ^2 distances between the columns and between the columns and the column center

    r(Drows) χ^2 distances between the rows and between the rows and the row center

estat inertia saves in r():

Matrices

    r(Q)   matrix of (squared) inertia (or χ^2) contributions

estat profiles saves in r():

Matrices

    r(Pcolumns)   column profiles (columns normalized to 1)

    r(Prows)   row profiles (rows normalized to 1)

estat tables saves in r():

Matrices

    r(Fit)   fitted (reconstructed) values

    r(Fit0)   fitted (reconstructed) values, assuming independence of row and column variables

    r(Obs)   correspondence table

Methods and Formulas

All postestimation commands listed above are implemented as ado-files. See [MV] **ca** for methods and formulas.

References

Van der Heijden, P. G. M. and J. de Leeuw. 1985. Correspondence analysis used complementary to loglinear analysis. *Psychometrika* 50: 429–447.

Weesie, J. 1997. sg68: Goodness-of-fit statistics for multinomial distributions. *Stata Technical Bulletin* 36: 26–28. Reprinted in *Stata Technical Bulletin Reprints*, vol. 6, pp. 183–186.

See [MV] **ca** for additional references.

Also See

Complementary:	[MV] **ca**,
	[R] **estimates**
Background:	[R] **estat**, [R] **predict**

Title

canon — Canonical correlations

Syntax

canon (*varlist*₁) (*varlist*₂) $\left[\,if\,\right]$ $\left[\,in\,\right]$ $\left[\,weight\,\right]$ $\left[\,,\,options\,\right]$

options	description
Model	
coefmatrix	output matrices of raw coefficients
stdcoef	output matrices of standardized coefficients
lc(#)	calculate the linear combinations for canonical correlation #
first(#)	calculate the linear combinations for the first # canonical correlations
noconstant	do not subtract means when calculating correlations
Reporting	
level(#)	set confidence level; default is level(95)
test(*numlist*)	display significance tests for the specified canonical correlations
notests	do not display tests
format(%*fmt*)	numerical format for coefficient matrices; default is format(%8.4f)

by and statsby may be used with canon; see [U] **11.1.10 Prefix commands**.
aweights and fweights are allowed; see [U] **11.1.6 weight**.
See [MV] **canon postestimation** for features available after estimation.

Description

canon estimates canonical correlations and provides the coefficients for calculating the appropriate linear combinations corresponding to those correlations.

canon typed without arguments redisplays previous estimation results.

Options

⌐ Model ⌐

coefmatrix specifies that the first part of the output contain the raw coefficients of the canonical correlations in matrix form. The default is to present the raw coefficients, standard errors, and confidence intervals in the standard estimation table. coefmatrix may not be specified with stdcoef.

stdcoef specifies that the first part of the output contain the standard coefficients of the canonical correlations in matrix form. The default is to present the raw coefficients, standard errors, and confidence intervals in the standard estimation table. stdcoef may not be specified with coefmatrix.

lc(#) specifies that linear combinations for canonical correlation # be calculated. By default, all are calculated.

first(*#*) specifies that linear combinations for the first *#* canonical correlations be calculated. By default, all are calculated.

noconstant specifies that means not be subtracted when calculating correlations.

level(*#*) specifies the confidence level, as a percentage, for confidence intervals. The default is level(95) or as set by set level; see [U] **20.6 Specifying the width of confidence intervals**. These "confidence intervals" are the result of an approximate calculation; see the technical note later in this entry.

test(*numlist*) specifies that significance tests of the canonical correlations in the *numlist* be displayed. Because of the nature of significance testing, if there are 3 canonical correlations, test(1) will test the significance of all 3 correlations, test(2) will test the significance of canonical correlations 2 and 3, and test(3) will test the significance of the third canonical correlation alone.

notests specifies that significance tests of the canonical correlation not be displayed.

format(*%fmt*) specifies the display format for numbers in coefficient matrices; see [D] **format**. format(%8.4f) is the default.

Remarks

Canonical correlations attempt to describe the relationships between two sets of variables. Given two sets of variables, $\mathbf{X} = (x_1, x_2, \ldots, x_K)$ and $\mathbf{Y} = (y_1, y_2, \ldots, y_L)$, the goal is to find linear combinations of $\mathbf{X}$ and $\mathbf{Y}$ so that the correlation between the linear combinations is as high as possible. That is, letting $\widehat{x}_1$ and $\widehat{y}_1$ be the linear combinations,

$$\widehat{x}_1 = \beta_{11}x_1 + \beta_{12}x_2 + \cdots + \beta_{1K}x_K$$
$$\widehat{y}_1 = \gamma_{11}y_1 + \gamma_{12}y_2 + \cdots + \gamma_{1L}y_L$$

you wish to find the maximum correlation between $\widehat{x}_1$ and $\widehat{y}_1$ as functions of the β's and the γ's. The second canonical correlation coefficient is defined as the ordinary correlation between

$$\widehat{x}_2 = \beta_{21}x_1 + \beta_{22}x_2 + \cdots + \beta_{2K}x_K \quad \text{and}$$
$$\widehat{y}_2 = \gamma_{21}y_1 + \gamma_{22}y_2 + \cdots + \gamma_{2L}y_L$$

This correlation is maximized subject to the constraints that $\widehat{x}_1$ and $\widehat{x}_2$, along with $\widehat{y}_1$ and $\widehat{y}_2$, are orthogonal and that $\widehat{x}_1$ and $\widehat{y}_2$, along with $\widehat{x}_2$ and $\widehat{y}_1$, are also orthogonal. The third and further correlations are defined similarly. There are $m = \min(K, L)$ such correlations.

Canonical correlation analysis originated with the work of Hotelling (1935, 1936). For an introduction, see Rencher (2002) or Johnson and Wichern (1992).

▷ Example 1

Consider two scientists trying to describe how "big" a car is. The first scientist takes physical measurements—the length, weight, headroom, and trunk space—whereas the second takes mechanical measurements—the engine displacement, mileage rating, gear ratio, and turning circle. Can they agree on a conceptual framework?

```
. use http://www.stata-press.com/data/r9/auto
(1978 Automobile Data)
```

```
. canon (length weight headroom trunk) (displ mpg gear_ratio turn)
Linear combinations for canonical correlations        Number of obs =        74
```

	Coef.	Std. Err.	t	P>|t|	[95% Conf.	Interval]
u1						
length	.0094779	.0060748	1.56	0.123	-.0026292	.021585
weight	.0010162	.0001615	6.29	0.000	.0006943	.0013381
headroom	.0351132	.0641755	0.55	0.586	-.0927884	.1630148
trunk	-.0022823	.0158555	-0.14	0.886	-.0338823	.0293176
v1						
displacement	.0053704	.0009541	5.63	0.000	.0034688	.007272
mpg	-.0461481	.0107324	-4.30	0.000	-.0675377	-.0247585
gear_ratio	.0329583	.1598716	0.21	0.837	-.2856654	.3515821
turn	.0793927	.0158975	4.99	0.000	.0477091	.1110762
u2						
length	.1441402	.0498307	2.89	0.005	.0448277	.2434526
weight	-.003663	.0013249	-2.76	0.007	-.0063036	-.0010225
headroom	-.370143	.5264215	-0.70	0.484	-1.419299	.6790136
trunk	-.0342739	.1300601	-0.26	0.793	-.2934833	.2249355
v2						
displacement	-.0125498	.0078267	-1.60	0.113	-.0281483	.0030487
mpg	-.0412777	.0880361	-0.47	0.641	-.2167333	.134178
gear_ratio	1.02798	1.311402	0.78	0.436	-1.58564	3.6416
turn	.3112667	.1304043	2.39	0.020	.0513713	.5711621
u3						
length	.0329362	.2837201	0.12	0.908	-.5325172	.5983897
weight	-.0009583	.0075437	-0.13	0.899	-.0159929	.0140764
headroom	1.536107	2.997275	0.51	0.610	-4.437454	7.509669
trunk	-.2135492	.7405205	-0.29	0.774	-1.689404	1.262306
v3						
displacement	.0191297	.0445624	0.43	0.669	-.0696831	.1079425
mpg	.0683398	.5012493	0.14	0.892	-.9306487	1.067328
gear_ratio	3.659567	7.466703	0.49	0.626	-11.22155	18.54068
turn	.0033034	.7424803	0.00	0.996	-1.476458	1.483065
u4						
length	.0211708	.4027108	0.05	0.958	-.7814306	.8237722
weight	.000683	.0107075	0.06	0.949	-.020657	.0220231
headroom	-.0440331	4.254316	-0.01	0.992	-8.522872	8.434806
trunk	-.32528	1.051091	-0.31	0.758	-2.4201	1.769541
v4						
displacement	-.0005364	.0632517	-0.01	0.993	-.1265968	.125524
mpg	.2478095	.7114704	0.35	0.729	-1.170149	1.665768
gear_ratio	-1.031143	10.5982	-0.10	0.923	-22.15332	20.09103
turn	.2240488	1.053872	0.21	0.832	-1.876316	2.324413

```
                                   (Standard errors estimated conditionally)
Canonical correlations:
  0.9476  0.3400  0.0634  0.0447
```

```
Tests of significance of all canonical correlations
                         Statistic      df1      df2           F     Prob>F
       Wilks' lambda      .0897314       16  202.271     15.1900     0.0000 a
       Pillai's trace      1.01956       16      276      5.9009     0.0000 a
Lawley-Hotelling trace     8.93344       16      258     36.0129     0.0000 a
   Roy's largest root     8.79667        4       69    151.7426     0.0000 u
```

e = exact, a = approximate, u = upper bound on F

By default, `canon` uses a standard estimation table to report the linear combinations corresponding to the canonical correlations, reports the canonical correlations, and finally, reports the tests of significance of all canonical correlations. The two views on car size are closely related: the best linear combination of the physical measurements is correlated at almost 0.95 with the best linear combination of the mechanical measurements. All the tests are significant.

If we are interested in seeing the standardized coefficients instead of the raw coefficients, we can use the `stdcoef` option on replay, which gives the standardized coefficients in matrix form. We specify the `notests` option to suppress the display of tests this time.

```
. use http://www.stata-press.com/data/r9/auto, clear
(1978 Automobile Data)
. canon, stdcoef notests

Canonical correlation analysis                  Number of obs =       74

Standardized coefficients for the first variable set
                          1          2          3          4
         length      0.2110     3.2095     0.7334     0.4714
         weight      0.7898    -2.8469    -0.7448     0.5308
       headroom      0.0297    -0.3131     1.2995    -0.0373
          trunk     -0.0098    -0.1466    -0.9134    -1.3914


Standardized coefficients for the second variable set
                          1          2          3          4
   displacement      0.4932    -1.1525     1.7568    -0.0493
            mpg     -0.2670    -0.2388     0.3954     1.4337
      gear_ratio     0.0150     0.4691     1.6698    -0.4705
           turn      0.3493     1.3694     0.0145     0.9857
```

```
Canonical correlations:
  0.9476   0.3400   0.0634   0.0447
```

To see the raw coefficients in matrix form rather than in the standard estimation table, we could use the `coefmatrix` option to `canon` (or on replay).

◁

❏ Technical Note

`canon` reports standard errors for the coefficients in the linear combinations; most other software does not. You should view these standard errors as lower bounds for the true standard errors. It is based on the assumption that the coefficients for one set of measurements are correct for the purpose of calculating the coefficients and standard errors of the other relationship based on a linear regression.

❏

Saved Results

canon saves in e():

Scalars
e(N)	number of observations
e(df_r)	residual degrees of freedom
e(df)	degrees of freedom
e(n_lc)	the linear combination calculated
e(n_cc)	number of canonical correlations calculated

Macros
e(cmd)	canon
e(wexp)	weight expression
e(wtype)	weight type
e(predict)	program used to implement predict
e(estat_cmd)	program used to implement estat

Matrices
e(b)	coefficient vector
e(V)	variance–covariance matrix of the estimators
e(stat_#)	statistics for canonical correlation #
e(stat_m)	statistics for overall model
e(canload11)	canonical loadings for *varlist*$_1$
e(canload22)	canonical loadings for *varlist*$_2$
e(canload12)	correlation between *varlist*$_1$ and the canonical variates from *varlist*$_2$
e(canload21)	correlation between *varlist*$_2$ and the canonical variates from *varlist*$_1$
e(rawcoef_var1)	raw coefficients for *varlist*$_1$
e(rawcoef_var2)	raw coefficients for *varlist*$_2$
e(stdcoef_var1)	standardized coefficients for *varlist*$_1$
e(stdcoef_var2)	standardized coefficients for *varlist*$_2$
e(ccorr)	canonical correlation coefficients
e(corr_var1)	correlation matrix for *varlist*$_1$
e(corr_var2)	correlation matrix for *varlist*$_2$
e(corr_mixed)	correlation matrix between *varlist*$_1$ and *varlist*$_2$

Functions
e(sample)	marks estimation sample

Methods and Formulas

canon is implemented as an ado-file.

Let the correlation matrix between the two sets of variables be

$$\begin{pmatrix} \mathbf{A} & \mathbf{B} \\ \mathbf{B}' & \mathbf{C} \end{pmatrix}$$

That is, $\mathbf{A}$ is the correlation matrix of the first set of variables with themselves, $\mathbf{C}$ is the correlation matrix of the second set of variables with themselves, and $\mathbf{B}$ contains the cross-correlations.

The squared canonical correlations are then the eigenvalues of $\mathbf{V} = \mathbf{B}'\mathbf{A}^{-1}\mathbf{B}\mathbf{C}^{-1}$ or $\mathbf{W} = \mathbf{B}\mathbf{C}^{-1}\mathbf{B}'\mathbf{A}^{-1}$ (either will work), which are both nonsymmetric matrices (Wilks 1962, 587–592). The corresponding left eigenvectors are the linear combinations for the two sets of variables in standardized (variance 1) form. These eigenvectors, or linear combinations of the two sets of variables, in standardized form are the canonical variates.

To calculate standard errors in this form, assume that the left eigenvectors of $\mathbf{V}$ are fixed, and write $\mathbf{V} = (\mathbf{v}_1, \mathbf{v}_2, \ldots, \mathbf{v}_m)$. The left eigenvector of $\mathbf{W}$ corresponding to $\mathbf{v}_k$ is proportional to $\mathbf{v}_k \mathbf{B}' \mathbf{C}^{-1}$, which has variance $(1 - r_k^2) \mathbf{C}^{-1}$, where r_k is the corresponding canonical correlation. These results are then scaled to have mean 0 and variance 1 and are in terms of the original scale of the variables.

If the eigenvalues are $r_1, r_2, \ldots, r_m$ where m is the number of canonical correlations, we test the hypothesis that there is no (linear) relationship between the two variable sets. This is equivalent to the statement that none of the correlations $r_1, r_2, \ldots, r_m$ is significant.

Wilks' (1932) lambda statistic is

$$\Lambda_1 = \prod_{i=1}^{m}(1 - r_i^2)$$

and is a likelihood-ratio statistic. This statistic is distributed as the Wilks Λ-distribution. Rejection of the null hypothesis is for small values of Λ_1.

Pillai's (1955) trace for canonical correlations is

$$V^{(m)} = \sum_{i=1}^{m} r_i^2$$

and the Lawley–Hotelling trace (Lawley 1938 and Hotelling 1951) is

$$U^{(m)} = \sum_{i=1}^{m} \frac{r_i^2}{1 - r_i^2}$$

Roy's (1939) largest root is given by

$$\theta = r_1^2$$

Rencher (2002) has tables providing critical values for these statistics and discussion on significance testing for canonical correlations.

Canonical loadings, the correlation between a variable set and its corresponding canonical variate set, are calculated by `canon` and used in [MV] **canon postestimation**.

For a note about Harold Hotelling, see [MV] **hotelling**.

Acknowledgment

Significance testing of canonical correlations is based on the `cancor` package originally written by Philip B. Ender, UCLA.

References

Hotelling, H. 1935. The most predictable criterion. *Journal of Educational Psychology* 26: 139–142.

——. 1936. Relations between two sets of variates. *Biometrika* 28: 321–377.

——. 1951 A generalized t^2 test and measure of multivariate dispersion. *Proceedings of the Second Berkeley Symposium on Mathematical Statistics and Probability* 1: 23–41.

Johnson, R. and D. Wichern. 1992. *Applied Multivariate Statistical Analysis.* 3rd ed. Englewood Cliffs, NJ: Prentice Hall.

Lawley, D. N. 1938. A generalization of Fisher's z-test. *Biometrika* 30: 180–187.

Pillai, K. C. S. 1955. Some new test criteria in multivariate analysis. *Annals of Mathematical Statistics* 26: 117–121.

Rencher, A. C. 2002. *Methods of Multivariate Analysis.* 2nd ed. New York: Wiley.

Roy, S. N. 1939. p-Statistics or some generalizations in analysis of variance appropriate to multivariate problems. *Sankhyā* 4: 381–396

Wilks, S. S. 1932. Certain generalizations in the analysis of variance. *Biometrika* 24: 471–494.

——. 1962. *Mathematical Statistics.* New York: Wiley.

Also See

Complementary:	[MV] **canon postestimation**
Related:	[MV] **factor**, [MV] **pca**,
	[R] **correlate**, [R] **mvreg**, [R] **pcorr**, [R] **regress**
Background:	[U] **11.1.10 Prefix commands**,
	[U] **20 Estimation and postestimation commands**

Title

canon postestimation — Postestimation tools for canon

Description

The following postestimation commands are of special interest after `canon`:

command	description
estat correlations	show correlation matrices
estat loadings	show loading matrices

For information about these commands, see below.

In addition, the following standard postestimation commands are available:

command	description
estat	VCE and estimation sample summary
estimates	cataloging estimation results
lincom	point estimates, standard errors, testing, and inference for linear combinations of coefficients
nlcom	point estimates, standard errors, testing, and inference for nonlinear combinations of coefficients
predict	predictions, residuals, influence statistics, and other diagnostic measures
predictnl	point estimates, standard errors, testing, and inference for generalized predictions
test	Wald tests for simple and composite linear hypotheses
testnl	Wald tests of nonlinear hypotheses

See the corresponding entries in the *Stata Base Reference Manual* for details.

Special-interest postestimation commands

estat correlations displays the correlation matrices calculated by `canon` for *varlist*$_1$ and *varlist*$_2$ and between the two lists.

estat loadings displays the canonical loadings computed by `canon`.

(*Continued on next page*)

Syntax for predict

predict [*type*] *newvar* [*if*] [*in*], *statistic** [<u>c</u>orrelation(#)]

statistic*	description
Main	
u	calculate linear combination of *varlist*$_1$
v	calculate linear combination of *varlist*$_2$
stdu	calculate standard error of the linear combination of *varlist*$_1$
stdv	calculate standard error of the linear combination of *varlist*$_2$

* There is no default statistic; you must specify one *statistic* from the list.

These statistics are available both in and out of sample; type predict ... if e(sample) ... if wanted only for the estimation sample.

Options for predict

      Main

u and v calculate the linear combinations of *varlist*$_1$ and *varlist*$_2$, respectively. For the first canonical correlation, u and v are the linear combinations having maximal correlation. For the second canonical correlation, specified in predict with the correlation(2) option, u and v have maximal correlation subject to the constraints that u is orthogonal to the u from the first canonical correlation, and v is orthogonal to the v from the first canonical correlation. The third and higher correlations are defined similarly.

stdu and stdv calculate the standard errors of the respective linear combinations.

correlation(#) specifies the canonical correlation for which the requested statistic is to be computed. The default value for correlation() is 1. If the lc() option to canon was used to calculate a particular canonical correlation, then only this canonical correlation is in the estimation results. You can obtain estimates for it either by specifying correlation(1) or by omitting the correlation() option entirely.

Syntax for estat

Display the correlation matrices

estat <u>corr</u>elations [, <u>f</u>ormat(% *fmt*)]

Display the canonical loadings

estat <u>loa</u>dings [, <u>f</u>ormat(% *fmt*)]

Option for estat

      Main

format(% *fmt*) specifies the display format for numbers in matrices; see [D] **format**. format(% 8.4f) is the default.

Remarks

In addition to the coefficients presented by `canon` in computing canonical correlations, several other matrices may be of interest.

▷ Example 1

Recall from `canon` the example of two scientists trying to describe how "big" a car is. One took physical measurements—the length, weight, headroom and trunk space—whereas the second took mechanical measurements—engine displacement, mileage rating, gear ratio, and turning radius. We discovered that these two views are closely related, with the best linear combination of the two types of measurements, the largest canonical correlation, at 0.9476. We can prove that the first canonical correlation is correct by calculating the two linear combinations and then calculating the ordinary correlation.

```
. use http://www.stata-press.com/data/r9/auto
(1978 Automobile Data)
. quietly canon (length weight headroom trunk) (displ mpg gear_ratio turn)
. predict physical, u corr(1)
. predict mechanical, v corr(1)
. correlate mechanical physical
(obs=74)
```

	mechan~l	physical
mechanical	1.0000	
physical	0.9476	1.0000

```
. drop mechanical physical
```

◁

▷ Example 2

Researchers are often interested in the canonical loadings, the correlations between the original variable lists and their canonical variates, The canonical loadings are used to interpret the canonical variates. However, as shown in the technical note later in this entry, Rencher (1988; 1992; 1998, section 8.6.3) has shown that there is no information in these correlations about how one variable list contributes jointly to canonical correlation with the other. Loadings are still frequently discussed, and `estat loadings` reports these as well as the "cross-loadings" or correlations between *varlist*$_1$ and the canonical variates for *varlist*$_2$ and the correlations between *varlist*$_2$ and the canonical variates for *varlist*$_1$. The loadings and cross-loadings are all computed by `canon`.

```
. estat loadings
Canonical loadings for variable list 1
```

	1	2	3	4
length	0.9664	0.2481	0.0361	-0.0566
weight	0.9972	-0.0606	-0.0367	0.0235
headroom	0.5140	-0.1295	0.7134	-0.4583
trunk	0.6941	0.0644	-0.0209	-0.7167

```
Canonical loadings for variable list 2
                          1          2          3          4

       displacement    0.9404    -0.3091     0.1050     0.0947
                mpg   -0.8569    -0.1213     0.1741     0.4697
          gear_ratio  -0.7945     0.3511     0.4474    -0.2129
               turn    0.9142     0.3286    -0.0345     0.2345


Correlation between variable list 1 and canonical variates from list 2
                          1          2          3          4

             length    0.9158     0.0844     0.0023    -0.0025
             weight    0.9449    -0.0206    -0.0023     0.0011
           headroom    0.4871    -0.0440     0.0452    -0.0205
              trunk    0.6577     0.0219    -0.0013    -0.0320


Correlation between variable list 2 and canonical variates from list 1
                          1          2          3          4

       displacement    0.8912    -0.1051     0.0067     0.0042
                mpg   -0.8120    -0.0413     0.0110     0.0210
          gear_ratio  -0.7529     0.1194     0.0284    -0.0095
               turn    0.8663     0.1117    -0.0022     0.0105
```

```
. mat load2 = r(canload22)
```

◁

▷ Example 3

In example 2, we saved the loading matrix for *varlist*$_2$, containing the mechanical variables, and we wish to verify that it is correct. We predict the canonical variates for *varlist*$_2$ and then find the canonical correlations between the canonical variates and the original mechanical variables as a means of getting the correlation matrices, which we then display using estat correlations. The mixed correlation matrix is the same as the loading matrix that we saved.

```
. predict mechanical1, v corr(1)
. predict mechanical2, v corr(2)
. predict mechanical3, v corr(3)
. predict mechanical4, v corr(4)
. quietly canon (mechanical1-mechanical4) (displ mpg gear_ratio turn)
. estat correlation
Correlations for variable list 1
                 mechan~1   mechan~2   mechan~3   mechan~4

    mechanical1    1.0000
    mechanical2   -0.0000     1.0000
    mechanical3   -0.0000     0.0000     1.0000
    mechanical4   -0.0000    -0.0000    -0.0000     1.0000
```

```
Correlations for variable list 2
                     displa~t        mpg  gear_r~o       turn

     displacement      1.0000
              mpg     -0.7056     1.0000
        gear_ratio    -0.8289     0.6162    1.0000
             turn      0.7768    -0.7192   -0.6763     1.0000
```

```
Correlations between variable lists 1 and 2
                     mechan~1   mechan~2   mechan~3   mechan~4

     displacement      0.9404    -0.3091     0.1050     0.0947
              mpg     -0.8569    -0.1213     0.1741     0.4697
        gear_ratio    -0.7945     0.3511     0.4474    -0.2129
             turn      0.9142     0.3286    -0.0345     0.2345
```

```
. matlist load2, format(%8.4f) border(bottom)
                          1          2          3          4

     displacement      0.9404    -0.3091     0.1050     0.0947
              mpg     -0.8569    -0.1213     0.1741     0.4697
        gear_ratio    -0.7945     0.3511     0.4474    -0.2129
             turn      0.9142     0.3286    -0.0345     0.2345
```

◁

❏ Technical Note

estat loadings reports the canonical loadings, or correlations between a *varlist* and its corresponding canonical variates. It is widely claimed that the loadings provide a more valid interpretation of the canonical variates. Rencher (1988; 1992; 1998, section 8.6.3) has shown that a weighted sum of the correlations between a $x_j \in$ *varlist*$_1$ and the canonical variates from *varlist*$_1$ is equal to the squared multiple correlation between x_j and the variables in *varlist*$_2$. The correlations do not give new information on the importance of a given variable in the context of the others. Rencher (2002) notes, "The researcher who uses these correlations for interpretation is unknowingly reducing the multivariate setting to a univariate one."

❏

Saved Results

estat correlations saves in r():

Matrices
    r(corr_var1)   correlations for *varlist*$_1$
    r(corr_var2)   correlations for *varlist*$_2$
    r(corr_mixed)   correlations between *varlist*$_1$ and *varlist*$_2$

estat loadings saves in r():

Matrices
    r(canload11)   canonical loadings for *varlist*$_1$
    r(canload22)   canonical loadings for *varlist*$_2$
    r(canload21)   correlations between *varlist*$_2$ and the canonical variates for *varlist*$_1$
    r(canload12)   correlations between *varlist*$_1$ and the canonical variates for *varlist*$_2$

Methods and Formulas

All postestimation commands listed above are implemented as ado-files.

References

Rencher, A. C. 1988. On the use of correlations to interpret canonical functions. *Biometrika* 75: 363–365.

——. 1992. Interpretation of canonical discriminant functions, canonical variates and principal components. *American Statistician* 46: 217–225.

——. 1998. *Multivariate Statistical Inference and Applications.* New York: Wiley.

——. 2002. *Methods of Multivariate Analysis.* 2nd ed. New York: Wiley.

Also See

Complementary:	[MV] **canon**,
	[R] **adjust**, [R] **estimates**, [R] **hausman**, [R] **lincom**, [R] **nlcom**,
	[R] **predictnl**, [R] **test**, [R] **testnl**
Background:	[R] **estat**, [R] **predict**

Title

cluster — Introduction to cluster-analysis commands

Syntax

Cluster analysis of data

    cluster *subcommand* ...

Cluster analysis of a dissimilarity matrix

    clustermat *subcommand* ...

Description

Stata's cluster-analysis routines provide several hierarchical and partition clustering methods, postclustering summarization methods, and cluster-management tools. This entry presents an overview of cluster analysis, the `cluster` and `clustermat` commands (also see [MV] **clustermat**), and Stata's cluster-analysis management tools. The hierarchical clustering methods may be applied to the data using the `cluster` command or to a user supplied dissimilarity matrix using the `clustermat` command.

The `cluster` command has the following *subcommand*s, which are detailed in their respective manual entries.

Partition-clustering methods for observations

kmeans	[MV] **cluster kmeans**	Kmeans cluster analysis
kmedians	[MV] **cluster kmedians**	Kmedians cluster analysis

Hierarchical clustering methods for observations

singlelinkage	[MV] **cluster singlelinkage**	Single-linkage cluster analysis
averagelinkage	[MV] **cluster averagelinkage**	Average-linkage cluster analysis
completelinkage	[MV] **cluster completelinkage**	Complete-linkage cluster analysis
waveragelinkage	[MV] **cluster waveragelinkage**	Weighted-average linkage cluster analysis
medianlinkage	[MV] **cluster medianlinkage**	Median-linkage cluster analysis
centroidlinkage	[MV] **cluster centroidlinkage**	Centroid-linkage cluster analysis
wardslinkage	[MV] **cluster wardslinkage**	Ward's linkage cluster analysis

Postclustering commands

stop	[MV] **cluster stop**	Cluster-analysis stopping rules
dendrogram	[MV] **cluster dendrogram**	Dendrograms for hierarchical cluster analysis
generate	[MV] **cluster generate**	Generate summary or grouping variables from a cluster analysis

User utilities

notes	[MV] **cluster notes**	Place notes in cluster analysis
dir	[MV] **cluster utility**	Directory list of cluster analyses
list	[MV] **cluster utility**	List cluster analyses
drop	[MV] **cluster utility**	Drop cluster analyses
rename	[MV] **cluster utility**	Rename cluster analyses
renamevar	[MV] **cluster utility**	Rename cluster-analysis variables

Programmer utilities

	[MV] **cluster programming subroutines**	Add cluster-analysis routines
query	[MV] **cluster programming utilities**	Obtain cluster-analysis attributes
set	[MV] **cluster programming utilities**	Set cluster-analysis attributes
delete	[MV] **cluster programming utilities**	Delete cluster-analysis attributes
parsedistance	[MV] **cluster programming utilities**	Parse (dis)similarity measure names
measures	[MV] **cluster programming utilities**	Compute (dis)similarity measures

The clustermat command has the following *subcommand*s, which are detailed along with the related cluster command manual entries. Also see [MV] **clustermat**.

Hierarchical clustering methods for matrices

singlelinkage	[MV] **cluster singlelinkage**	Single-linkage cluster analysis
averagelinkage	[MV] **cluster averagelinkage**	Average-linkage cluster analysis
completelinkage	[MV] **cluster completelinkage**	Complete-linkage cluster analysis
waveragelinkage	[MV] **cluster waveragelinkage**	Weighted-average linkage cluster analysis
medianlinkage	[MV] **cluster medianlinkage**	Median-linkage cluster analysis
centroidlinkage	[MV] **cluster centroidlinkage**	Centroid-linkage cluster analysis
wardslinkage	[MV] **cluster wardslinkage**	Ward's linkage cluster analysis

In addition, the clustermat stop postclustering command has similar syntax as the cluster stop command; see [MV] **cluster stop**. For the remaining postclustering commands and user utilities, you may specify either cluster or clustermat—it does not matter which.

If you are new to Stata's cluster-analysis commands, we recommend that you first read this entry and then read:

[MV] *measure_option*	Option for similarity and dissimilarity measures
[MV] **clustermat**	Cluster analysis of a dissimilarity matrix
[MV] **cluster kmeans**	Kmeans cluster analysis
[MV] **cluster singlelinkage**	Single-linkage cluster analysis
[MV] **cluster dendrogram**	Dendrograms for hierarchical cluster analysis
[MV] **cluster stop**	Cluster-analysis stopping rules
[MV] **cluster generate**	Generate summary or grouping variables from a cluster analysis

Remarks

Remarks are presented under the headings

Introduction to cluster analysis

Cluster analysis attempts to determine the natural groupings (or clusters) of observations. Sometimes this is called "classification", but this term is used by others to mean discriminant analysis, which is related but is not the same. To avoid confusion, we will use "cluster analysis" or "clustering" when referring to finding groups in data. It is difficult (maybe impossible) to define cluster analysis. Kaufman and Rousseeuw (1990) start their book by saying, "Cluster analysis is the art of finding groups in data." Everitt, Landau, and Leese (2001, 6) use the terms "cluster", "group", and "class" and say, concerning a formal definition for these terms, "In fact it turns out that such formal definition is not only difficult but may even be misplaced."

Everitt, Landau, and Leese (2001) and Gordon (1999) provide examples of the use of cluster analysis, such as in refining or redefining diagnostic categories in psychiatry, detecting similarities in artifacts by archaeologists to study the spatial distribution of artifact types, discovering hierarchical relationships in the field of taxonomy, and identifying sets of similar cities so that one city from each class can be sampled in a market research task. In addition, the activity that is now called "data mining" relies extensively on cluster-analysis methods.

We view cluster analysis as an exploratory data-analysis technique. According to Everitt, "Many cluster-analysis techniques have taken their place alongside other exploratory data-analysis techniques as tools of the applied statistician. The term exploratory is important here since it explains the largely absent 'p-value', ubiquitous in many other areas of statistics. ...Clustering methods are intended largely for generating rather than testing hypotheses" (1993, 10).

It has been said that there are as many cluster-analysis methods as there are people performing cluster analysis. This is a gross understatement! There exist infinitely more ways to perform a cluster analysis than people who perform them.

There are several general types of cluster-analysis methods, each having numerous specific methods. Additionally, most cluster-analysis methods allow a variety of distance measures for determining the similarity or dissimilarity between observations. Some of the measures do not meet the requirements to be called a distance metric, so we use the more general term "dissimilarity measure" in place of distance. Similarity measures may be used in place of dissimilarity measures. There are an infinite number of similarity and dissimilarity measures. For instance, there are an infinite number of Minkowski distance metrics, with the familiar Euclidean, absolute-value, and maximum-value distances being special cases.

In addition to cluster method and dissimilarity measure choice, those performing a cluster analysis might decide to perform data transformations and/or variable selection before clustering. Then they might need to determine how many clusters there really are in the data, which you can do using stopping rules. There is a surprisingly large number of stopping rules mentioned in the literature. For example, Milligan and Cooper (1985) compare 30 different stopping rules.

Looking at all of these choices, you can see why there are more cluster-analysis methods than people performing cluster analysis.

Stata's cluster-analysis system

Stata's `cluster` and `clustermat` commands were designed to allow you to keep track of the various cluster analyses performed on your data. The main clustering subcommands `singlelinkage`, `averagelinkage`, `completelinkage`, `waveragelinkage`, `medianlinkage`, `centroidlinkage`, `wardslinkage`, `kmeans`, and `kmedians` create named Stata cluster objects that keep track of the variables these methods create and hold other identifying information for the cluster analysis. These cluster objects become part of your dataset. They are saved with your data when your data are `saved` and are retrieved when you again `use` your dataset; see [D] **use** and [D] **save**.

Postcluster-analysis subcommands are available with the `cluster` and `clustermat` commands so that you can examine the created clusters. Cluster-management tools are provided that allow you to add information to the cluster objects and to manipulate them as needed. The main clustering subcommands, postclustering subcommands, and cluster-management tools are discussed in the following sections.

Stata's clustering methods fall into two general types: partition and hierarchical. These two types are discussed below. There exist other types, such as fuzzy partition (where observations can belong to more than one group). Stata's `cluster` command is designed so that programmers can extend it by adding more methods; see [MV] **cluster programming subroutines** and [MV] **cluster programming utilities** for details.

❏ Technical Note

If you are familiar with Stata's large array of estimation commands, be careful to distinguish between cluster analysis (the `cluster` command) and the `cluster()` option allowed with many estimation commands. Cluster analysis finds groups in data. The `cluster()` option allowed with various estimation commands indicates that the observations are independent across the groups defined by the option but are not necessarily independent within those groups. A grouping variable produced by the `cluster` command will seldom satisfy the assumption behind the use of the `cluster()` option.

❏

Data transformations and variable selection

Stata's `cluster` command does not have any built-in data transformations, but since Stata has full data-management and statistical capabilities, you can use other Stata commands to transform your data before calling the `cluster` command. In some cases, standardization of the variables is important to keep a variable with high variability from dominating the cluster analysis. In other cases, standardization of variables acts to hide the true groupings present in the data. The decision to standardize or perform other data transformations depends on the type of data and the nature of the groups.

Data transformations (such as standardization of variables) and the variables selected for use in clustering can also have a great impact on the groupings that are discovered. These and other cluster-analysis data issues are covered in Milligan and Cooper (1988) and Schaffer and Green (1996) and many of the cluster-analysis texts, including Anderberg (1973); Gordon (1999); Everitt, Landau, and Leese (2001); and Späth (1980).

Similarity and dissimilarity measures

A variety of similarity and dissimilarity measures have been implemented for Stata's clustering commands for both continuous variables and binary variables. For information about these measures, see [MV] *measure_option*.

Partition cluster-analysis methods

Partition methods break the observations into a distinct number of nonoverlapping groups. Stata has implemented two partition methods, kmeans and kmedians.

One of the more commonly used partition clustering methods is called kmeans cluster analysis. In kmeans clustering, the user specifies the number of clusters, k, to create using an iterative process. Each observation is assigned to the group whose mean is closest, and then based on that categorization, new group means are determined. These steps continue until no observations change groups. The algorithm begins with k seed values, which act as the k group means. There are many ways to specify the beginning seed values. See [MV] **cluster kmeans** for the details of the `cluster kmeans` command.

A variation of kmeans clustering is kmedians clustering. The same process is followed in kmedians as in kmeans, except that medians, instead of means, are computed to represent the group centers at each step; see [MV] **cluster kmedians** for details.

These partition clustering methods will generally be quicker and will allow larger datasets than the hierarchical clustering methods outlined next. However, if you wish to examine clustering to various numbers of clusters, you will need to execute `cluster` numerous times with the partition methods. Clustering to various numbers of groups using a partition method typically does not produce clusters that are hierarchically related. If this is important for your application, consider using one of the hierarchical methods.

Hierarchical cluster-analysis methods

Hierarchical clustering creates hierarchically related sets of clusters. Hierarchical clustering methods are generally of two types: agglomerative or divisive.

Agglomerative hierarchical clustering methods begin with each observation being considered as a separate group (N groups each of size 1). The closest two groups are combined ($N - 1$ groups, one of size 2 and the rest of size 1), and this process continues until all observations belong to the same group. This process creates a hierarchy of clusters.

In addition to choosing the similarity or dissimilarity measure to use in comparing two observations, you can choose what to compare between groups that contain more than one observation. The method used to compare groups is called a linkage method. Stata's `cluster` and `clustermat` commands provide several hierarchical agglomerative linkage methods, which are discussed in the next section.

Unlike hierarchical agglomerative clustering, divisive hierarchical clustering begins with all observations belonging to one group. This group is then split in some fashion to create two groups. One of these two groups is then split to create three groups, each of which is split to create four groups, and so on, until all observations are in their own separate group. Stata does not currently have any divisive hierarchical clustering commands. There are relatively few mentioned in the literature, and they tend to be particularly time-consuming to compute.

To appreciate the underlying computational complexity of both agglomerative and divisive hierarchical clustering, consider the following information paraphrased from Kaufman and Rousseeuw (1990). The first step of an agglomerative algorithm considers $N(N-1)/2$ possible fusions of observations to find the closest pair. This number grows quadratically with N. For divisive hierarchical clustering, the first step would be to find the best split into two nonempty subsets, and if all possibilities were considered, it would amount to $2^{(N-1)} - 1$ comparisons. This number grows exponentially with N.

Agglomerative methods

Stata's `cluster` and `clustermat` commands provide the following hierarchical agglomerative linkage methods: single linkage, complete linkage, average linkage, Ward's method, centroid linkage, median linkage, and weighted-average linkage. There are others mentioned in the literature, but these are the best-known methods.

Single-linkage clustering computes the similarity or dissimilarity between two groups as the similarity or dissimilarity between the closest pair of observations between the two groups. Complete-linkage clustering, on the other hand, uses the farthest pair of observations between the two groups to determine the similarity or dissimilarity of the two groups. Average-linkage clustering uses the average similarity or dissimilarity of observations between the groups as the measure between the two groups. Ward's method joins the two groups that result in the minimum increase in the error sum of squares. The other linkage methods provide alternatives to these basic linkage methods.

The `cluster singlelinkage` and `clustermat singlelinkage` commands implement single-linkage hierarchical agglomerative clustering; see [MV] **cluster singlelinkage** for details. Single-linkage clustering suffers (or benefits, depending on your point of view) from what is called chaining. Since the closest points between two groups determine the next merger, long, thin clusters can result. If this chaining feature is not what you desire, consider using one of the other methods, such as complete linkage or average linkage. Single-linkage clustering is faster and uses less memory than the other linkage methods due to special properties of the method that can be exploited computationally.

Complete-linkage hierarchical agglomerative clustering is implemented by the `cluster completelinkage` and `clustermat completelinkage` commands; see [MV] **cluster completelinkage** for details. Complete-linkage clustering is at the other extreme from single-linkage clustering. Complete linkage produces spatially compact clusters, so it is not the best method for recovering elongated cluster structures. Several sources, including Kaufman and Rousseeuw (1990), discuss the chaining of single linkage and the clumping of complete linkage.

Kaufman and Rousseeuw (1990) indicate that average linkage works well for many situations and is reasonably robust. The `cluster averagelinkage` and `clustermat averagelinkage` commands provide average-linkage clustering; see [MV] **cluster averagelinkage**.

Ward (1963) presented a general hierarchical clustering approach where groups were joined so as to maximize an objective function. He used an error-sum-of-squares objective function to illustrate. Ward's method of clustering became synonymous with using the error-sum-of-squares criteria. Kaufman and Rousseeuw (1990) indicate that Ward's method does well with groups that are multivariate normal and spherical but does not do as well if the groups are of different sizes or have unequal numbers of observations. The `cluster wardslinkage` and `clustermat wardslinkage` commands provide Ward's linkage clustering; see [MV] **cluster wardslinkage**.

At each step of the clustering, centroid linkage merges the groups whose means are closest. The centroid of a group is the component-wise mean and can be interpreted as the center of gravity for the group. Centroid linkage differs from average linkage in that centroid linkage is concerned with the distance between the means of the groups, while average linkage looks at the average distance between the points of the two groups. The `cluster centroidlinkage` and `clustermat centroidlinkage` commands provide centroid-linkage clustering; see [MV] **cluster centroidlinkage**.

Weighted-average linkage and median linkage are variations on average linkage and centroid linkage, respectively. In both cases, the difference is in how groups of unequal size are treated when merged. In average linkage and centroid linkage, the number of elements of each group is factored into the computation, giving correspondingly larger influence to the larger group. These two methods are called unweighted because each observation carries the same weight. In weighted-average linkage and median linkage, the two groups are given equal weighting in determining the combined group, regardless of the number of observations in each group. These two methods are said to be weighted since observations from groups with few observations carry more weight than observations from groups with many observations. The `cluster waveragelinkage` and `clustermat waveragelinkage` commands provide weighted-average linkage clustering; see [MV] **cluster waveragelinkage**. The `cluster medianlinkage` and `clustermat medianlinkage` commands provide median linkage clustering; see [MV] **cluster medianlinkage**.

Lance and Williams' recurrence formula

Lance and Williams (1967) developed a recurrence formula that defines, as special cases, most of the well-known hierarchical clustering methods, including all the hierarchical clustering methods found in Stata. Anderberg (1973), Jain and Dubes (1988), Kaufman and Rousseeuw (1990), Gordon (1999), Everitt, Landau, and Leese (2001), and Rencher (2002) discuss the Lance and Williams formula and how most popular hierarchical clustering methods are contained within it.

Using the notation of Everitt, Landau, and Leese (2001, 61), the Lance and Williams recurrence formula is

$$d_{k(ij)} = \alpha_i d_{ki} + \alpha_j d_{kj} + \beta d_{ij} + \gamma |d_{ki} - d_{kj}|$$

where d_{ij} is the distance (or dissimilarity) between cluster i and cluster j; $d_{k(ij)}$ is the distance (or dissimilarity) between cluster k and the new cluster formed by joining clusters i and j; and α_i, α_j, β, and γ are parameters that are set based on the particular hierarchical cluster-analysis method.

The recurrence formula allows, at each new level of the hierarchical clustering, the dissimilarity between the newly formed group and the rest of the groups to be computed from the dissimilarities of the current grouping. This can result in a large computational savings compared with recomputing at each step in the hierarchy from the observation-level data. This feature of the recurrence formula allows `clustermat` to operate on a similarity or dissimilarity matrix instead of the data.

The following table shows the values of α_i, α_j, β, and γ for the hierarchical clustering methods implemented in Stata. n_i, n_j, and n_k are the number of observations in group i, j, and k, respectively.

(Continued on next page)

Clustering method	α_i	α_j	β	γ
Single linkage	$\frac{1}{2}$	$\frac{1}{2}$	0	$-\frac{1}{2}$
Complete linkage	$\frac{1}{2}$	$\frac{1}{2}$	0	$\frac{1}{2}$
Average linkage	$\frac{n_i}{n_i + n_j}$	$\frac{n_j}{n_i + n_j}$	0	0
Weighted-average linkage	$\frac{1}{2}$	$\frac{1}{2}$	0	0
Centroid linkage	$\frac{n_i}{n_i + n_j}$	$\frac{n_j}{n_i + n_j}$	$-\alpha_i \alpha_j$	0
Median linkage	$\frac{1}{2}$	$\frac{1}{2}$	$-\frac{1}{4}$	0
Ward's linkage	$\frac{n_i + n_k}{n_i + n_j + n_k}$	$\frac{n_j + n_k}{n_i + n_j + n_k}$	$\frac{-n_k}{n_i + n_j + n_k}$	0

For information on the use of various similarity and dissimilarity measures in hierarchical clustering, see the next two sections.

Dissimilarity transformations and the Lance and Williams formula

The Lance and Williams formula, which is used as the basis for computing hierarchical clustering in Stata, is designed for use with dissimilarity measures. Before performing hierarchical clustering, Stata transforms similarity measures, both continuous and binary, to dissimilarities. After cluster analysis, Stata transforms the fusion values (heights at which the various groups join in the hierarchy) back to similarities.

Stata's `cluster` command uses

$$\text{dissimilarity} = 1 - \text{similarity}$$

to transform from a similarity to a dissimilarity measure and back again. Stata's similarity measures range from either 0 to 1 or -1 to 1. The resulting dissimilarities range from 1 down to 0 and from 2 down to 0, respectively.

For continuous data, Stata provides both the L2 and L2squared dissimilarity measures, as well as both the L(#) and Lpower(#) dissimilarity measures. Why have both a L2 and L2squared dissimilarity measure, and why have both a L(#) and Lpower(#) dissimilarity measure?

For single- and complete-linkage hierarchical clustering (and for kmeans and kmedians partition clustering), there is no need for the additional L2squared and Lpower(#) dissimilarities. The same cluster solution is obtained using L2 and L2squared (or L(#) and Lpower(#)), except the resulting heights in the dendrogram are raised to a power.

However, for the other hierarchical clustering methods, there is a difference. For some of these other hierarchical clustering methods, the natural default for dissimilarity measure is L2squared. For instance, the traditional Ward's (1963) method is obtained using the L2squared dissimilarity option.

Warning concerning similarity or dissimilarity choice

With hierarchical centroid, median, Ward's, and weighted-average linkage clustering, Lance and Williams (1967), Anderberg (1973), Jain and Dubes (1988), Kaufman and Rousseeuw (1990), Everitt, Landau, and Leese (2001), and Gordon (1999) give various levels of warnings about using many of the similarity and dissimilarity measures ranging from saying that you should never use anything other than the default squared Euclidean distance (or Euclidean distance) to saying that the results may lack a useful interpretation.

The second example in [MV] **cluster wardslinkage** illustrates part of the basis for this warning. The simple matching coefficient is used on binary data. The range of the fusion values for the resulting hierarchy is not between 1 and 0, as you would expect for the matching coefficient. The conclusions from the cluster analysis, however, agree well with the results obtained in other ways.

Stata does not restrict your choice of similarity or dissimilarity. If you are not familiar with these hierarchical clustering methods, use the default dissimilarity measure.

Synonyms

Cluster-analysis methods have been developed by researchers in many different disciplines. Since researchers did not always know what was happening in other fields, many synonyms for the different hierarchical cluster-analysis methods exist.

Blashfield and Aldenderfer (1978) provide a table of equivalent terms. Jain and Dubes (1988) and Day and Edelsbrunner (1984) also mention some of the synonyms and use various acronyms. Here is a list of synonyms:

Single linkage
    Nearest-neighbor method
    Minimum method
    Hierarchical analysis
    Space-contracting method
    Elementary linkage analysis
    Connectedness method

Complete linkage
    Furthest-neighbor method
    Maximum method
    Compact method
    Space-distorting method
    Space-dilating method
    Rank-order typal analysis
    Diameter analysis

Average linkage
    Arithmetic-average clustering
    Unweighted pair-group method using
       arithmetic averages
    UPGMA
    Unweighted clustering
    Group-average method
    Unweighted group mean
    Unweighted pair-group method

Weighted-average linkage
    Weighted pair-group method using
       arithmetic averages
    WPGMA
    Weighted group-average method

Centroid linkage
    Unweighted centroid method
    Unweighted pair-group centroid method
    UPGMC
    Nearest-centroid sorting

Median linkage
    Gower's method
    Weighted centroid method
    Weighted pair-group centroid method
    WPGMC
    Weighted pair method
    Weighted group method

Ward's method
    Minimum-variance method
    Error-sum-of-squares method
    Hierarchical grouping to minimize tr(W)
    HGROUP

Reversals

Unlike the other hierarchical methods implemented in Stata, centroid linkage (see [MV] **cluster centroidlinkage**) and median linkage (see [MV] **cluster medianlinkage**) can (and often do) produce reversals or crossovers; see Anderberg (1973), Jain and Dubes (1988), Gordon (1999), and Rencher (2002). Normally, the dissimilarity or clustering criterion increases monotonically as the agglomerative hierarchical clustering progresses from many to few clusters. (For similarity measures, it monotonically decreases.) In other words, the dissimilarity value at which $k + 1$ clusters form will be larger than the value at which k clusters form. When the dissimilarity does not increase monotonically through the levels of the hierarchy, it is said to have reversals or crossovers.

The word *crossover*, in this context, comes from the appearance of the resulting dendrogram (see [MV] **cluster dendrogram**). In a hierarchical clustering without reversals, the dendrogram branches extend in one direction (increasing dissimilarity measure). With reversals, some of the branches reverse and go in the opposite direction, causing the resulting dendrogram to be drawn with crossing lines (crossovers).

When reversals happen, Stata still produces correct results. You can still generate grouping variables (see [MV] **cluster generate**) and compute stopping rules (see [MV] **cluster stop**). However, the cluster dendrogram command will not draw a dendrogram with reversals; see [MV] **cluster dendrogram**. In all but the simplest cases, dendrograms with reversals are almost impossible to interpret visually.

Hierarchical cluster analysis applied to a dissimilarity matrix

What if you want to perform a cluster analysis using a similarity or dissimilarity measure that Stata does not provide? What if you want to cluster variables instead of observations? The clustermat command gives you the flexibility to do either; see [MV] **clustermat**.

User-supplied dissimilarities

There are situations where the dissimilarity between objects is evaluated subjectively (perhaps on a scale from 1 to 10 by a rater). These dissimilarities may be entered in a matrix and passed to the clustermat command in order to perform hierarchical clustering. Likewise, if Stata does not offer the dissimilarity measure you desire, you may compute the dissimilarities yourself and place them in a matrix and then use clustermat to perform the cluster analysis. [MV] **clustermat** illustrates both of these situations.

Clustering variables instead of observations

Sometimes, instead of clustering observations, the desire is to cluster variables. The cluster command clusters observations. One approach to clustering variables in Stata is to use xpose (see [D] **xpose**) to transpose the variables and observations and then to use cluster. Another approach is to use the matrix dissimilarity command with the variables option (see [MV] **matrix dissimilarity**) to produce a dissimilarity matrix for the variables. This matrix is then passed to clustermat to obtain the hierarchical clustering. See [MV] **clustermat**.

Postclustering commands

Stata's cluster stop and clustermat stop commands are used to determine the number of clusters. Two stopping rules are provided, the Caliński and Harabasz (1974) pseudo-F index and the Duda and Hart (1973) Je(2)/Je(1) index with associated pseudo-T-squared. You can easily add stopping rules to the cluster stop command; see [MV] **cluster stop** for details.

The `cluster dendrogram` command presents the dendrogram (cluster tree) after a hierarchical cluster analysis; see [MV] **cluster dendrogram**. Options allow you to view the top portion of the tree or the portion of the tree associated with a group. These options are important with larger datasets since the full dendrogram cannot be presented.

The `cluster generate` command produces grouping variables after hierarchical clustering; see [MV] **cluster generate**. These variables can then be used in other Stata commands, such as those that tabulate, summarize, and provide graphs. For instance, you might use `cluster generate` to create a grouping variable. You then might use the `pca` command (see [MV] **pca**) to obtain the first two principal components of the data. You could follow that with a graph (see *Stata Graphics Reference Manual*) to plot the principal components, using the grouping variable from the `cluster generate` command to control the point labeling of the graph. This would allow you to get one type of view into the clustering behavior of your data.

Cluster-management tools

You may add notes to your cluster analysis with the `cluster notes` command; see [MV] **cluster notes**. This command also allows you to view and delete notes attached to the cluster analysis.

The `cluster dir` and `cluster list` commands allow you to list the cluster objects and attributes currently defined for your dataset. `cluster drop` lets you remove a cluster object. See [MV] **cluster utility** for details.

Cluster objects are referenced by name. If no name is provided, many of the `cluster` commands will, by default, use the cluster object from the most recently performed cluster analysis. The `cluster use` command tells Stata which cluster object to use. You can change the name attached to a cluster object with the `cluster rename` command and the variables associated with a cluster analysis with the `cluster renamevar` command. See [MV] **cluster utility** for details.

You can exercise fine control over the attributes that are stored with a cluster object; see [MV] **cluster programming utilities**.

References

Anderberg, M. R. 1973. *Cluster Analysis for Applications.* New York: Academic Press.

Blashfield, R. K. and M. S. Aldenderfer. 1978. The literature on cluster analysis. *Multivariate Behavioral Research* 13: 271–295.

Caliński, T. and J. Harabasz. 1974. A dendrite method for cluster analysis. *Communications in Statistics* 3: 1–27.

Day, W. H. E. and H. Edelsbrunner. 1984. Efficient algorithms for agglomerative hierarchical clustering methods. *Journal of Classification* 1: 7–24.

Duda, R. O. and P. E. Hart. 1973. *Pattern Classification and Scene Analysis.* New York: Wiley.

Everitt, B. S. 1993. *Cluster Analysis.* 3rd ed. London: Arnold.

Everitt, B. S., S. Landau, and M. Leese. 2001. *Cluster Analysis.* 4th ed. London: Arnold.

Gordon, A. D. 1999. *Classification.* 2nd ed. Boca Raton, FL: CRC.

Jain, A. K. and R. C. Dubes. 1988. *Algorithms for Clustering Data.* Englewood Cliffs, NJ: Prentice Hall.

Kaufman, L. and P. J. Rousseeuw. 1990. *Finding Groups in Data.* New York: Wiley.

Lance, G. N. and W. T. Williams. 1967. A general theory of classificatory sorting strategies: 1. Hierarchical systems. *Computer Journal* 9: 373–380.

Milligan, G. W. and M. C. Cooper. 1985. An examination of procedures for determining the number of clusters in a dataset. *Psychometrika* 50: 159–179.

———. 1988. A study of standardization of variables in cluster analysis. *Journal of Classification* 5: 181–204.

Rencher, A. C. 2002. *Methods of Multivariate Analysis.* 2nd ed. New York: Wiley.

Rohlf, F. J. 1982. Single-link clustering algorithms. In *Handbook of Statistics,* Vol. 2, ed. P. R. Krishnaiah and L. N. Kanal, 267–284. Amsterdam: North-Holland.

Schaffer, C. M. and P. E. Green. 1996. An empirical comparison of variable standardization methods in cluster analysis. *Multivariate Behavioral Research* 31: 149–167.

Sibson, R. 1973. SLINK: An optimally efficient algorithm for the single-link cluster method. *Computer Journal* 16: 30–34.

Späth, H. 1980. *Cluster Analysis Algorithms for Data Reduction and Classification of Objects.* Chichester, UK: Ellis Horwood.

Ward, J. H., Jr. 1963. Hierarchical grouping to optimize an objective function. *Journal of the American Statistical Association* 58: 236–244.

Also See

Complementary:	[MV] **cluster averagelinkage**, [MV] **cluster centroidlinkage**, [MV] **cluster completelinkage**, [MV] **cluster dendrogram**, [MV] **cluster generate**, [MV] **cluster kmeans**, [MV] **cluster kmedians**, [MV] **cluster medianlinkage**, [MV] **cluster notes**, [MV] **cluster singlelinkage**, [MV] **cluster stop**, [MV] **cluster utility**, [MV] **cluster wardslinkage**, [MV] **cluster waveragelinkage**
Related:	[MV] **clustermat**, [MV] **cluster programming subroutines**, [MV] **cluster programming utilities**

Title

> **clustermat** — Introduction to clustermat commands

Syntax

clustermat *method matname* ...

method	hierarchical clustering method
<u>s</u>inglelinkage	single linkage
<u>a</u>veragelinkage	average linkage
<u>c</u>ompletelinkage	complete linkage
<u>wav</u>eragelinkage	weighted average linkage
<u>med</u>ianlinkage	median linkage
<u>c</u>entroidlinkage	centroid linkage
<u>wards</u>linkage	Ward's linkage

See [MV] **cluster singlelinkage**, [MV] **cluster averagelinkage**, [MV] **cluster completelinkage**,
   [MV] **cluster waveragelinkage**, [MV] **cluster medianlinkage**, [MV] **cluster centroidlinkage**,
   and [MV] **cluster wardslinkage**.

clustermat stop has similar syntax to cluster stop; see [MV] **cluster stop**. For the remaining postclustering subcommands and user utilities, you may specify either cluster or clustermat—it does not matter which.

Description

clustermat performs hierarchical cluster analysis on the dissimilarity matrix *matname*. clustermat is part of the cluster suite of commands; see [MV] **cluster**. All of Stata's hierarchical clustering methods are allowed with clustermat. The partition clustering methods (kmeans and kmedians) are not allowed since they require the data.

See [MV] **cluster** for a listing of all the cluster and clustermat commands. The cluster dendrogram command (see [MV] **cluster dendrogram**) will display the resulting dendrogram, the clustermat stop command (see [MV] **cluster stop**) will help in determining the number of groups, and the cluster generate command (see [MV] **cluster generate**) will produce grouping variables. Other useful cluster subcommands include: notes, dir, list, drop, use, rename, and renamevar; see [MV] **cluster notes** and [MV] **cluster utility**.

Remarks

If you are clustering observations using one of the similarity or dissimilarity measures provided by Stata, the cluster command is what you need. If, however, you already have a dissimilarity matrix or can produce one for a dissimilarity measure that Stata does not provide, or if you want to cluster variables instead of observations, the clustermat command is what you need.

▷ Example 1

   Table 6 of Kaufman and Rousseeuw (1990) provides a subjective dissimilarity matrix between 11
sciences. Fourteen postgraduate economics students from different parts of the world gave subjective
dissimilarities between these 11 sciences on a scale from 0 (identical) to 10 (very different). The final
dissimilarity matrix was obtained by averaging the results from the fourteen students.

   We begin by creating a label variable and a shorter version of the label variable corresponding to
the 11 sciences. Then we create a row vector containing the lower triangle of the dissimilarity matrix.

```
. clear
. input str13 science
           science
  1. Astronomy
  2. Biology
  3. Chemistry
  4. Computer sci.
  5. Economics
  6. Geography
  7. History
  8. Mathematics
  9. Medicine
 10. Physics
 11. Psychology
 12. end
. gen str4 shortsci = substr(science,1,4)
. matrix input D = (
 0.00
 7.86 0.00
 6.50 2.93 0.00
 5.00 6.86 6.50 0.00
 8.00 8.14 8.21 4.79 0.00
 4.29 7.00 7.64 7.71 5.93 0.00
 8.07 8.14 8.71 8.57 5.86 3.86 0.00
 3.64 7.14 4.43 1.43 3.57 7.07 9.07 0.0
 8.21 2.50 2.93 6.36 8.43 7.86 8.43 6.29 0.00
 2.71 5.21 4.57 4.21 8.36 7.29 8.64 2.21 5.07 0.00
 9.36 5.57 7.29 7.21 6.86 8.29 7.64 8.71 3.79 8.64 0.00 )
```

   There are several ways we could have stored the dissimilarity information in a matrix. To avoid
entering both the upper and lower triangle of the matrix, we entered the dissimilarities as a row
vector containing the lower triangular entries of the dissimilarity matrix, including the diagonal of
zeros (although there are options that would allow us to omit the diagonal of zeros). Notice that we
typed `matrix input D = ...` instead of `matrix D = ...` so that we could omit the commas between
entries; see [P] **matrix define**.

   We now perform a complete-linkage cluster analysis on these dissimilarities. The `name()` option
names the cluster analysis. We will name it `complink`. The `shape(lower)` option is what signals
that the dissimilarity matrix is stored as a row vector containing the lower triangle of the dissimilarity
matrix, including the diagonal of zeros. The `add` option indicates that the resulting cluster information
should be added to the existing dataset. In our case, the existing dataset consists of the `science` label
variable and the shortened version `shortsci`. See [MV] **cluster completelinkage** for details concerning
these options. The short labels are passed to `cluster dendrogram` so that we can see which subjects
were most-closely related when viewing the dendrogram; see [MV] **cluster dendrogram**.

```
. clustermat completelinkage D, shape(lower) add name(complink)
. cluster dendrogram complink, labels(shortsci)
                 title(Complete-linkage clustering)
                 ytitle("Subjective dissimilarity"
                     "0=Same, 10=Very different")
```

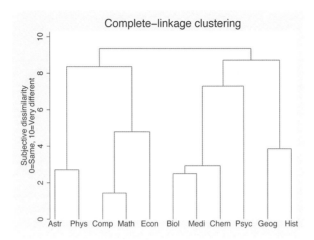

From the dendrogram, we see that mathematics and computer science were deemed most similar and that the economists most closely related their field of study to those two disciplines.

◁

▷ Example 2

Stata does not provide the Bray and Curtis (1957) dissimilarity measure first described by Odum (1950). Using the same notation as found in [MV] *measure_option*, the Bray–Curtis dissimilarity between observations i and j is

$$\frac{\sum_{a=1}^{p} |x_{ia} - x_{ja}|}{\sum_{a=1}^{p} (x_{ia} + x_{ja})}$$

Stata does not provide this measure because of the numerous cases where the measure is undefined (due to dividing by zero). However, when the data are positive the Bray–Curtis dissimilarity is well behaved.

Even though Stata does not automatically provide this measure, it is easy to obtain it and then use it with clustermat to perform hierarchical clustering. The numerator of the Bray–Curtis dissimilarity measure is the L1 (absolute value) distance. We use the matrix dissimilarity command (see [MV] **matrix dissimilarity**) to obtain the L1 dissimilarity matrix, and then divide the elements of that matrix by the appropriate values to obtain the Bray–Curtis dissimilarity.

Fisher (1936) presented data, originally from Anderson (1935), on three species of iris. Measurements of the length and width of the sepal and petal were obtained for 50 samples of each of the three iris species. We obtained the data from Morrison (2005). Here we demonstrate average-linkage clustering of these 150 observations.

```
. use http://www.stata-press.com/data/r9/iris, clear
(Iris data)

. summarize seplen sepwid petlen petwid
```

Variable	Obs	Mean	Std. Dev.	Min	Max
seplen	150	5.843333	.8280661	4.3	7.9
sepwid	150	3.057333	.4358663	2	4.4
petlen	150	3.758	1.765298	1	6.9
petwid	150	1.199333	.7622377	.1	2.5

```
. matrix dissimilarity irisD = seplen sepwid petlen petwid, L1

. egen rtot = rowtotal(seplen sepwid petlen petwid)

. forvalues a = 1/150 {
  2.          forvalues b = 1/150 {
  3.                  mat irisD['a','b'] = irisD['a','b']/(rtot['a']+rtot['b'])
  4.          }
  5. }

. matlist irisD[1..5,1..5]
```

	obs1	obs2	obs3	obs4	obs5
obs1	0				
obs2	.035533	0			
obs3	.0408163	.026455	0		
obs4	.0510204	.026455	.0212766	0	
obs5	.0098039	.035533	.0408163	.0510204	0

The egen rowtotal() function provided the row totals used in the denominator of the Bray–Curtis dissimilarity measure; see [D] **egen**. We listed the dissimilarities between the first five observations.

We now compute the average-linkage cluster analysis on these 150 observations (see [MV] **cluster averagelinkage**) and examine the Caliński and Harabasz pseudo-F index and the Duda and Hart Je(2)/Je(1) index (cluster stopping rules; see [MV] **cluster stop**) to try to determine the number of clusters.

```
. clustermat averagelink irisD, name(iris) add

. clustermat stop, variables(seplen sepwid petlen petwid)
```

Number of clusters	Calinski/ Harabasz pseudo-F
2	502.82
3	299.96
4	201.58
5	332.89
6	288.61
7	244.61
8	252.39
9	223.28
10	268.47
11	241.51
12	232.61
13	233.46
14	255.84
15	273.96

```
. clustermat stop, variables(seplen sepwid petlen petwid) rule(duda)
```

Number of clusters	Je(2)/Je(1)	Duda/Hart pseudo T-squared
1	0.2274	502.82
2	0.8509	17.18
3	0.8951	5.63
4	0.4472	116.22
5	0.6248	28.23
6	0.9579	2.55
7	0.5438	28.52
8	0.8843	5.10
9	0.5854	40.37
10	0.0000	.
11	0.8434	6.68
12	0.4981	37.28
13	0.5526	25.91
14	0.6342	16.15
15	0.6503	3.23

The stopping rules are not conclusive in this case. From the Duda and Hart pseudo T-squared (small values) you might best conclude that there are 3, 6, or 8 natural clusters. The Caliński and Harabasz pseudo-F (large values) indicates that there might be 2, 3 or 5 groups.

With the iris data, we know the three species. Let's compare the average-linkage hierarchical cluster solutions with the actual species. The cluster generate command (see [MV] **cluster generate**) will generate grouping variables for our hierarchical cluster analysis.

```
. cluster generate g = groups(2/6)
. tabulate g2 species
```

g2	Iris seto	Iris vers	Iris virg	Total
1	50	0	0	50
2	0	50	50	100
Total	50	50	50	150

```
. tabulate g3 species
```

g3	Iris seto	Iris vers	Iris virg	Total
1	50	0	0	50
2	0	46	50	96
3	0	4	0	4
Total	50	50	50	150

```
. tabulate g4 species
```

g4	Iris seto	Iris vers	Iris virg	Total
1	49	0	0	49
2	1	0	0	1
3	0	46	50	96
4	0	4	0	4
Total	50	50	50	150

```
. tabulate g5 species
```

	Iris seto	Iris vers	Iris virg	Total
g5		Iris species		
1	49	0	0	49
2	1	0	0	1
3	0	45	15	60
4	0	1	35	36
5	0	4	0	4
Total	50	50	50	150

```
. tabulate g6 species
```

	Iris seto	Iris vers	Iris virg	Total
g6		Iris species		
1	41	0	0	41
2	8	0	0	8
3	1	0	0	1
4	0	45	15	60
5	0	1	35	36
6	0	4	0	4
Total	50	50	50	150

The 2-group cluster solution splits Iris setosa from Iris versicolor and Iris virginica. The 3- and 4-group cluster solutions appear to split off some outlying observations from the two main groups. The 5-group solution finally splits the majority of Iris virginica from the Iris versicolor, but leaves some overlap.

Though this is not shown here, cluster solutions that better match the known species can be found using dissimilarity measures other than Bray–Curtis.

◁

▷ Example 3

The cluster command clusters observations. If you want to cluster variables, you have two choices. You can use xpose (see [D] **xpose**) to transpose the variables and observations, or you can use matrix dissimilarity with the variables option (see [MV] **matrix dissimilarity**) and then use clustermat.

In example 2 of [MV] **cluster kmeans**, we introduce the women's club data. 30 women were asked 35 yes/no questions. In [MV] **cluster kmeans** and [MV] **cluster kmedians**, our interest was in clustering the 30 women for placement at luncheon tables. Here our interest is in understanding the relationship between the 35 variables. Which questions produced similar response patterns from the 30 women?

```
. use http://www.stata-press.com/data/r9/wclub

. describe
```

Contains data from http://www.stata-press.com/data/r9/wclub.dta
```
  obs:            30
 vars:            35                          1 May 2005 16:56
 size:         1,170 (99.9% of memory free)
```

variable name	storage type	display format	value label	variable label
bike	byte	%8.0g		enjoy bicycle riding Y/N
bowl	byte	%8.0g		enjoy bowling Y/N
swim	byte	%8.0g		enjoy swimming Y/N
jog	byte	%8.0g		enjoy jogging Y/N
hock	byte	%8.0g		enjoy watching hockey Y/N
foot	byte	%8.0g		enjoy watching football Y/N
base	byte	%8.0g		enjoy baseball Y/N
bask	byte	%8.0g		enjoy basketball Y/N
arob	byte	%8.0g		participate in aerobics Y/N
fshg	byte	%8.0g		enjoy fishing Y/N
dart	byte	%8.0g		enjoy playing darts Y/N
clas	byte	%8.0g		enjoy classical music Y/N
cntr	byte	%8.0g		enjoy country music Y/N
jazz	byte	%8.0g		enjoy jazz music Y/N
rock	byte	%8.0g		enjoy rock and roll music Y/N
west	byte	%8.0g		enjoy reading western novels Y/N
romc	byte	%8.0g		enjoy reading romance novels Y/N
scif	byte	%8.0g		enjoy reading sci. fiction Y/N
biog	byte	%8.0g		enjoy reading biographies Y/N
fict	byte	%8.0g		enjoy reading fiction Y/N
hist	byte	%8.0g		enjoy reading history Y/N
cook	byte	%8.0g		enjoy cooking Y/N
shop	byte	%8.0g		enjoy shopping Y/N
soap	byte	%8.0g		enjoy watching soap operas Y/N
sew	byte	%8.0g		enjoy sewing Y/N
crft	byte	%8.0g		enjoy craft activities Y/N
auto	byte	%8.0g		enjoy automobile mechanics Y/N
pokr	byte	%8.0g		enjoy playing poker Y/N
brdg	byte	%8.0g		enjoy playing bridge Y/N
kids	byte	%8.0g		have children Y/N
hors	byte	%8.0g		have a horse Y/N
cat	byte	%8.0g		have a cat Y/N
dog	byte	%8.0g		have a dog Y/N
bird	byte	%8.0g		have a bird Y/N
fish	byte	%8.0g		have a fish Y/N

Sorted by:

The `matrix dissimilarity` command allows us to compute the Jaccard similarity measure (the Jaccard option), comparing variables (the `variables` option) instead of observations, saving one minus the Jaccard measure (the `dissim(oneminus)` option) as a dissimilarity matrix.

```
. matrix dissimilarity clubD = , variables Jaccard dissim(oneminus)

. matlist clubD[1..5,1..5]
```

	bike	bowl	swim	jog	hock
bike	0				
bowl	.7333333	0			
swim	.5625	.625	0		
jog	.6	.8235294	.5882353	0	
hock	.8461538	.6	.8	.8571429	0

We pass the clubD matrix to clustermat and ask for a single-linkage cluster analysis. We need to specify the clear option in order to replace the 30 observations currently in memory with the 35 observations containing the cluster results. Using the labelvar() option, we also ask for a label variable, question, to be created from the clubD matrix row names. To visualize the resulting cluster analysis, we call cluster dendrogram; see [MV] **cluster dendrogram**.

```
. clustermat singlelink clubD, name(club) clear labelvar(question)
obs was 0, now 35

. describe

Contains data
  obs:            35
  vars:            4
  size:          630 (99.9% of memory free)
```

variable name	storage type	display format	value label	variable label
club_id	byte	%8.0g		
club_ord	byte	%8.0g		
club_hgt	double	%10.0g		
question	str4	%9s		

```
Sorted by:
    Note:  dataset has changed since last saved

. cluster dendrogram club, labels(question)
                  xlabel(, angle(90) labsize(*.75))
                  title(Single-linkage clustering)
                  ytitle(1 - Jaccard similarity, suffix)
```

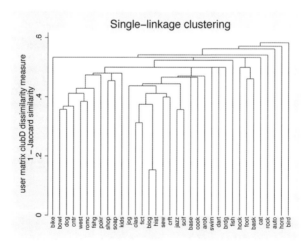

From these 30 women, we see that the biog (enjoy reading biographies) and hist (enjoy reading history) questions were most closely related. auto (enjoy automobile mechanics), hors (have a horse), and bird (have a bird) seem to be the least related to the other variables. These three variables, in turn, merge last into the super group containing the remaining variables.

The variables of this dataset are clustered using each of the available hierarchical clustering methods. See example 3 of [MV] **cluster averagelinkage**, [MV] **cluster centroidlinkage**, [MV] **cluster completelinkage**, [MV] **cluster medianlinkage**, [MV] **cluster singlelinkage**, [MV] **cluster wardslinkage**, and [MV] **cluster waveragelinkage**.

◁

References

Anderson, E. 1935. The irises of the Gaspe Peninsula. *Bulletin of the American Iris Society* 59: 2–5.

Bray, R. J. and J. T. Curtis. 1957. An ordination of the upland forest communities of southern Wisconsin. *Ecological Monographs* 27: 325–349.

Fisher, R. A. 1936. The use of multiple measurements in taxonomic problems. *Annals of Eugenics* 7: 179–188.

Kaufman, L. and P. J. Rousseeuw. 1990. *Finding Groups in Data.* New York: Wiley.

Morrison, D. F. 2005. *Multivariate Statistical Methods.* 4th ed. Belmont, CA: Duxbury.

Odum, E. P. 1950. Bird populations of the Highlands (North Carolina) plateau in relation to plant succession and avian invasion. *Ecology* 31: 587–605.

Also See

Complementary:	[MV] **cluster averagelinkage**, [MV] **cluster centroidlinkage**,
	[MV] **cluster completelinkage**, [MV] **cluster dendrogram**,
	[MV] **cluster generate**, [MV] **cluster medianlinkage**, [MV] **cluster notes**,
	[MV] **cluster singlelinkage**, [MV] **cluster stop**, [MV] **cluster utility**,
	[MV] **cluster wardslinkage**, [MV] **cluster waveragelinkage**
Related:	[MV] **cluster programming subroutines**,
	[MV] **cluster programming utilities**
Background:	[MV] **cluster**

Title

cluster averagelinkage — Average-linkage cluster analysis

Syntax

Cluster analysis of data

cluster averagelinkage [*varlist*] [*if*] [*in*] [, *cluster_options*]

Cluster analysis of a dissimilarity matrix

clustermat averagelinkage *matname* [*if*] [*in*] [, *clustermat_options*]

cluster_options	description
Main	
measure(*measure*)	similarity or dissimilarity measures; see *Options* for available measures; default is L2
name(*clname*)	name of resulting cluster analysis
Advanced	
generate(*stub*)	prefix for generated variables; default prefix is *clname*

clustermat_options	description
Main	
shape(*shape*)	shape (storage method) of *matname*
add	add cluster information to data currently in memory
clear	replace data in memory with cluster information
labelvar(*varname*)	place dissimilarity matrix row names in *varname*
name(*clname*)	name of resulting cluster analysis
Advanced	
force	perform clustering after fixing *matname* problems
generate(*stub*)	prefix for generated variables; default prefix is *clname*

shape	*matname* is stored as a
full	square symmetric matrix; the default
lower	vector of rowwise lower triangle (with diagonal)
llower	vector of rowwise strict lower triangle (no diagonal)
upper	vector of rowwise upper triangle (with diagonal)
uupper	vector of rowwise strict upper triangle (no diagonal)

Description

cluster averagelinkage performs hierarchical agglomerative average-linkage cluster analysis. See [MV] **cluster** for a general discussion of cluster analysis and a description of the other cluster commands.

clustermat averagelinkage performs hierarchical agglomerative average-linkage cluster analysis on the dissimilarity matrix *matname*. See [MV] **clustermat** for a general discussion of cluster analysis of dissimilarity matrices and a description of the other clustermat commands.

After cluster averagelinkage or clustermat averagelinkage, the cluster dendrogram command (see [MV] **cluster dendrogram**) displays the resulting dendrogram, the cluster stop or clustermat stop commands (see [MV] **cluster stop**) help determine the number of groups, and the cluster generate command (see [MV] **cluster generate**) produces grouping variables.

Options for cluster averagelinkage

‾‾| Main |‾‾‾

measure(*measure*) is one of the similarity or dissimilarity measures allowed by Stata. This option is not case sensitive. See [MV] *measure_option* for a discussion of these measures.

The available measures designed for continuous data are L2 (synonym Euclidean), which is the default; L2squared; L1 (synonyms absolute, cityblock, and manhattan); Linfinity (synonym maximum); L(#); Lpower(#); Canberra; correlation; and angular (synonym angle).

The available measures designed for binary data are matching, Jaccard, Russell, Hamann, Dice, antiDice, Sneath, Rogers, Ochiai, Yule, Anderberg, Kulczynski, Gower2, and Pearson.

name(*clname*) specifies the name to attach to the resulting cluster analysis. If name() is not specified, Stata finds an available cluster name, displays it for your reference, and attaches the name to your cluster analysis.

‾‾| Advanced |‾‾‾

generate(*stub*) provides a prefix for the variable names created by cluster averagelinkage. By default, the variable-name prefix will be the name specified in name(). Three variables are created and attached to the cluster-analysis results with the suffixes _id, _ord, and _hgt. Users generally will not need to access these variables directly.

Options for clustermat averagelinkage

‾‾| Main |‾‾‾

shape(*shape*) specifies the storage mode of *matname*, the matrix of dissimilarities. The following shapes are allowed:

full specifies that *matname* is an $n \times n$ symmetric matrix.

lower specifies that *matname* is a row or column vector of length $n(n+1)/2$, with the rowwise lower triangle of the dissimilarity matrix including the diagonal of zeros.

$$D_{11}\ D_{21}\ D_{22}\ D_{31}\ D_{32}\ D_{33}\ \dots\ D_{n1}\ D_{n2}\ \dots\ D_{nn}$$

llower specifies that *matname* is a row or column vector of length $n(n-1)/2$, with the rowwise lower triangle of the dissimilarity matrix excluding the diagonal.

$$\mathrm{D}_{21} \; \mathrm{D}_{31} \; \mathrm{D}_{32} \; \mathrm{D}_{41} \; \mathrm{D}_{42} \; \mathrm{D}_{43} \; \ldots \; \mathrm{D}_{n1} \; \mathrm{D}_{n2} \; \ldots \; \mathrm{D}_{n,n-1}$$

upper specifies that *matname* is a row or column vector of length $n(n+1)/2$, with the rowwise upper triangle of the dissimilarity matrix including the diagonal of zeros.

$$\mathrm{D}_{11} \; \mathrm{D}_{12} \; \ldots \; \mathrm{D}_{1n} \; \mathrm{D}_{22} \; \mathrm{D}_{23} \; \ldots \; \mathrm{D}_{2n} \; \mathrm{D}_{33} \; \mathrm{D}_{34} \; \ldots \; \mathrm{D}_{3n} \; \ldots \; \mathrm{D}_{nn}$$

uupper specifies that *matname* is a row or column vector of length $n(n-1)/2$, with the rowwise upper triangle of the dissimilarity matrix excluding the diagonal.

$$\mathrm{D}_{12} \; \mathrm{D}_{13} \; \ldots \; \mathrm{D}_{1n} \; \mathrm{D}_{23} \; \mathrm{D}_{24} \; \ldots \; \mathrm{D}_{2n} \; \mathrm{D}_{34} \; \mathrm{D}_{35} \; \ldots \; \mathrm{D}_{3n} \; \ldots \; \mathrm{D}_{n-1,n}$$

add specifies that clustermat's results be added to the dataset currently in memory. The number of observations (selected observations based on the if and in qualifiers) must equal the number of rows and columns of *matname*. Either clear or add is required if a dataset is currently in memory.

clear drops all the variables and cluster solutions in the current dataset in memory (even if that dataset has changed since the data were last saved) before generating clustermat's results. Either clear or add is required if a dataset is currently in memory.

labelvar(*varname*) specifies the name of a new variable to be created containing the row names of matrix *matname*.

name(*clname*) specifies the name to attach to the resulting cluster analysis. If name() is not specified, Stata finds an available cluster name, displays it for your reference, and attaches the name to your cluster analysis.

⌐ Advanced ⌐

force allows computations to continue when *matname* is nonsymmetric or has nonzeros on the diagonal. By default, clustermat will complain and exit when it encounters these conditions. force specifies that clustermat operate on the symmetric matrix $(matname * matname')/2$, with any nonzero diagonal entries treated as if they were zero.

generate(*stub*) provides a prefix for the variable names created by clustermat. By default, the variable-name prefix is the name specified in name(). Three variables are created and attached to the cluster-analysis results with the suffixes _id, _ord, and _hgt. Users generally will not need to access these variables directly.

Remarks

An example using the default L2 (Euclidean) distance on continuous data and an example using the matching coefficient on binary data illustrate the cluster averagelinkage command. A third example illustrates the use of clustermat averagelinkage in clustering variables instead of observations. These are the same datasets used as examples in [MV] **cluster centroidlinkage**, [MV] **cluster completelinkage**, [MV] **cluster medianlinkage**, [MV] **cluster singlelinkage**, [MV] **cluster wardslinkage**, and [MV] **cluster waveragelinkage** so that you can compare the results from using different hierarchical clustering methods.

▷ Example 1

As explained in the first example of [MV] **cluster singlelinkage**, as the senior data analyst for a small biotechnology firm, you are given a dataset with 4 chemical laboratory measurements on 50 different samples of a particular plant gathered from the rain forest. The head of the expedition that gathered the samples thinks, based on information from the natives, that an extract from the plant might reduce the negative side effects associated with your company's best-selling nutritional supplement.

While the company chemists and botanists continue exploring the possible uses of the plant and plan future experiments, the head of product development asks you to look at the preliminary data and report anything that might be helpful to the researchers.

While all 50 plants are supposed to be of the same type, you decide to perform a cluster analysis to see if there are subgroups or anomalies among them. Using single-linkage clustering, you discover an anomaly in the data. You now wish to see if you discover the same thing using average-linkage clustering with the default Euclidean distance.

You first call `cluster averagelinkage` and use the `name()` option to attach the name L2alnk to the resulting cluster analysis. The `cluster list` command (see [MV] **cluster utility**) is then applied to list the components of your cluster analysis. The `cluster dendrogram` command then graphs the dendrogram; see [MV] **cluster dendrogram**. As described in the [MV] **cluster singlelinkage** example, the `labels()` option is used, instead of the default action of showing the observation number, to identify which laboratory technician produced the data.

```
. use http://www.stata-press.com/data/r9/labtech
. cluster averagelinkage x1 x2 x3 x4, name(L2alnk)
. cluster list L2alnk
L2alnk  (type: hierarchical,  method: average,  dissimilarity: L2)
     vars: L2alnk_id (id variable)
           L2alnk_ord (order variable)
           L2alnk_hgt (height variable)
    other: range: 0 .
           cmd: cluster averagelinkage x1 x2 x3 x4, name(L2alnk)
           varlist: x1 x2 x3 x4
. cluster dendrogram L2alnk, labels(labtech) xlabel(, angle(90) labsize(*.75))
```

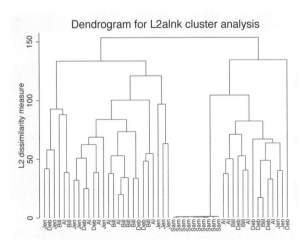

As with single-linkage clustering, you see that the samples analyzed by Sam, the lab technician, cluster together closely (dissimilarity measures near zero) and are separated from the rest of the data by a large dissimilarity gap (the long vertical line going up from Sam's cluster to eventually combine with other observations). When you examined the data, you discovered that Sam's data are all between zero and one, while the other four technicians have data that range from zero up to near 150. It appears that Sam has made a mistake.

If you compare the dendrogram from this average-linkage clustering with those from single-linkage clustering and complete-linkage clustering, you will notice that the y-axis range falls between that of these two other methods. This is a property of these linkage methods. With average linkage, the average of the (dis)similarities between the two groups determines the distance between the groups. This is in contrast to the smallest distance and largest distance, which define single-linkage and complete-linkage clustering.

◁

▷ Example 2

This example analyzes the same data that was introduced in the second example of [MV] **cluster singlelinkage**. The sociology professor of your graduate-level class gives, as homework, a dataset containing 30 observations on 60 binary variables, with the assignment to tell him something about the 30 subjects represented by the observations.

In addition to examining single-linkage clustering of these data, you decide to see what average-linkage clustering shows. As with the single-linkage clustering, you pick the simple matching binary coefficient to measure the similarity between groups. The name() option is used to attach the name alink to the cluster analysis. cluster list displays the details; see [MV] **cluster utility**. cluster tree, which is a synonym for cluster dendrogram, then displays the cluster tree (dendrogram); see [MV] **cluster dendrogram**.

```
. use http://www.stata-press.com/data/r9/homework, clear
. cluster a a1-a60, measure(match) name(alink)
. cluster list alink
alink  (type: hierarchical,  method: average,  similarity: matching)
       vars: alink_id (id variable)
             alink_ord (order variable)
             alink_hgt (height variable)
      other: range: 1 0
             cmd: cluster averagelinkage a1-a60, measure(match) name(alink)
             varlist: a1 a2 a3 a4 a5 a6 a7 a8 a9 a10 a11 a12 a13 a14 a15 a16 a17
                 a18 a19 a20 a21 a22 a23 a24 a25 a26 a27 a28 a29 a30 a31 a32
                 a33 a34 a35 a36 a37 a38 a39 a40 a41 a42 a43 a44 a45 a46 a47
                 a48 a49 a50 a51 a52 a53 a54 a55 a56 a57 a58 a59 a60
```

. cluster tree

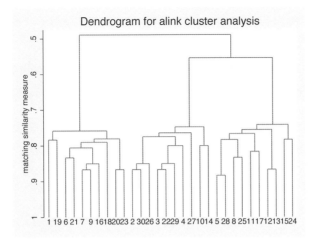

Since, by default, Stata uses the most-recently performed cluster analysis, you do not need to type the cluster name when calling `cluster tree`.

As with single-linkage clustering, the dendrogram from average-linkage clustering seems to indicate the presence of 3 groups among the 30 observations. Later you receive another variable called `truegrp` that identifies the groups that the teacher believes are in the data. You use the `cluster generate` command (see [MV] **cluster generate**) to create a grouping variable, based on your average-linkage clustering, to compare with `truegrp`. You do a cross-tabulation of `truegrp` and `agrp3`, your grouping variable, to see if your conclusions match those of the teacher.

. cluster gen agrp3 = group(3)

. table agrp3 truegrp

agrp3	truegrp		
	1	2	3
1		10	
2	10		
3			10

Other than the numbers arbitrarily assigned to the three groups, your teacher's conclusions and the results from the average-linkage clustering are in complete agreement.

◁

▷ Example 3

The `wclub` dataset contains answers from 30 women to 35 yes/no questions. The variables are described in example 3 of [MV] **clustermat**. We are interested in seeing how average-linkage clustering will cluster the 35 variables (instead of the observations).

We use the `matrix dissimilarity` command to produce a dissimilarity matrix equal to one minus the Jaccard similarity; see [MV] **matrix dissimilarity**.

. use http://www.stata-press.com/data/r9/wclub

. matrix dissimilarity clubD = , variables Jaccard dissim(oneminus)

```
. clustermat averagelink clubD, name(clubavg) clear labelvar(question)
obs was 0, now 35

. cluster dendrogram clubavg, labels(question)
                 xlabel(, angle(90) labsize(*.75))
                 title(Average-linkage clustering)
                 ytitle(1 - Jaccard similarity, suffix)
```

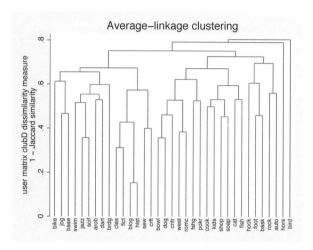

From these 30 women, we see that the `biog` (enjoy reading biographies) and `hist` (enjoy reading history) questions were most closely related. `hors` (have a horse) and `bird` (have a bird) seem to be the least related to the other variables. These two variables, in turn, merge last into the super group containing the remaining variables.

◁

❑ Technical Note

    `cluster averagelinkage` requires more memory and more execution time than `cluster singlelinkage` does. With a large number of observations, the execution time may be significant.

❑

Methods and Formulas

[MV] **cluster** discusses hierarchical clustering and places average-linkage clustering in this general framework. It compares average linkage to single linkage and complete linkage.

Conceptually, hierarchical agglomerative average-linkage clustering proceeds as follows. The N observations start out as N separate groups, each of size one. The two closest observations are merged into one group, producing $N-1$ total groups. The closest two groups are then merged so that there are $N-2$ total groups. This process continues until all the observations are merged into one large group, producing a hierarchy of groupings from one group to N groups. For average-linkage clustering, the "closest two groups" are determined by the average (dis)similarity between the observations of the two groups.

The average-linkage clustering algorithm produces two variables that together act as a pointer representation of a dendrogram. To this, Stata adds a third variable used to restore the sort order, as needed, so that the two variables of the pointer representation remain valid. The first variable of the pointer representation gives the order of the observations. The second variable has one less element and gives the height in the dendrogram at which the adjacent observations in the order-variable join.

See [MV] *measure_option* for the details and formulas of the available *measure*s, which include (dis)similarity measures for continuous and binary data.

Also See

Complementary:	[MV] **cluster dendrogram**, [MV] **cluster generate**, [MV] **cluster notes**, [MV] **cluster stop**, [MV] **cluster utility**
Related:	[MV] **cluster centroidlinkage**, [MV] **cluster completelinkage**, [MV] **cluster medianlinkage**, [MV] **cluster singlelinkage**, [MV] **cluster wardslinkage**, [MV] **cluster waveragelinkage**
Background:	[MV] **cluster**, [MV] **clustermat**

Title

> **cluster centroidlinkage** — Centroid-linkage cluster analysis

Syntax

Cluster analysis of data

> cluster <u>cent</u>roidlinkage [*varlist*] [*if*] [*in*] [, *cluster_options*]

Cluster analysis of a dissimilarity matrix

> clustermat <u>cent</u>roidlinkage *matname* [*if*] [*in*] [, *clustermat_options*]

cluster_options	description
Main	
<u>me</u>asure(*measure*)	similarity or dissimilarity measures; see *Options* for available measures; default is L2squared
<u>n</u>ame(*clname*)	name of resulting cluster analysis
Advanced	
<u>g</u>enerate(*stub*)	prefix for generated variables; default prefix is *clname*

clustermat_options	description
Main	
<u>sh</u>ape(*shape*)	shape (storage method) of *matname*
add	add cluster information to data currently in memory
clear	replace data in memory with cluster information
<u>l</u>abelvar(*varname*)	place dissimilarity matrix row names in *varname*
<u>n</u>ame(*clname*)	name of resulting cluster analysis
Advanced	
force	perform clustering after fixing *matname* problems
<u>g</u>enerate(*stub*)	prefix for generated variables; default prefix is *clname*

shape	*matname* is stored as a
<u>f</u>ull	square symmetric matrix; the default
<u>l</u>ower	vector of rowwise lower triangle (with diagonal)
<u>ll</u>ower	vector of rowwise strict lower triangle (no diagonal)
<u>u</u>pper	vector of rowwise upper triangle (with diagonal)
<u>uu</u>pper	vector of rowwise strict upper triangle (no diagonal)

Description

cluster centroidlinkage performs hierarchical agglomerative centroid-linkage cluster analysis. See [MV] **cluster** for a general discussion of cluster analysis and a description of the other cluster commands.

clustermat centroidlinkage performs hierarchical agglomerative centroid-linkage cluster analysis on the dissimilarity matrix *matname*. See [MV] **clustermat** for a general discussion of cluster analysis of dissimilarity matrices and a description of the other clustermat commands.

After cluster centroidlinkage or cluster centroidlinkage, the cluster dendrogram command (see [MV] **cluster dendrogram**) displays the resulting dendrogram, the cluster stop or clustermat stop commands (see [MV] **cluster stop**) help determine the number of groups, and the cluster generate command (see [MV] **cluster generate**) produces grouping variables.

Options for cluster centroidlinkage

⌐ Main ⌐

measure(*measure*) is one of the similarity or dissimilarity measures allowed by Stata. This option is not case sensitive. See [MV] *measure_option* for a discussion of these measures.

The available measures designed for continuous data are L2 (synonym Euclidean); L2squared, which is the default for cluster centroidlinkage; L1 (synonyms absolute, cityblock, and manhattan); Linfinity (synonym maximum); L(#); Lpower(#); Canberra; correlation; and angular (synonym angle).

The available measures designed for binary data are matching, Jaccard, Russell, Hamann, Dice, antiDice, Sneath, Rogers, Ochiai, Yule, Anderberg, Kulczynski, Gower2, and Pearson.

Several authors advise using the L2squared *measure* exclusively with centroid linkage. See *(Dis)similarity transformations and the Lance and Williams formula* and *Warning concerning (dis)similarity choice* in [MV] **cluster** for details.

name(*clname*) specifies the name to attach to the resulting cluster analysis. If name() is not specified, Stata finds an available cluster name, displays it for your reference, and attaches the name to your cluster analysis.

⌐ Advanced ⌐

generate(*stub*) provides a prefix for the variable names created by cluster centroidlinkage. By default, the variable-name prefix is the name specified in name(). Three variables are created and attached to the cluster-analysis results with the suffixes _id, _ord, and _hgt. Users generally will not need to access these variables directly.

Centroid linkage can produce reversals or crossovers; see [MV] **cluster** for details. When reversals happen, cluster centroidlinkage also creates a fourth variable with the suffix _pht. This is a pseudoheight variable that is used by some of the postclustering commands to properly interpret the _hgt variable.

Options for clustermat centroidlinkage

shape(*shape*) specifies the storage mode of *matname*, the matrix of dissimilarities. The following shapes are allowed:

full specifies that *matname* is an $n \times n$ symmetric matrix.

lower specifies that *matname* is a row or column vector of length $n(n+1)/2$, with the rowwise lower triangle of the dissimilarity matrix including the diagonal of zeros.

$$D_{11}\ D_{21}\ D_{22}\ D_{31}\ D_{32}\ D_{33}\ \ldots\ D_{n1}\ D_{n2}\ \ldots\ D_{nn}$$

llower specifies that *matname* is a row or column vector of length $n(n-1)/2$, with the rowwise lower triangle of the dissimilarity matrix excluding the diagonal.

$$D_{21}\ D_{31}\ D_{32}\ D_{41}\ D_{42}\ D_{43}\ \ldots\ D_{n1}\ D_{n2}\ \ldots\ D_{n,n-1}$$

upper specifies that *matname* is a row or column vector of length $n(n+1)/2$, with the rowwise upper triangle of the dissimilarity matrix including the diagonal of zeros.

$$D_{11}\ D_{12}\ \ldots\ D_{1n}\ D_{22}\ D_{23}\ \ldots\ D_{2n}\ D_{33}\ D_{34}\ \ldots\ D_{3n}\ \ldots\ D_{nn}$$

uupper specifies that *matname* is a row or column vector of length $n(n-1)/2$, with the rowwise upper triangle of the dissimilarity matrix excluding the diagonal.

$$D_{12}\ D_{13}\ \ldots\ D_{1n}\ D_{23}\ D_{24}\ \ldots\ D_{2n}\ D_{34}\ D_{35}\ \ldots\ D_{3n}\ \ldots\ D_{n-1,n}$$

add specifies that clustermat's results be added to the dataset currently in memory. The number of observations (selected observations based on the if and in qualifiers) must equal the number of rows and columns of matname. Either clear or add is required if a dataset is currently in memory.

clear drops all the variables and cluster solutions in the current dataset in memory (even if that dataset has changed since the data were last saved) before generating clustermat's results. Either clear or add is required if a dataset is currently in memory.

labelvar(*varname*) specifies the name of a new variable to be created containing the row names of matrix *matname*.

name(*clname*) specifies the name to attach to the resulting cluster analysis. If name() is not specified, Stata finds an available cluster name, displays it for your reference, and attaches the name to your cluster analysis.

force allows computations to continue when *matname* is nonsymmetric or has nonzeros on the diagonal. By default, clustermat will complain and exit when it encounters these conditions. force specifies that clustermat operate on the symmetric matrix $(matname * matname')/2$, with any nonzero diagonal entries treated as if they were zero.

generate(*stub*) provides a prefix for the variable names created by cluster centroidlinkage. By default, the variable-name prefix will be the name specified in name(). Three variables are created and attached to the cluster-analysis results with the suffixes _id, _ord, and _hgt. Users generally will not need to access these variables directly.

Centroid linkage can produce reversals or crossovers; see [MV] **cluster** for details. When reversals happen, cluster centroidlinkage also creates a fourth variable with the suffix _pht. This is a pseudoheight variable that is used by some of the postclustering commands to properly interpret the _hgt variable.

Remarks

An example using the default L2squared (squared Euclidean) distance and L2 (Euclidean) distance on continuous data and an example using the matching coefficient on binary data illustrate the cluster centroidlinkage command. A third example illustrates the use of clustermat centroidlinkage in clustering variables instead of observations. These are the same datasets introduced in [MV] **cluster singlelinkage**, which are used as examples for all the hierarchical clustering methods so that you can compare the results from different hierarchical clustering methods.

▷ Example 1

As explained in the first example of [MV] **cluster singlelinkage**, as the senior data analyst for a small biotechnology firm, you are given a dataset with 4 chemical laboratory measurements on 50 different samples of a particular plant gathered from the rain forest. The head of the expedition that gathered the samples thinks, based on information from the natives, that an extract from the plant might reduce the negative side effects associated with your company's best-selling nutritional supplement.

While the company chemists and botanists continue exploring the possible uses of the plant and plan future experiments, the head of product development asks you to look at the preliminary data and report anything that might be helpful to the researchers.

While all 50 of the plants are supposed to be of the same type, you decide to perform a cluster analysis to see if there are subgroups or anomalies among them. Single-linkage clustering helped you discover an anomaly in the data. You now wish to see if you discover the same thing using centroid-linkage clustering with the default squared Euclidean distance and with Euclidean distance.

You first call cluster centroidlinkage, letting the distance default to L2squared (squared Euclidean distance), and use the name() option to attach the name cent to the resulting cluster analysis. You then use the cluster list command (see [MV] **cluster utility**) to list the components of your cluster analysis.

```
. use http://www.stata-press.com/data/r9/labtech
. cluster centroidlinkage x1 x2 x3 x4, name(cent)
. cluster list cent
cent  (type: hierarchical,  method: centroid,  dissimilarity: L2squared)
        vars: cent_id (id variable)
              cent_ord (order variable)
              cent_hgt (real_height variable)
              cent_pht (pseudo_height variable)
       other: range: 0 .
              cmd: cluster centroidlinkage x1 x2 x3 x4, name(cent)
              varlist: x1 x2 x3 x4
```

You repeat the same process, this time using L2 (Euclidean distance) and giving it the name L2cent.

```
. cluster centroidlinkage x1 x2 x3 x4, name(L2cent) measure(L2)
. cluster list L2cent
L2cent  (type: hierarchical,  method: centroid,  dissimilarity: L2)
      vars: L2cent_id (id variable)
            L2cent_ord (order variable)
            L2cent_hgt (real_height variable)
            L2cent_pht (pseudo_height variable)
     other: range: 0 .
            cmd: cluster centroidlinkage x1 x2 x3 x4, name(L2cent) measure(L2)
            varlist: x1 x2 x3 x4
```

You wish to use the `cluster dendrogram` command to graph the dendrogram (see [MV] **cluster dendrogram**), but since this particular cluster analysis produces reversals, Stata refuses to draw the dendrogram.

You decide to use the `cluster generate` command (see [MV] **cluster generate**) to produce grouping variables for 2 to 10 groups for each of the two cluster analyses. You also wish to examine the cross-tabulation of each of these generated groups against a variable that identifies which laboratory technician produced the data. (For the sake of brevity, only one cross-tabulation is shown for each cluster analysis.)

```
. cluster gen gc = groups(2/10), name(cent)
. cluster gen gL2c = groups(2/10), name(L2cent)
. table labtech gc6
```

			gc6			
labtech	1	2	3	4	5	6
Al	1	5			4	
Bill	2	5			3	
Deb	1	4			4	1
Jen	2	3	3		2	
Sam				10		

```
. table labtech gL2c2
```

	gL2c2	
labtech	1	2
Al	10	
Bill	10	
Deb	10	
Jen	10	
Sam		10

The samples analyzed by Sam appear to be clustered together more strongly than the samples analyzed by the other technicians. The reason for this phenomenon is not as obvious from this analysis as it was when we viewed the dendrogram from the single-linkage clustering (see [MV] **cluster singlelinkage**).

◁

▷ Example 2

This example analyzes the same data that was introduced in the second example of [MV] **cluster singlelinkage**. The sociology professor of your graduate-level class gives, as homework, a dataset containing 30 observations on 60 binary variables, with the assignment to tell him something about the 30 subjects represented by the observations.

In addition to examining single-linkage clustering of these data, you decide to see what centroid-linkage clustering shows. As with the single-linkage clustering, you pick the simple matching binary coefficient to measure the similarity between groups. The name() option is used to attach the name centlink to the cluster analysis. cluster list displays the details; see [MV] **cluster utility**.

```
. use http://www.stata-press.com/data/r9/homework
. cluster cent a1-a60, measure(match) name(centlink)
. cluster list centlink
centlink  (type: hierarchical,  method: centroid,  similarity: matching)
      vars: centlink_id (id variable)
            centlink_ord (order variable)
            centlink_hgt (real_height variable)
            centlink_pht (pseudo_height variable)
     other: range: 1 0
            cmd: cluster centroidlinkage a1-a60, measure(match) name(centlink)
            varlist: a1 a2 a3 a4 a5 a6 a7 a8 a9 a10 a11 a12 a13 a14 a15 a16 a17
                a18 a19 a20 a21 a22 a23 a24 a25 a26 a27 a28 a29 a30 a31 a32
                a33 a34 a35 a36 a37 a38 a39 a40 a41 a42 a43 a44 a45 a46 a47
                a48 a49 a50 a51 a52 a53 a54 a55 a56 a57 a58 a59 a60
```

You attempt to use the cluster dendrogram command to display the dendrogram, but since this particular cluster analysis produced reversals, cluster dendrogram refuses to produce the dendrogram. You realize that with reversals, the resulting dendrogram would not be easy to interpret anyway.

You decide to compare the three-group solution from this centroid-linkage clustering with the variable called truegrp provided by the teacher. You use the cluster generate command (see [MV] **cluster generate**) to create a grouping variable, based on your centroid clustering, to compare with truegrp.

```
. cluster gen centgrp3 = group(3)
. table centgrp3 truegrp
```

centgrp3	truegrp 1	2	3
1		10	
2	10		
3			10

Other than the numbers arbitrarily assigned to the three groups, your teacher's conclusions and the results from the three-group centroid-linkage clustering are in agreement.

◁

▷ Example 3

The wclub dataset contains answers from 30 women to 35 yes/no questions. The variables are described in example 3 of [MV] **clustermat**. We are interested in seeing how centroid-linkage clustering will cluster the 35 variables (instead of the observations).

We use the `matrix dissimilarity` command to produce a dissimilarity matrix equal to one minus the Jaccard similarity; see [MV] **matrix dissimilarity**.

```
. use http://www.stata-press.com/data/r9/wclub
. matrix dissimilarity clubD = , variables Jaccard dissim(oneminus)
. clustermat centroid clubD, name(clubcent) clear labelvar(question)
obs was 0, now 35
```

We cannot draw the dendrogram due to reversals, but we can examine some of the groupings produced by the cluster analysis.

```
. cluster generate g = groups(2/4)
. table g2
```

g2	Freq.
1	34
2	1

```
. table g3
```

g3	Freq.
1	33
2	1
3	1

```
. table g4
```

g4	Freq.
1	32
2	1
3	1
4	1

```
. list g2 g3 g4 question if g4 > 1
```

	g2	g3	g4	question
5.	2	3	4	hock
31.	1	1	2	hors
34.	1	2	3	bird

`hors` (have a horse), `bird` (have a bird), and `hock` (enjoy watching hockey) seem to be the least related to the other variables. These three variables, in turn, merge last into the super group containing the remaining variables.

◁

❏ Technical Note

`cluster centroidlinkage` requires more memory and more execution time than `cluster singlelinkage`. With a large number of observations, the execution time may be significant.

❏

Methods and Formulas

[MV] **cluster** discusses hierarchical clustering and places centroid-linkage clustering in this general framework. Conceptually, hierarchical agglomerative clustering proceeds as follows. The N observations start out as N separate groups, each of size one. The two closest observations are merged into one group, producing $N - 1$ total groups. The closest two groups are then merged so that there are $N - 2$ total groups. This process continues until all the observations are merged into one large group, producing a hierarchy of groupings from one group to N groups. The difference between the various hierarchical-linkage methods depends on how they define "closest" when comparing groups. Centroid linkage merges the groups whose means are closest.

The centroid-linkage clustering algorithm produces two variables that together act as a pointer representation of a dendrogram. To this, Stata adds a third variable used to restore the sort order, as needed, so that the two variables of the pointer representation remain valid. The first variable of the pointer representation gives the order of the observations. The second variable has one less element and gives the height in the dendrogram at which the adjacent observations in the order-variable join. When reversals happen, which they often do, a fourth variable, called a pseudoheight, is produced. This is used by postclustering commands with the height variable to properly interpret the ordering of the hierarchy.

See [MV] *measure_option* for the details and formulas of the available *measure*s, which include (dis)similarity measures for continuous and binary data. See [MV] **cluster** for a warning concerning (dis)similarity measure choice.

Also See

Complementary:	[MV] **cluster dendrogram**, [MV] **cluster generate**, [MV] **cluster notes**, [MV] **cluster stop**, [MV] **cluster utility**
Related:	[MV] **cluster averagelinkage**, [MV] **cluster completelinkage**, [MV] **cluster medianlinkage**, [MV] **cluster singlelinkage**, [MV] **cluster wardslinkage**, [MV] **cluster waveragelinkage**
Background:	[MV] **cluster**, [MV] **clustermat**

Title

cluster completelinkage — Complete-linkage cluster analysis

Syntax

Cluster analysis of data

> cluster <u>c</u>ompletelinkage [*varlist*] [*if*] [*in*] [, *cluster_options*]

Cluster analysis of a dissimilarity matrix

> clustermat <u>c</u>ompletelinkage *matname* [*if*] [*in*] [, *clustermat_options*]

cluster_options	description
Main	
<u>mea</u>sure(*measure*)	similarity or dissimilarity measures; see *Options* for available measures; default is L2
<u>n</u>ame(*clname*)	name of resulting cluster analysis
Advanced	
<u>gen</u>erate(*stub*)	prefix for generated variables; default prefix is *clname*

clustermat_options	description
Main	
<u>sh</u>ape(*shape*)	shape (storage method) of *matname*
add	add cluster information to data currently in memory
clear	replace data in memory with cluster information
<u>labe</u>lvar(*varname*)	place dissimilarity matrix row names in *varname*
<u>n</u>ame(*clname*)	name of resulting cluster analysis
Advanced	
force	perform clustering after fixing *matname* problems
<u>gen</u>erate(*stub*)	prefix for generated variables; default prefix is *clname*

shape	*matname* is stored as a
<u>full</u>	square symmetric matrix; the default
<u>lower</u>	vector of rowwise lower triangle (with diagonal)
<u>ll</u>ower	vector of rowwise strict lower triangle (no diagonal)
<u>upper</u>	vector of rowwise upper triangle (with diagonal)
<u>uu</u>pper	vector of rowwise strict upper triangle (no diagonal)

Description

cluster completelinkage performs hierarchical agglomerative complete-linkage cluster analysis, which is also known (among other names) as the furthest-neighbor technique. See [MV] **cluster** for a general discussion of cluster analysis and a description of the other cluster commands.

clustermat completelinkage performs hierarchical agglomerative complete-linkage cluster analysis on the dissimilarity matrix *matname*. See [MV] **clustermat** for a general discussion of cluster analysis of dissimilarity matrices and a description of the other clustermat commands.

After cluster complelinkage or clustermat completelinkage, the cluster dendrogram command (see [MV] **cluster dendrogram**) displays the resulting dendrogram, the cluster stop or clustermat stop commands (see [MV] **cluster stop**) help determine the number of groups, and the cluster generate command (see [MV] **cluster generate**) produces grouping variables.

Options for cluster completelinkage

⌐ Main └

measure(*measure*) is one of the similarity or dissimilarity measures allowed by Stata. This option is not case sensitive. See [MV] *measure_option* for a discussion of these measures.

The available measures designed for continuous data are L2 (synonym Euclidean), which is the default; L2squared; L1 (synonyms absolute, cityblock, and manhattan); Linfinity (synonym maximum); L(#); Lpower(#); Canberra; correlation; and angular (synonym angle).

The available measures designed for binary data are matching, Jaccard, Russell, Hamann, Dice, antiDice, Sneath, Rogers, Ochiai, Yule, Anderberg, Kulczynski, Gower2, and Pearson.

name(*clname*) specifies the name to attach to the resulting cluster analysis. If name() is not specified, Stata finds an available cluster name, displays it for your reference, and attaches the name to your cluster analysis.

⌐ Advanced └

generate(*stub*) provides a prefix for the variable names created by cluster completelinkage. By default, the variable-name prefix will be the name specified in name(). Three variables are created and attached to the cluster-analysis results with the suffixes _id, _ord, and _hgt. Users generally will not need to access these variables directly.

Options for clustermat completelinkage

⌐ Main └

shape(*shape*) specifies the storage mode of *matname*, the matrix of dissimilarities. The following shapes are allowed:

full specifies that *matname* is an $n \times n$ symmetric matrix.

lower specifies that *matname* is a row or column vector of length $n(n+1)/2$, with the rowwise lower triangle of the dissimilarity matrix including the diagonal of zeros.

$$D_{11} \ D_{21} \ D_{22} \ D_{31} \ D_{32} \ D_{33} \ \ldots \ D_{n1} \ D_{n2} \ \ldots \ D_{nn}$$

llower specifies that *matname* is a row or column vector of length $n(n-1)/2$, with the rowwise lower triangle of the dissimilarity matrix excluding the diagonal.

$$D_{21}\ D_{31}\ D_{32}\ D_{41}\ D_{42}\ D_{43}\ \ldots\ D_{n1}\ D_{n2}\ \ldots\ D_{n,n-1}$$

upper specifies that *matname* is a row or column vector of length $n(n+1)/2$, with the rowwise upper triangle of the dissimilarity matrix including the diagonal of zeros.

$$D_{11}\ D_{12}\ \ldots\ D_{1n}\ D_{22}\ D_{23}\ \ldots\ D_{2n}\ D_{33}\ D_{34}\ \ldots\ D_{3n}\ \ldots\ D_{nn}$$

uupper specifies that *matname* is a row or column vector of length $n(n-1)/2$, with the rowwise upper triangle of the dissimilarity matrix excluding the diagonal.

$$D_{12}\ D_{13}\ \ldots\ D_{1n}\ D_{23}\ D_{24}\ \ldots\ D_{2n}\ D_{34}\ D_{35}\ \ldots\ D_{3n}\ \ldots\ D_{n-1,n}$$

add specifies that **clustermat**'s results be added to the dataset currently in memory. The number of observations (selected observations based on the **if** and **in** qualifiers) must equal the number of rows and columns of *matname*. Either **clear** or **add** is required if a dataset is currently in memory.

clear drops all the variables and cluster solutions in the current dataset in memory (even if that dataset has changed since the data were last saved) before generating **clustermat**'s results. Either **clear** or **add** is required if a dataset is currently in memory.

labelvar(*varname*) specifies the name of a new variable to be created containing the row names of matrix *matname*.

name(*clname*) specifies the name to attach to the resulting cluster analysis. If **name()** is not specified, Stata finds an available cluster name, displays it for your reference, and attaches the name to your cluster analysis.

⌐ Advanced ⌐

force allows computations to continue when *matname* is nonsymmetric or has nonzeros on the diagonal. By default, **clustermat** will complain and exit when it encounters these conditions. **force** specifies that **clustermat** operate on the symmetric matrix $(matname * matname')/2$, with any nonzero diagonal entries treated as if they were zero.

generate(*stub*) provides a prefix for the variable names created by **clustermat**. By default, the variable-name prefix is the name specified in **name()**. Three variables are created and attached to the cluster-analysis results with the suffixes **_id**, **_ord**, and **_hgt**. Users generally will not need to access these variables directly.

Remarks

An example using the default L2 (Euclidean) distance on continuous data and an example using the **matching** coefficient on binary data illustrate the **cluster completelinkage** command. A third example illustrates the use of **clustermat completelinkage** in clustering variables instead of observations. These are the same datasets used as examples in [MV] **cluster averagelinkage**, [MV] **cluster centroidlinkage**, [MV] **cluster medianlinkage**, [MV] **cluster singlelinkage**, [MV] **cluster wardslinkage**, and [MV] **cluster waveragelinkage** so that you can compare the results from using different hierarchical clustering methods.

▷ Example 1

As explained in the first example of [MV] **cluster singlelinkage**, as the senior data analyst for a small biotechnology firm, you are given a dataset with 4 chemical laboratory measurements on 50 different samples of a particular plant gathered from the rain forest. The head of the expedition that gathered the samples thinks, based on information from the natives, that an extract from the plant might reduce the negative side effects associated with your company's best-selling nutritional supplement.

While the company chemists and botanists continue exploring the possible uses of the plant and plan future experiments, the head of product development asks you to look at the preliminary data and to report anything that might be helpful to the researchers.

While all 50 of the plants are supposed to be of the same type, you decide to perform a cluster analysis to see if there are subgroups or anomalies among them. Single-linkage clustering helped you discover an anomaly in the data. You now wish to see if you discover the same thing using complete-linkage clustering with the default Euclidean distance.

You first call `cluster completelinkage` and use the `name()` option to attach the name L2clnk to the resulting cluster analysis. You then use the `cluster list` command (see [MV] **cluster utility**) to list the components of your cluster analysis. You use the `cluster dendrogram` command to graph the dendrogram; see [MV] **cluster dendrogram**. As described in the [MV] **cluster singlelinkage** example, you use the `labels()` option, instead of the default action of showing the observation number, to identify which laboratory technician produced the data.

```
. use http://www.stata-press.com/data/r9/labtech
. cluster completelinkage x1 x2 x3 x4, name(L2clnk)
. cluster list L2clnk
L2clnk (type: hierarchical,  method: complete,  dissimilarity: L2)
     vars: L2clnk_id (id variable)
           L2clnk_ord (order variable)
           L2clnk_hgt (height variable)
    other: range: 0 .
           cmd: cluster completelinkage x1 x2 x3 x4, name(L2clnk)
           varlist: x1 x2 x3 x4
. cluster dendrogram L2clnk, labels(labtech) xlabel(, angle(90) labsize(*.75))
```

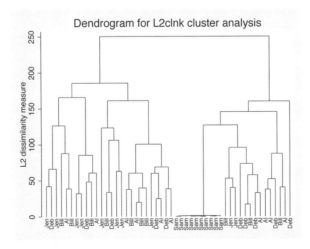

Dendrogram for L2clnk cluster analysis

As with single-linkage clustering, you see that the samples analyzed by Sam, the lab technician, cluster together closely (dissimilarity measures near zero) and are separated from the rest of the data by a large dissimilarity gap (the long vertical line going up from Sam's cluster to eventually combine with other observations). When you examined the data, you discovered that Sam's data are all between zero and one, while the other four technicians have data that range from zero up to near 150. It appears that Sam has made a mistake.

If you compare the dendrogram from this complete-linkage clustering with those from single-linkage clustering and average-linkage clustering, you will notice that the vertical lines at the top of the tree are relatively longer and the y-axis range is larger. This is a property of these linkage methods. The distance between groups is larger for complete linkage, since with complete linkage, the distance between two groups is the distance between their farthest members.

◁

▷ Example 2

This example analyzes the same data that was introduced in the second example of [MV] **cluster singlelinkage**. The sociology professor of your graduate-level class gives, as homework, a dataset containing 30 observations on 60 binary variables, with the assignment to tell him something about the 30 subjects represented by the observations.

In addition to examining single-linkage clustering of these data, you decide to see what complete-linkage clustering shows. As with the single-linkage clustering, you pick the simple matching binary coefficient to measure the similarity between groups. The `name()` option is used to attach the name `clink` to the cluster analysis. `cluster list` displays the details; see [MV] **cluster utility**. `cluster tree`, which is a synonym for `cluster dendrogram`, then displays the cluster tree (dendrogram); see [MV] **cluster dendrogram**.

```
. use http://www.stata-press.com/data/r9/homework

. cluster c a1-a60, measure(match) name(clink)

. cluster list clink
clink  (type: hierarchical,  method: complete,  similarity: matching)
      vars: clink_id (id variable)
            clink_ord (order variable)
            clink_hgt (height variable)
     other: range: 1 0
            cmd: cluster completelinkage a1-a60, measure(match) name(clink)
            varlist: a1 a2 a3 a4 a5 a6 a7 a8 a9 a10 a11 a12 a13 a14 a15 a16 a17
                 a18 a19 a20 a21 a22 a23 a24 a25 a26 a27 a28 a29 a30 a31 a32
                 a33 a34 a35 a36 a37 a38 a39 a40 a41 a42 a43 a44 a45 a46 a47
                 a48 a49 a50 a51 a52 a53 a54 a55 a56 a57 a58 a59 a60
```

```
. cluster tree
```

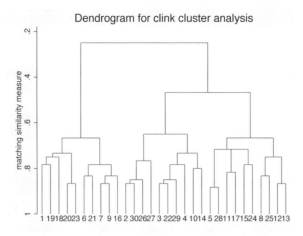

Dendrogram for clink cluster analysis

Since, by default, Stata uses the most-recently performed cluster analysis, you do not need to type the cluster name when calling `cluster tree`.

As with single-linkage clustering, the dendrogram from complete-linkage clustering seems to indicate the presence of 3 groups among the 30 observations. Later you receive another variable called `truegrp` that identifies the groups that the teacher believes are in the data. You use the `cluster generate` command (see [MV] **cluster generate**) to create a grouping variable, based on your complete-linkage clustering, to compare with `truegrp`. You do a cross-tabulation of `truegrp` and `cgrp3`, your grouping variable, to see if your conclusions match those of the teacher.

```
. cluster gen cgrp3 = group(3)
. table cgrp3 truegrp
```

	truegrp		
cgrp3	1	2	3
1		10	
2	10		
3			10

Other than the numbers arbitrarily assigned to the three groups, your teacher's conclusions and the results from the complete-linkage clustering are in complete agreement.

◁

▷ Example 3

The `wclub` dataset contains answers from 30 women to 35 yes/no questions. The variables are described in example 3 of [MV] **clustermat**. We are interested in seeing how complete-linkage clustering will cluster the 35 variables (instead of the observations).

We use the `matrix dissimilarity` command to produce a dissimilarity matrix equal to one minus the Jaccard similarity; see [MV] **matrix dissimilarity**.

```
. use http://www.stata-press.com/data/r9/wclub
. matrix dissimilarity clubD = , variables Jaccard dissim(oneminus)
```

```
. clustermat completelink clubD, name(clubcomp) clear labelvar(question)
obs was 0, now 35

. cluster dendrogram clubcomp, labels(question)
               xlabel(, angle(90) labsize(*.75))
               title(Complete-linkage clustering)
               ytitle(1 - Jaccard similarity, suffix)
```

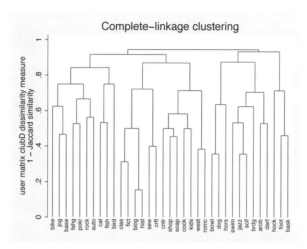

From these 30 women, we see that the biog (enjoy reading biographies) and hist (enjoy reading history) questions were most closely related. If we were to cut the dendrogram to produce four groups, the smallest of these groups would contain the variables: foot (enjoy watching football), bask (enjoy basketball), and hock (enjoy watching hockey).

◁

❏ Technical Note

cluster completelinkage requires more memory and more execution time than cluster singlelinkage. With a large number of observations, the execution time may be significant.

❏

Methods and Formulas

[MV] **cluster** discusses hierarchical clustering and places complete-linkage clustering in this general framework. It compares complete linkage to single linkage and average linkage.

Conceptually, hierarchical agglomerative complete-linkage clustering proceeds as follows. The N observations start out as N separate groups each of size one. The two closest observations are merged into one group, producing $N-1$ total groups. The closest two groups are then merged so that there are $N-2$ total groups. This process continues until all the observations are merged into one large group, producing a hierarchy of groupings from one group to N groups. For complete-linkage clustering, the "closest two groups" are determined by the farthest observations between the two groups.

The complete-linkage clustering algorithm produces two variables that together act as a pointer representation of a dendrogram. To this, Stata adds a third variable used to restore the sort order, as needed, so that the two variables of the pointer representation remain valid. The first variable of the pointer representation gives the order of the observations. The second variable has one less element and gives the height in the dendrogram at which the adjacent observations in the order-variable join.

See [MV] *measure_option* for the details and formulas of the available *measures*, which include (dis)similarity measures for continuous and binary data.

Also See

Complementary:	[MV] **cluster dendrogram**, [MV] **cluster generate**, [MV] **cluster notes**, [MV] **cluster stop**, [MV] **cluster utility**
Related:	[MV] **cluster averagelinkage**, [MV] **cluster centroidlinkage**, [MV] **cluster medianlinkage**, [MV] **cluster singlelinkage**, [MV] **cluster wardslinkage**, [MV] **cluster waveragelinkage**
Background:	[MV] **cluster**, [MV] **clustermat**

Title

cluster dendrogram — Dendrograms for hierarchical cluster analysis

Syntax

cluster dendrogram [*clname*] [*if*] [*in*] [, *options*]

options	description
Main	
quick	do not center parent branches
labels(*varname*)	name of variable containing leaf labels
cutnumber(#)	display top # branches only
cutvalue(#)	display branches above # (dis)similarity measure only
showcount	display number of observations for each branch
countprefix(*string*)	prefix the branch count with *string*; default is "n="
countsuffix(*string*)	suffix the branch count with *string*; default is empty string
countinline	put branch count in-line with branch label
vertical	orient dendrogram vertically (default)
horizontal	orient dendrogram horizontally
Plot	
line_options	affect rendition of the plotted lines
Add plot	
addplot(*plot*)	add other plots to the dendrogram
Y-Axis, X-axis, Title, Caption, Legend, Overall	
twoway_options	any options other than by() documented in [G] *twoway_options*

Note: cluster tree is a synonym for cluster dendrogram.

In addition to the restrictions imposed by if and in, the observations are automatically restricted to those that were used in the cluster analysis.

Description

cluster dendrogram produces dendrograms (also called cluster trees) for a hierarchical clustering. See [MV] **cluster** for a discussion of cluster analysis, hierarchical clustering, and the available cluster commands.

Dendrograms graphically present the information concerning which observations are grouped together at various levels of (dis)similarity. At the bottom of the dendrogram, each observation is considered its own cluster. Vertical lines extend up for each observation, and at various (dis)similarity values, these lines are connected to the lines from other observations with a horizontal line. The observations continue to combine until, at the top of the dendrogram, all observations are grouped together.

The height of the vertical lines and the range of the (dis)similarity axis give visual clues about the strength of the clustering. Long vertical lines indicate more distinct separation between the groups. Long vertical lines at the top of the dendrogram indicate that the groups represented by those lines are well separated from one another. Shorter lines indicate groups that are not as distinct.

Options

> Main

quick switches to a different style of dendrogram in which the vertical lines go straight up from the observations, instead of the default action of being recentered after each merge of observations in the dendrogram hierarchy. Some people prefer this representation, and it is quicker to render.

labels(*varname*) specifies that *varname* be used in place of observation numbers for labeling the observations at the bottom of the dendrogram.

cutnumber(*#*) displays only the top *#* branches of the dendrogram. With large dendrograms, the lower levels of the tree can become too crowded. With cutnumber(), you can limit your view to the upper portion of the dendrogram. Also see the cutvalue() and labcutn options.

cutvalue(*#*) displays only those branches of the dendrogram that are above the *#* (dis)similarity measure. With large dendrograms, the lower levels of the tree can become too crowded. With cutvalue(), you can limit your view to the upper portion of the dendrogram. Also see the cutnumber() and labcutn options.

showcount requests that the number of observations associated with each branch be displayed below the branches. showcount is most useful with cutnumber() and cutvalue() since, otherwise, the number of observations for each branch is one. When this option is specified, a label for each branch in constructed using a prefix string, the branch count, and a suffix string.

countprefix(*string*) specifies the prefix string for the branch count label. The default is countprefix(n=). This option implies the showcount option.

countsuffix(*string*) specifies the suffix string for the branch count label. The default is an empty string. This option implies the showcount option.

countinline requests that the branch count be put in-line with the corresponding branch label. The branch count is placed below the branch label by default. This option implies the showcount option.

vertical and horizontal specify whether the x- and y-coordinates are to be swapped before plotting—vertical (the default) does not swap the coordinates, whereas horizontal does.

> Plot

line_options affect the rendition of the lines; see [G] ***line_options***.

> Add plot

addplot(*plot*) allows adding additional graph twoway plots to the graph; see help [G] ***addplot_option***.

> Y-Axis, X-Axis, Title, Caption, Legend, Overall, By

twoway_options are any of the options documented in [G] ***twoway_options***, except by(). These include options for titling the graph (see [G] ***title_options***) and saving the graph to disk (see [G] ***saving_option***).

(*Continued on next page*)

Remarks

Examples of the cluster dendrogram command can be found in [MV] **cluster singlelinkage**, [MV] **cluster completelinkage**, [MV] **cluster averagelinkage**, [MV] **cluster wardslinkage**, [MV] **cluster waveragelinkage**, [MV] **cluster stop**, and [MV] **cluster generate**. Here we illustrate some of the additional options available with cluster dendrogram.

▷ Example 1

In the first example of [MV] **cluster completelinkage**, the dendrogram for the complete-linkage clustering of 50 observations on 4 variables was illustrated with the following dendrogram:

```
. use http://www.stata-press.com/data/r9/labtech
. cluster completelinkage x1 x2 x3 x4, name(L2clnk)
. cluster dendrogram L2clnk, labels(labtech) xlabel(, angle(90) labsize(*.75))
```

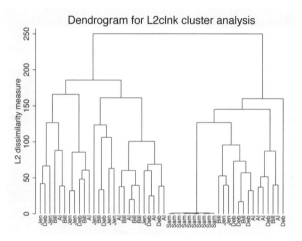

The same dendrogram can be rendered in a slightly different format using the quick option:

```
. cluster dendrogram L2clnk, quick labels(labtech)
        xlabel(, angle(90) labsize(*.75))
```

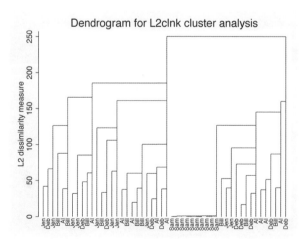

Some people prefer this style of dendrogram. The option name `quick` comes from the fact that this style of dendrogram is quicker to render.

You can use the `if` and `in` conditions to restrict the dendrogram to the observations for one subgroup. This is usually accomplished with the `cluster generate` command, which creates a grouping variable; see [MV] **cluster generate**.

Here we show the third of three groups in the dendrogram by first generating the grouping variable for three groups and then using `if` in the command for `cluster dendrogram` to restrict it to the third of those three groups.

```
. cluster gen g3 = group(3)
. cluster tree if g3==3
```

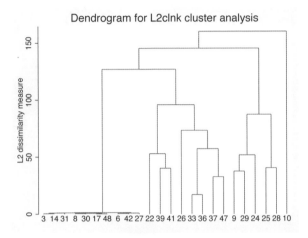

Dendrogram for L2clnk cluster analysis

Since we find it easier to type, we used the synonym `tree` instead of `dendrogram`. We did not specify the cluster name, allowing it to default to the most-recently performed cluster analysis. We also omitted the `labels()` and `xlabel()` options, which bring us back to the default action of showing, horizontally, the observation numbers.

This example has only 50 observations. When there are a large number of observations, the dendrogram can become too crowded. You will need to limit which part of the dendrogram you display. One way to view only part of the dendrogram is to use `if` and `in` to limit to one particular group, as we did above.

The other way to limit your view of the dendrogram is to specify that you only wish to view the top portion of the tree. The `cutnumber()` and `cutvalue()` options allow you to do this:

(Continued on next page)

```
. cluster tree, cutn(15) showcount
```

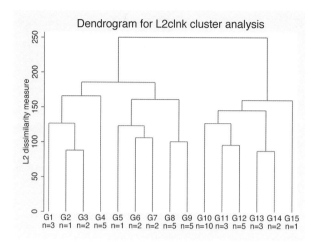

We limited our view to the top 15 branches of the dendrogram with `cutn(15)`. By default, the 15 branches were labeled G1 through G15. The `showcount` option provided, below these branch labels, the number of observations in each of the 15 groups.

The `cutvalue()` option provides another way to limit the view to the top branches of the dendrogram. With this option, you specify the similarity or dissimilarity value at which to trim the tree.

```
. cluster tree, cutvalue(75.3)
        countprefix("(") countsuffix(" obs)") countinline
        ylabel(, angle(0)) horizontal
```

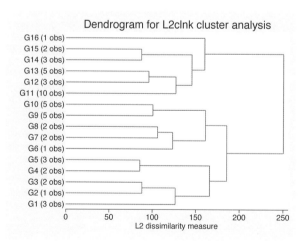

This time, we limited the dendrogram to those branches with dissimilarity greater than 75.3 by using the `cutvalue(75.3)` option. There were 16 branches (groups) that met that restriction. We used the `countprefix()` and `countsuffix()` options to display the number of observations in each branch as "(# obs)" instead of "n=#". The `countinline` option puts the branch counts in-line with the branch labels. We specified the `horizontal` option and the `angle(0)` suboption of `ylabel()` to get a horizontal dendrogram with horizontal branch labels.

◁

❑ Technical Note

Programmers can control the graphical procedure executed when `cluster dendrogram` is called. This will be helpful to programmers adding new hierarchical clustering methods that require a different dendrogram algorithm. See [MV] **cluster programming subroutines** for details.

❑

Also See

Complementary:	[MV] **cluster averagelinkage**, [MV] **cluster centroidlinkage**,
	[MV] **cluster completelinkage**, [MV] **cluster generate**,
	[MV] **cluster medianlinkage**, [MV] **cluster programming subroutines**,
	[MV] **cluster singlelinkage**, [MV] **cluster stop**,
	[MV] **cluster wardslinkage**, [MV] **cluster waveragelinkage**
Background:	[MV] **cluster**, [MV] **clustermat**

Title

> **cluster generate** — Generate summary or grouping variables from a cluster analysis

Syntax

Generate grouping variables for specified number of clusters

> cluster generate { *newvarname* | *stub* } = groups(*numlist*) [, *options*]

Generate grouping variable by cutting the dendrogram

> cluster generate *newvarname* = cut(#) [, name(*clname*)]

options	description
name(*clname*)	name of cluster analysis to use in producing new variables
ties(error)	produce error message in the case of ties; default
ties(skip)	ignore requests that result in ties
ties(fewer)	produce results for largest number of groups smaller than your request
ties(more)	produce results for smallest number of groups larger than your request

Description

The `cluster generate` command generates summary or grouping variables from a hierarchical cluster analysis. What is produced depends on the function. See [MV] **cluster** for information on available cluster-analysis commands.

The `groups`(*numlist*) function generates grouping variables, giving the grouping for the specified numbers of clusters from a hierarchical cluster analysis. If a single number is given, *newvarname* is produced with group numbers going from 1 to the number of clusters requested. If more than one number is specified, a new variable is generated for each number using the provided *stub* name appended with the number. For instance,

```
cluster gen xyz = groups(5/7), name(myclus)
```

creates variables xyz5, xyz6, and xyz7, giving the fifth, sixth, and seventh groups obtained from the cluster analysis named myclus.

The `cut`(#) function generates a grouping variable corresponding to cutting the dendrogram (see [MV] **cluster dendrogram**) of a hierarchical cluster analysis at the specified (dis)similarity value.

Additional `cluster generate` functions may be added; see [MV] **cluster programming subroutines**.

Options

name(*clname*) specifies the name of the cluster analysis to use in producing the new variables. The default is the name of the cluster analysis last performed, which can be reset using the `cluster use` command; see [MV] **cluster utility**.

ties(error | skip | fewer | more) indicates what to do with the groups() function in the case of ties. A hierarchical cluster analysis has ties when multiple groups are generated at a particular (dis)similarity value. For example, you might have the case where you can uniquely create two, three, and four groups, but the next possible grouping produces eight groups due to ties.

ties(error), the default, produces an error message and does not generate the requested variables.

ties(skip) specifies that the offending requests be ignored. No error message is produced, and only the requests that produce unique groupings will be honored. With multiple values specified in the groups() function, ties(skip) allows the processing of those that produce unique groupings and ignores the rest.

ties(fewer) produces the results for the largest number of groups less than or equal to your request. In the example above with groups(6) and using ties(fewer), you would get the same result that you would by using groups(4).

ties(more) produces the results for the smallest number of groups greater than or equal to your request. In the example above with groups(6) and using ties(more), you would get the same result that you would by using groups(8).

Remarks

Examples of the use of the groups() function of cluster generate can be found in [MV] **cluster singlelinkage**, [MV] **cluster completelinkage**, [MV] **cluster averagelinkage**, [MV] **cluster centroidlinkage**, [MV] **cluster medianlinkage**, [MV] **cluster wardslinkage**, [MV] **cluster waveragelinkage**, [MV] **cluster stop**, and [MV] **cluster dendrogram**. Additional examples of the groups() and cut() functions of cluster generate are provided here.

You may find it easier to understand these functions by looking at a dendrogram from a hierarchical cluster analysis. The cluster dendrogram command produces dendrograms (cluster trees) from a hierarchical cluster analysis; see [MV] **cluster dendrogram**.

▷ Example 1

The first example of [MV] **cluster completelinkage** performs a complete-linkage cluster analysis on 50 observations with 4 variables. The dendrogram presented there was labeled at the bottom by the name of the laboratory technician responsible for the data. Here we reproduce that same dendrogram but use the default action, placing observation numbers as labels. We then use the groups() function of cluster generate to produce a grouping variable, splitting the data into two groups.

```
. use http://www.stata-press.com/data/r9/labtech
. cluster completelinkage x1 x2 x3 x4, name(L2clnk)
. cluster dendrogram L2clnk, xlabel(, angle(90) labsize(*.75))
. cluster generate g2 = group(2), name(L2clnk)
. codebook g2
```

g2	(unlabeled)

type:	numeric (byte)		
range:	[1,2]	units:	1
unique values:	2	missing .:	0/50
tabulation:	Freq. Value		
	26 1		
	24 2		

```
. by g2, sort: summarize x*
```

```
-> g2 = 1
    Variable |       Obs        Mean    Std. Dev.        Min         Max
    ---------+-------------------------------------------------------------
          x1 |        26        91.5    37.29432        17.4         143
          x2 |        26    74.58077    41.19319         4.8       142.1
          x3 |        26    101.0077    36.95704        16.3       147.9
          x4 |        26    71.77308    43.04107         6.6       146.1
```

```
-> g2 = 2
    Variable |       Obs        Mean    Std. Dev.        Min         Max
    ---------+-------------------------------------------------------------
          x1 |        24        18.8    23.21742           0          77
          x2 |        24    30.05833    37.66979           0       143.6
          x3 |        24    18.54583    21.68215          .2        69.7
          x4 |        24    41.89167    43.62025          .1       130.9
```

The group() function of cluster generate created a grouping variable named g2, with ones indicating the 26 observations that belong to the left main branch of the dendrogram and twos indicating the 24 observations that belong to the right main branch of the dendrogram. The summary of the x variables used in the cluster analysis for each group shows that the second group is characterized by lower values.

We could have obtained the same grouping variable by using the cut() function of cluster generate.

```
. cluster gen g2cut = cut(200)
. table g2 g2cut
```

```
        |    g2cut
    g2  |    1       2
--------+---------------
    1   |   26
    2   |           24
--------+---------------
```

Looking at the y-axis of the dendrogram, we decide to cut the tree at the dissimilarity value of 200. We did not need to specify the name() option because this was the latest cluster analysis performed, which is the default. The table output shows that we obtained the same result with cut(200) as with group(2) for this example.

How many groups are produced if we cut the tree at the value 105.2?

```
. cluster gen z = cut(105.2)

. codebook z, tabulate(20)
```

---

z   (unlabeled)

```
              type:  numeric (byte)
             range:  [1,11]                          units:  1
     unique values:  11                          missing .:  0/50
        tabulation:  Freq.  Value
                        3   1
                        3   2
                        5   3
                        1   4
                        2   5
                        2   6
                       10   7
                       10   8
                        8   9
                        5   10
                        1   11
```

The codebook command shows that the result of cutting the dendrogram at the value 105.2 produced eleven groups ranging in size from one to ten observations.

The group() function of cluster generate may be used to create multiple grouping variables with a single call. Here we create the grouping variables for groups of size 3 to 12:

```
. cluster gen gp = gr(3/12)

. summarize gp*
```

Variable	Obs	Mean	Std. Dev.	Min	Max
gp3	50	2.26	.8033095	1	3
gp4	50	3.14	1.030356	1	4
gp5	50	3.82	1.438395	1	5
gp6	50	3.84	1.461897	1	6
gp7	50	3.96	1.603058	1	7
gp8	50	4.24	1.911939	1	8
gp9	50	5.18	2.027263	1	9
gp10	50	5.94	2.385415	1	10
gp11	50	6.66	2.781939	1	11
gp12	50	7.24	3.197959	1	12

In this case, we used abbreviations for generate and group(). The group() function takes a numlist; see [U] **11.1.8 numlist**. We specified 3/12, indicating the numbers 3 to 12. gp, the stub name we provide, is appended with the number as the variable name for each group variable produced.

◁

▷ Example 2

The second example of [MV] **cluster singlelinkage** shows the following dendrogram from the single-linkage clustering of 30 observations on 60 variables. In that example, we used the group() function of cluster generate to produce a grouping variable for three groups. What happens when we try to obtain four groups from this clustering?

```
. use http://www.stata-press.com/data/r9/homework
. cluster singlelinkage a1-a60, measure(matching)
cluster name: _cl_1
. cluster tree
```

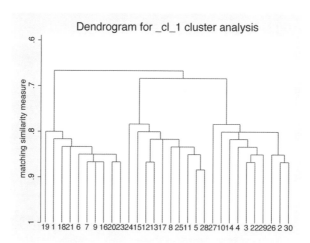

Dendrogram for _cl_1 cluster analysis

```
. cluster gen g4 = group(4)
cannot create 4 groups due to ties
r(198);
```

Stata complains that it cannot create four groups from this cluster analysis.

The `ties()` option gives us control over this situation. We just need to decide whether we want more groups or fewer groups than we asked for when faced with ties. We demonstrate both ways.

```
. cluster gen more4 = gr(4), ties(more)
. cluster gen less4 = gr(4), ties(fewer)
. summarize more4 less4
```

Variable	Obs	Mean	Std. Dev.	Min	Max
more4	30	2.933333	1.638614	1	5
less4	30	2	.8304548	1	3

For this cluster analysis, `ties(more)` with `group(4)` produces five groups, while `ties(fewer)` with `group(4)` produces three groups.

The `ties(skip)` option is convenient when we want to produce a range of grouping variables.

```
. cluster gen group = gr(4/20), ties(skip)
. summarize group*
```

Variable	Obs	Mean	Std. Dev.	Min	Max
group5	30	2.933333	1.638614	1	5
group9	30	4.866667	2.622625	1	9
group13	30	7.066667	3.92106	1	13
group18	30	9.933333	5.419844	1	18

With this cluster analysis, the only unique groupings available are 5, 9, 13, and 18 within the range 4 to 20.

Also See

Complementary:	[MV] **cluster averagelinkage**, [MV] **cluster centroidlinkage**,
	[MV] **cluster completelinkage**, [MV] **cluster dendrogram**,
	[MV] **cluster medianlinkage**, [MV] **cluster programming subroutines**,
	[MV] **cluster singlelinkage**, [MV] **cluster stop**,
	[MV] **cluster wardslinkage**, [MV] **cluster waveragelinkage**
Related:	[D] **egen**, [D] **generate**
Background:	[MV] **cluster**, [MV] **clustermat**

Title

> **cluster kmeans** — Kmeans cluster analysis

Syntax

> cluster <u>k</u>means [*varlist*] [*if*] [*in*] , k(*#*) [*options*]

options	description
Main	
* k(*#*)	perform cluster analysis resulting in *#* groups
<u>measure</u>(*measure*)	(dis)similarity measures; see *Options* for available measures; default is L2
<u>name</u>(*clname*)	name of resulting cluster analysis
Options	
<u>start</u>(*start_option*)	obtain *k* initial group centers using *start_option*; see *Options* for details
keepcenters	append the *k* final group means to the data
Advanced	
<u>generate</u>(*groupvar*)	name of grouping variable
<u>iterate</u>(*#*)	maximum number of iterations; default is iterate(10000)

* k(*#*) is required.

Description

cluster kmeans performs kmeans partition cluster analysis. See [MV] **cluster** for a general discussion of cluster analysis and a description of the other cluster commands. See [MV] **cluster kmedians** for an alternative that uses medians instead of means.

Options

> ⌐ Main ⌐

k(*#*) is required and specifies that *#* groups be formed by the cluster analysis.

measure(*measure*) is one of the similarity or dissimilarity measures allowed by Stata. This option is not case sensitive. See [MV] *measure_option* for a discussion of these measures.

The available measures designed for continuous data are L2 (synonym <u>Euc</u>lidean), which is the default; L2squared; L1 (synonyms <u>abs</u>olute, <u>city</u>block, and <u>man</u>hattan); <u>Linf</u>inity (synonym <u>max</u>imum); L(*#*); <u>Lpow</u>er(*#*); <u>Can</u>berra; <u>corr</u>elation; and <u>ang</u>ular (synonym <u>ang</u>le).

The available measures designed for binary data are <u>matc</u>hing, <u>Jacc</u>ard, <u>Russ</u>ell, Hamann, Dice, antiDice, Sneath, Rogers, Ochiai, Yule, <u>And</u>erberg, <u>Kulc</u>zynski, Gower2, and Pearson.

name(*clname*) specifies the name to attach to the resulting cluster analysis. If name() is not specified, Stata finds an available cluster name, displays it for your reference, and attaches the name to your cluster analysis.

start(*start_option*) indicates how the k initial group centers are to be obtained. The available *start_options* are <u>kr</u>andom[(*seed#*)], <u>f</u>irstk[, <u>ex</u>clude], <u>l</u>astk[, <u>ex</u>clude], <u>r</u>andom[(*seed#*)], <u>pr</u>andom[(*seed#*)], <u>everykth</u>, <u>segments</u>, and <u>group</u>(*varname*).

krandom[(*seed#*)], the default, specifies that k unique observations be chosen at random, from among those to be clustered, as starting centers for the k groups. Optionally, a random-number seed may be specified to cause the command set seed *seed#* (see [D] **generate**) to be applied before the k random observations are chosen.

firstk[, exclude] specifies that the first k observations from among those to be clustered be used as the starting centers for the k groups. With the exclude option, these first k observations are not included among the observations to be clustered.

lastk[, exclude] specifies that the last k observations from among those to be clustered be used as the starting centers for the k groups. With the exclude option, these last k observations are not included among the observations to be clustered.

random[(*seed#*)] specifies that k random initial group centers be generated. The values are randomly chosen from a uniform distribution over the range of the data. Optionally, a random-number seed may be specified to cause the command set seed *seed#* (see [D] **generate**) to be applied before the k group centers are generated.

prandom[(*seed#*)] specifies that k partitions be formed randomly among the observations to be clustered. The group means from the k groups defined by this partitioning are to be used as the starting group centers. Optionally, a random-number seed may be specified to cause the command set seed *seed#* (see [D] **generate**) to be applied before the k partitions are chosen.

everykth specifies that k partitions be formed by assigning observations $1, 1+k, 1+2k, \ldots$ to the first group; assigning observations $2, 2+k, 2+2k, \ldots$ to the second group; and so on, to form k groups. The group means from these k groups are to be used as the starting group centers.

segments specifies that k nearly equal partitions be formed from the data. Approximately the first N/k observations are assigned to the first group, the second N/k observations are assigned to the second group, and so on. The group means from these k groups are to be used as the starting group centers.

group(*varname*) provides an initial grouping variable, *varname*, that defines k groups among the observations to be clustered. The group means from these k groups are to be used as the starting group centers.

keepcenters specifies that the group means from the k groups that are produced be appended to the data.

generate(*groupvar*) provides the name of the grouping variable to be created by cluster kmeans. By default, this will be the name specified in name().

iterate(#) specifies the maximum number of iterations to allow in the kmeans clustering algorithm. The default is iterate(10000).

Remarks

Two examples of using `cluster kmeans` are presented, one using continuous data and the other using binary data. See [MV] **cluster kmedians** to see these same two datasets examined using kmedians clustering.

▷ Example 1

You have measured the flexibility, speed, and strength of the 80 students in your physical education class. You want to split the class into four groups, based on their physical attributes, so that they can receive the mix of flexibility, strength, and speed training that will best help them improve.

Here is a summary of the data and a matrix graph showing the data:

```
. use http://www.stata-press.com/data/r9/physed
. summarize flex speed strength
```

Variable	Obs	Mean	Std. Dev.	Min	Max
flexibility	80	4.402625	2.788541	.03	9.97
speed	80	3.875875	3.121665	.03	9.79
strength	80	6.439875	2.449293	.05	9.57

```
. graph matrix flex speed strength
```

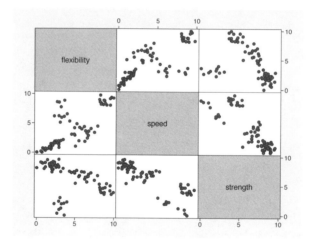

As you expected, based on what you saw the first day of class, the data indicate a wide range of levels of performance for the students. The graph seems to indicate that there are some distinct groups, which leads you to believe that your plan will work well.

You decide to perform a cluster analysis to create four groups, one for each of your class assistants. You have had good experience with kmeans clustering in the past and generally like the behavior of the absolute-value distance.

You don't really care what starting values are used in the cluster analysis, but you do want to be able to reproduce the same results if you ever decide to rerun your analysis. You decide to use the `krandom()` option to pick k of the observations at random as the initial group centers. You supply a random-number seed for reproducibility. You also add the `keepcenters` option so that the means of the four groups will be added to the bottom of your dataset.

```
. cluster k flex speed strength, k(4) name(g4abs) s(kr(385617)) mea(abs) keepcen
. cluster list g4abs
g4abs  (type: partition,  method: kmeans,  dissimilarity: L1)
      vars: g4abs (group variable)
     other: k: 4
            start: krandom(385617)
            range: 0 .
            cmd: cluster kmeans flex speed strength, k(4) name(g4abs)
                 start(kr(385617)) mea(abs) keepcen
            varlist: flexibility speed strength
. table g4abs
```

g4abs	Freq.
1	15
2	20
3	35
4	10

```
. list flex speed strength in 81/l
```

	flexib~y	speed	strength
81.	8.852	8.743333	4.358
82.	5.9465	3.4485	6.8325
83.	1.969429	1.144857	8.478857
84.	3.157	6.988	1.641

```
. drop in 81/l
(4 observations deleted)
. tabstat flex speed strength, by(g4abs) stat(min mean max)
Summary statistics: min, mean, max
  by categories of: g4abs
```

g4abs	flexib~y	speed	strength
1	8.12	8.05	3.61
	8.852	8.743333	4.358
	9.97	9.79	5.42
2	4.32	1.05	5.46
	5.9465	3.4485	6.8325
	7.89	5.32	7.66
3	.03	.03	7.38
	1.969429	1.144857	8.478857
	3.48	2.17	9.57
4	2.29	5.11	.05
	3.157	6.988	1.641
	3.99	8.87	3.02
Total	.03	.03	.05
	4.402625	3.875875	6.439875
	9.97	9.79	9.57

After looking at the last four observations (which are the group means since you specified keep-centers), you decide that what you really wanted to see was the minimum and maximum values and the mean for the four groups. You remove the last four observations and then use the tabstat command to view the desired statistics.

Group 1, with 15 students, is already doing very well in flexibility and speed but will need extra strength training. Group 2, with 20 students, needs to emphasize speed training but could use some improvement in the other categories as well. Group 3, the largest, with 35 students, has serious problems with both flexibility and speed, though they did very well in the strength category. Group 4, the smallest, with 10 students, needs help with flexibility and strength.

Since you like looking at graphs, you decide to view the matrix graph again but with group numbers used as plotting symbols.

```
. graph matrix flex speed strength, m(i) mlabel(g4abs) mlabpos(0)
```

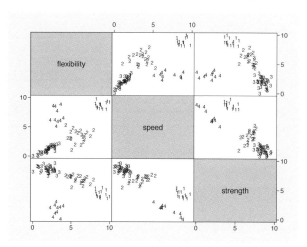

The groups, as shown in the graph, do appear reasonably distinct. However, you had hoped to have groups that were about the same size. You are curious what clustering to three or five groups would produce. For no good reason, you decide to use the first k observations as initial group centers for clustering to three groups and random numbers within the range of the data for clustering to five groups.

```
. cluster k flex speed strength, k(3) name(g3abs) start(firstk) measure(abs)
. cluster k flex speed strength, k(5) name(g5abs) start(random(33576))
> measure(abs)
. table g3abs g4abs, col
```

		g4abs			
g3abs	1	2	3	4	Total
1				10	10
2		18	35		53
3	15	2			17

```
. table g5abs g4abs, col
```

		g4abs			
g5abs	1	2	3	4	Total
1		20			20
2	15				15
3				6	6
4				4	4
5			35		35

With three groups, the unequal-group-size problem gets worse. With five groups, the smallest group gets split. Four groups seem like the best option for this class. You will try to help the assistant assigned to group 3 in dealing with the larger group.

◁

▷ Example 2

You have just started a women's club. Thirty women from throughout the community have sent in their requests to join. You have them fill out a questionnaire with 35 yes/no questions relating to sports, music, reading, and hobbies. Here is a description of the dataset:

```
. use http://www.stata-press.com/data/r9/wclub

. describe
Contains data from http://www.stata-press.com/data/r9/wclub.dta
  obs:            30
  vars:           35                           1 May 2005 16:56
  size:         1,170 (99.5% of memory free)
```

variable name	storage type	display format	value label	variable label
bike	byte	%8.0g		enjoy bicycle riding Y/N
bowl	byte	%8.0g		enjoy bowling Y/N
swim	byte	%8.0g		enjoy swimming Y/N
jog	byte	%8.0g		enjoy jogging Y/N
hock	byte	%8.0g		enjoy watching hockey Y/N
foot	byte	%8.0g		enjoy watching football Y/N
base	byte	%8.0g		enjoy baseball Y/N
bask	byte	%8.0g		enjoy basketball Y/N
arob	byte	%8.0g		participate in aerobics Y/N
fshg	byte	%8.0g		enjoy fishing Y/N
dart	byte	%8.0g		enjoy playing darts Y/N
clas	byte	%8.0g		enjoy classical music Y/N
cntr	byte	%8.0g		enjoy country music Y/N
jazz	byte	%8.0g		enjoy jazz music Y/N
rock	byte	%8.0g		enjoy rock and roll music Y/N
west	byte	%8.0g		enjoy reading western novels Y/N
romc	byte	%8.0g		enjoy reading romance novels Y/N
scif	byte	%8.0g		enjoy reading sci. fiction Y/N
biog	byte	%8.0g		enjoy reading biographies Y/N
fict	byte	%8.0g		enjoy reading fiction Y/N
hist	byte	%8.0g		enjoy reading history Y/N
cook	byte	%8.0g		enjoy cooking Y/N
shop	byte	%8.0g		enjoy shopping Y/N
soap	byte	%8.0g		enjoy watching soap operas Y/N
sew	byte	%8.0g		enjoy sewing Y/N
crft	byte	%8.0g		enjoy craft activities Y/N
auto	byte	%8.0g		enjoy automobile mechanics Y/N
pokr	byte	%8.0g		enjoy playing poker Y/N
brdg	byte	%8.0g		enjoy playing bridge Y/N
kids	byte	%8.0g		have children Y/N
hors	byte	%8.0g		have a horse Y/N
cat	byte	%8.0g		have a cat Y/N
dog	byte	%8.0g		have a dog Y/N
bird	byte	%8.0g		have a bird Y/N
fish	byte	%8.0g		have a fish Y/N

```
Sorted by:
```

Now you are trying to plan the first club meeting. You decide to have a lunch along with the business meeting that will officially organize the club and ratify its charter. You want the club to get off to a good start, so you worry about the best way to seat the guests. You decide to use kmeans clustering on the yes/no data from the questionnaires to put people with similar interests at the same tables.

You have five tables that can each seat up to eight comfortably. You request clustering to five groups and hope that the group sizes will fall under this table-size limit.

You really want people placed together who share positive interests, not who share dislikes. From among all the available binary similarity measures, you decide to use the Jaccard coefficient since it does not include jointly zero comparisons in its formula; see [MV] *measure_option*. The Jaccard coefficient is also easy to understand.

```
. cluster k bike-fish, k(5) measure(Jaccard) st(firstk) name(gr5)
. cluster list gr5
gr5 (type: partition,  method: kmeans,  similarity: Jaccard)
       vars: gr5 (group variable)
      other: k: 5
             start: firstk
             range: 1 0
             cmd: cluster k bike-fish, k(5) measure(Jaccard) st(firstk) name(gr5)
             varlist: bike bowl swim jog hock foot base bask arob fshg dart clas
                   cntr jazz rock west romc scif biog fict hist cook shop soap
                   sew crft auto pokr brdg kids hors cat dog bird fish

. table gr5
```

gr5	Freq.
1	7
2	7
3	5
4	5
5	6

You get lucky; the groups are reasonably close in size. You will seat yourself at one of the tables with only five people and your sister, who did not fill out a questionnaire, at the other table with only five people to make things as even as possible.

Now you wonder, what are the characteristics of these five groups? You decide to use the `tabstat` command to view the proportion answering yes to each question for each of the five groups.

```
. tabstat bike-fish, by(gr5) format(%4.3f)
Summary statistics: mean
  by categories of: gr5
```

gr5	bike	bowl	swim	jog	hock	foot
1	0.714	0.571	0.714	0.571	0.143	0.143
2	0.286	0.143	0.571	0.714	0.143	0.143
3	0.400	0.200	0.600	0.200	0.200	0.400
4	0.200	0.000	0.200	0.200	0.000	0.400
5	0.000	0.500	0.000	0.000	0.333	0.167
Total	0.333	0.300	0.433	0.367	0.167	0.233

gr5	base	bask	arob	fshg	dart	clas
1	0.429	0.571	0.857	0.429	0.571	0.429
2	0.571	0.286	0.714	0.429	0.857	0.857
3	0.600	0.400	0.000	0.800	0.200	0.000
4	0.200	0.600	0.400	0.000	0.000	0.800
5	0.167	0.333	0.000	0.500	0.167	0.000
Total	0.400	0.433	0.433	0.433	0.400	0.433

gr5	cntr	jazz	rock	west	romc	scif
1	0.857	0.571	0.286	0.714	0.571	0.286
2	0.571	0.857	0.429	0.143	0.143	0.857
3	0.200	0.200	0.600	0.000	0.000	0.200
4	0.200	0.400	0.400	0.200	0.400	0.000
5	0.833	0.167	0.667	0.500	0.667	0.000
Total	0.567	0.467	0.467	0.333	0.367	0.300

gr5	biog	fict	hist	cook	shop	soap
1	0.429	0.429	0.571	0.714	0.571	0.571
2	0.429	0.571	0.571	0.000	0.429	0.143
3	0.000	0.200	0.000	0.600	1.000	0.600
4	1.000	1.000	1.000	0.600	0.600	0.200
5	0.000	0.167	0.000	0.333	1.000	0.667
Total	0.367	0.467	0.433	0.433	0.700	0.433

gr5	sew	crft	auto	pokr	brdg	kids
1	0.429	0.571	0.143	0.571	0.429	0.714
2	0.143	0.714	0.429	0.286	0.714	0.143
3	0.400	0.200	0.600	1.000	0.200	0.600
4	0.800	0.800	0.000	0.000	0.000	1.000
5	0.000	0.000	0.333	0.667	0.000	0.500
Total	0.333	0.467	0.300	0.500	0.300	0.567

gr5	hors	cat	dog	bird	fish
1	0.571	0.571	1.000	0.286	0.429
2	0.143	0.571	0.143	0.429	0.143
3	0.000	0.200	0.200	0.400	0.800
4	0.000	0.400	0.000	0.000	0.200
5	0.167	0.167	0.833	0.167	0.167
Total	0.200	0.400	0.467	0.267	0.333

It appears that group 1 likes participating in most sporting activities, prefers country music, likes reading western and romance novels, enjoys cooking, and is more likely to have children and various animals, including horses.

Group 2 likes some sports (swimming, jogging, aerobics, baseball, and darts), prefers classical and jazz music, prefers science fiction (but also enjoys biography, fiction, and history), dislikes cooking, enjoys playing bridge, is not likely to have children, and is more likely to have a cat than any other animal.

Group 3 seems to enjoy swimming, baseball, and fishing (but dislikes aerobics), prefers rock and roll music (disliking classical), does not enjoy reading, prefers poker over bridge, and is more likely to own a fish than any other animal.

Group 4 dislikes many of the sports, prefers classical music, likes reading biographies, fiction, and history, enjoys sewing and crafts, dislikes card games, has children, and is not likely to have pets.

Group 5 dislikes sports, prefers country and rock and roll music, will pick up romance and western novels on occasion, dislikes sewing and crafts, prefers poker instead of bridge, and is most likely to have a dog.

◁

Methods and Formulas

Kmeans cluster analysis is discussed in most cluster-analysis books; see the references in [MV] **cluster**. [MV] **cluster** also provides a general discussion of cluster analysis, including kmeans clustering, and discusses the available `cluster` subcommands.

Kmeans clustering is an iterative procedure that partitions the data into k groups or clusters. The procedure begins with k initial group centers. Observations are assigned to the group with the closest center. The mean of the observations assigned to each of the groups is computed, and the process is repeated. These steps continue until all observations remain in the same group from the previous iteration.

To avoid endless loops, an observation will only be reassigned to a different group if it is closer to the other group center. In the case of a tied distance between an observation and two or more group centers, the observation is assigned to its current group if that is one of the closest, and to the lowest numbered group otherwise.

The `start()` option provides many ways to specify the beginning group centers. These include methods that specify the actual starting centers, as well as methods that specify initial partitions of the data from which the beginning centers are computed.

Some kmeans clustering algorithms recompute the group centers after each reassignment of an observation. Other kmeans clustering algorithms, including Stata's `cluster kmeans` command, recompute the group centers only after a complete pass through the data. A disadvantage of this method is that orphaned group centers—one that has no observations that are closest to it—can occur. The advantage of recomputing means only at the end of each pass through the data is that the sort order of the data does not potentially change your final result.

Stata deals with orphaned centers by finding the observations that are farthest from the centers and using them as new group centers. The observations are then reassigned to the closest groups, including these new centers.

Continuous or binary data are allowed with `cluster kmeans`. The mean of a group of binary observations for a variable is the proportion of ones for that group of observations and variable. The binary similarity measures can accommodate the comparison of a binary observation to a binary mean (proportion). See [MV] *measure_option* for details on this subject and for the formulas for all the available (dis)similarity measures.

Also See

Complementary:	[MV] **cluster notes**, [MV] **cluster stop**, [MV] **cluster utility**
Related:	[MV] **cluster kmedians**
Background:	[MV] **cluster**

Title

cluster kmedians — Kmedians cluster analysis

Syntax

cluster <u>kmed</u>ians [*varlist*] [*if*] [*in*] , k(#) [*options*]

options	description
Main	
* k(#)	perform cluster analysis resulting in # groups
<u>measure</u>(*measure*)	(dis)similarity measure; see *Options* for available measures; default is L2
<u>name</u>(*clname*)	name of resulting cluster analysis
Options	
<u>start</u>(*start_option*)	obtain *k* initial group centers using *start_option*; see *Options* for details
<u>keepcenters</u>	append the *k* final group medians to the data
Advanced	
<u>generate</u>(*groupvar*)	name of grouping variable
<u>iterate</u>(#)	maximum number of iterations; default is iterate(10000)

* k(#) is required.

Description

cluster kmedians performs kmedians partition cluster analysis. See [MV] **cluster** for a general discussion of cluster analysis and a description of the other cluster commands. See [MV] **cluster kmeans** for an alternative that uses means instead of medians.

Options

⌐ Main ⌐

k(#) is required and specifies that # groups be formed by the cluster analysis.

measure(*measure*) is one of the similarity or dissimilarity measures allowed by Stata. This option is not case sensitive. See [MV] *measure_option* for a discussion of these measures.

The available measures designed for continuous data are L2 (synonym <u>Euclidean</u>), which is the default; L2squared; L1 (synonyms <u>absolute</u>, cityblock, and <u>manhattan</u>); <u>Linfinity</u> (synonym <u>maximum</u>); L(#); Lpower(#); Canberra; <u>correlation</u>; and angular (synonym angle).

The available measures designed for binary data are <u>matching</u>, <u>Jaccard</u>, <u>Russell</u>, Hamann, Dice, antiDice, Sneath, Rogers, Ochiai, Yule, <u>Anderberg</u>, <u>Kulczynski</u>, Gower2, and Pearson.

name(*clname*) specifies the name to attach to the resulting cluster analysis. If name() is not specified, Stata finds an available cluster name, displays it for your reference, and attaches the name to your cluster analysis.

137

┌─ Options ┐

start(*start_option*) indicates how the k initial group centers are to be obtained. The available *start_option*s are <u>kr</u>andom$\left[(seed\#)\right]$, <u>f</u>irstk$\left[, \text{ } \underline{ex}\text{clude}\right]$, <u>l</u>astk$\left[, \text{ } \underline{ex}\text{clude}\right]$, <u>r</u>andom$\left[(seed\#)\right]$, <u>pr</u>andom$\left[(seed\#)\right]$, <u>every</u>kth, <u>seg</u>ments, and <u>gr</u>oup(*varname*).

krandom$\left[(seed\#)\right]$, the default, specifies that k unique observations be chosen at random, from among those to be clustered, as starting centers for the k groups. Optionally, a random-number seed may be specified to cause the command set seed *seed#* (see [D] **generate**) to be applied before the k random observations are chosen.

firstk$\left[, \text{ } exclude\right]$ specifies that the first k observations from among those to be clustered be used as the starting centers for the k groups. With the exclude option, these first k observations are not included among the observations to be clustered.

lastk$\left[, \text{ } exclude\right]$ specifies that the last k observations from among those to be clustered be used as the starting centers for the k groups. With the exclude option, these last k observations are not included among the observations to be clustered.

random$\left[(seed\#)\right]$ specifies that k random initial group centers be generated. The values are randomly chosen from a uniform distribution over the range of the data. Optionally, a random-number seed may be specified to cause the command set seed *seed#* (see [D] **generate**) to be applied before the k group centers are generated.

prandom$\left[(seed\#)\right]$ specifies that k partitions be formed randomly among the observations to be clustered. The group medians from the k groups defined by this partitioning are to be used as the starting group centers. Optionally, a random-number seed may be specified to cause the command set seed *seed#* (see [D] **generate**) to be applied before the k partitions are chosen.

everykth specifies that k partitions be formed by assigning observations $1, 1+k, 1+2k, \ldots$ to the first group; assigning observations $2, 2+k, 2+2k, \ldots$ to the second group; and so on, to form k groups. The group medians from these k groups are to be used as the starting group centers.

segments specifies that k nearly equal partitions be formed from the data. Approximately the first N/k observations are assigned to the first group, the second N/k observations are assigned to the second group, and so on. The group medians from these k groups are to be used as the starting group centers.

group(*varname*) provides an initial grouping variable, *varname*, that defines k groups among the observations to be clustered. The group medians from these k groups are to be used as the starting group centers.

keepcenters specifies that the group medians from the k groups that are produced be appended to the data.

┌─ Advanced ┐

generate(*groupvar*) provides the name of the grouping variable to be created by cluster kmedians. By default, this will be the name specified in name().

iterate(*#*) specifies the maximum number of iterations to allow in the kmedians clustering algorithm. The default is iterate(10000).

Remarks

The data from the two examples introduced in [MV] **cluster kmeans** are presented here to demonstrate the use of `cluster kmedians`. The first dataset contains continuous data, and the second dataset contains binary data.

▷ Example 1

You have measured the flexibility, speed, and strength of the 80 students in your physical education class. You want to split the class into four groups, based on their physical attributes, so that they can receive the mix of flexibility, strength, and speed training that will best help them improve.

The data are summarized and graphed in [MV] **cluster kmeans**. You previously performed kmeans clustering on these data to obtain three, four, and five groups.

```
. use http://www.stata-press.com/data/r9/physed
. cluster kmeans flex speed strength, k(4) name(g4abs) measure(abs)
            start(kr(385617))
. cluster kmeans flex speed strength, k(3) name(g3abs) measure(abs) start(firstk)
. cluster kmeans flex speed strength, k(5) name(g5abs) measure(abs)
            start(random(33576))
```

You now wish to see if kmedians clustering will produce the same grouping for this dataset. You specify four groups, absolute-value distance, and k random observations as beginning centers (but using a different random-number seed).

```
. cluster kmedians flex speed strength, k(4) name(kmed4) measure(abs)
> start(kr(11736))
. cluster list kmed4
kmed4  (type: partition,  method: kmedians,  dissimilarity: L1)
      vars: kmed4 (group variable)
     other: k: 4
            start: krandom(11736)
            range: 0 .
            cmd: cluster kmedians flex speed strength, k(4) name(kmed4)
                measure(abs) start(kr(11736))
            varlist: flexibility speed strength
```

```
. table g4abs kmed4
```

		kmed4		
g4abs	1	2	3	4
1		15		
2			20	
3	35			
4				10

Other than a difference in how the groups are numbered, kmedians clustering and kmeans clustering produced the same results for this dataset.

Now you want to see what happens with kmedians clustering for three groups and for five groups.

```
. cluster kmed flex speed strength, k(3) name(kmed3) measure(abs) start(lastk)
. cluster kmed flex speed strength, k(5) name(kmed5) measure(abs)
> start(prand(8723))
```

```
. cluster list kmed3 kmed5
kmed3  (type: partition,  method: kmedians,  dissimilarity: L1)
      vars: kmed3 (group variable)
     other: k: 3
            start: lastk
            range: 0 .
            cmd: cluster kmedians flex speed strength, k(3) name(kmed3)
                measure(abs) start(lastk)
            varlist: flexibility speed strength
kmed5  (type: partition,  method: kmedians,  dissimilarity: L1)
      vars: kmed5 (group variable)
     other: k: 5
            start: prandom(8723)
            range: 0 .
            cmd: cluster kmedians flex speed strength, k(5) name(kmed5)
                measure(abs) start(prand(8723))
            varlist: flexibility speed strength

. table g3abs kmed3, row
```

g3abs	kmed3 1	2	3
1	6		4
2	18	35	
3	2		15
Total	26	35	19

```
. table g5abs kmed5, row
```

g5abs	kmed5 1	2	3	4	5
1		20			
2	15				
3				6	
4				4	
5			20		15
Total	15	20	20	10	15

Kmeans and kmedians clustering produced different groups for three groups and five groups.

Since one of your concerns was having a better balance in the group sizes, you decide to look a little bit closer at the five-group solution produced by kmedians clustering.

```
. table g4abs kmed5, row col
```

g4abs	kmed5 1	2	3	4	5	Total
1	15					15
2		20				20
3			20		15	35
4				10		10
Total	15	20	20	10	15	80

```
. tabstat flex speed strength, by(kmed5) stat(min mean max)
```

Summary statistics: min, mean, max
  by categories of: kmed5

kmed5	flexib~y	speed	strength
1	8.12	8.05	3.61
	8.852	8.743333	4.358
	9.97	9.79	5.42
2	4.32	1.05	5.46
	5.9465	3.4485	6.8325
	7.89	5.32	7.66
3	1.85	1.18	7.38
	2.4425	1.569	8.2775
	3.48	2.17	9.19
4	2.29	5.11	.05
	3.157	6.988	1.641
	3.99	8.87	3.02
5	.03	.03	7.96
	1.338667	.5793333	8.747333
	2.92	.99	9.57
Total	.03	.03	.05
	4.402625	3.875875	6.439875
	9.97	9.79	9.57

A matrix graph with group numbers as plotting symbols helps you visualize the data.

```
. graph matrix flex speed strength, m(i) mlabel(kmed5) mlabpos(0)
```

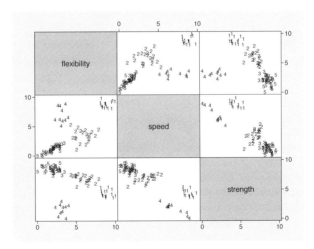

The five-group kmedians clustering split the group with 35 students from the four-group clustering into groups of size 20 and size 15. Looking at the output from `tabstat`, you see that this group was broken up so that the 15 slowest students are split apart from the other 20 (who are still slower than the remaining groups).

The characteristics of the five groups are as follows: Group 1, with 15 students, is already doing very well in flexibility and speed but will need extra strength training. Group 2, with 20 students, needs to emphasize speed training but could use some improvement in the other categories, as well. Group 3, which used to have 35 students, now has 20 students and has serious problems with both flexibility and speed, though they did very well in the strength category. Group 4, with 10 students, needs help with flexibility and strength. Group 5, which was split off of Group 3, has the 15 slowest students.

Even though the matrix graph showing the five groups does not indicate that groups 3 and 5 are very distinct, you decide to go with five groups anyway to even out the group sizes. You will take the slowest group and work with them directly, since they will need a lot of extra help, while your four class assistants will take care of the other four groups.

◁

▷ Example 2

As explained in the second example of [MV] **cluster kmeans**, you have just started a women's club. Thirty women from throughout the community have sent in their requests to join. You have them fill out a questionnaire with 35 yes/no questions relating to sports, music, reading, and hobbies. A description of the data is found in [MV] **cluster kmeans**.

In planning the first meeting of the club, you want to assign seats at the five lunch tables based on shared interests among the women. Kmeans clustering gave you five groups that are each about the right size and seem to make sense in terms of each groups common characteristics; see [MV] **cluster kmeans**. Now you want to see if kmedian clustering suggests a better solution.

As before, you select the Jaccard coefficient as the binary similarity measure and the first k observations as starting centers.

```
. use http://www.stata-press.com/data/r9/wclub, clear

. cluster kmeans bike-fish, k(5) measure(Jaccard) st(firstk) name(gr5)

. cluster kmed bike-fish, k(5) measure(Jaccard) st(firstk) name(kmedian5)

. cluster list kmedian5
kmedian5 (type: partition,  method: kmedians,  similarity: Jaccard)
      vars: kmedian5 (group variable)
     other: k: 5
            start: firstk
            range: 1 0
            cmd: cluster kmedians bike-fish, k(5) measure(Jaccard) st(firstk)
                 name(kmedian5)
            varlist: bike bowl swim jog hock foot base bask arob fshg dart clas
                 cntr jazz rock west romc scif biog fict hist cook shop soap
                 sew crft auto pokr brdg kids hors cat dog bird fish

. table gr5 kmedian5, row col
```

	kmedian5					
gr5	1	2	3	4	5	Total
1	7					7
2	1	6				7
3			5			5
4				5		5
5	1		1		4	6
Total	9	6	6	5	4	30

The resulting groups are similar. Three ladies are grouped differently with this kmedian clustering compared with the kmeans clustering. However, there is a more even distribution of women to the five groups of the kmeans clustering. Since the five lunch tables can seat only eight comfortably, and the kmedians clustering produces one group of size nine, you decide to stick with the groups produced by kmeans clustering.

◁

Methods and Formulas

Kmedians cluster analysis is a variation of the standard kmeans clustering discussed in most cluster analysis books; see the references in [MV] **cluster**. [MV] **cluster** also provides a general discussion of cluster analysis, including kmeans and kmedians clustering, and discusses the available `cluster` subcommands.

Kmedians clustering is an iterative procedure that partitions the data into k groups or clusters. The procedure begins with k initial group centers. Observations are assigned to the group with the closest center. The median of the observations assigned to each of the groups is computed, and the process is repeated. These steps continue until all observations remain in the same group from the previous iteration.

To avoid endless loops, an observation will only be reassigned to a different group if it is closer to the other group center. In the case of a tied distance between an observation and two or more group centers, the observation is assigned to its current group if that is one of the closest, or to the lowest numbered group otherwise.

The `start()` option provides many ways to specify the beginning group centers. These include methods that specify the actual starting centers, as well as methods that specify initial partitions of the data from which the beginning centers are computed.

Stata's `cluster kmedians` command recomputes the group centers only after a complete pass through the data. A disadvantage of this method is that orphaned group centers—one that has no observations that are closest to it—can occur. The advantage of recomputing means only at the end of each pass through the data, instead of after each reassignment, is that the sort order of the data does not potentially change your final result.

Stata deals with orphaned centers by finding the observations that are farthest from the centers and using them as new group centers. The observations are then reassigned to the closest groups, including these new centers.

Continuous or binary data are allowed with `cluster kmedians`. The median of a group of binary observations for a variable is almost always either zero or one. However, if there are an equal number of zeros and ones for a group, the median is 0.5, which is treated as a proportion (just as with kmeans clustering). The binary similarity measures can accommodate the comparison of a binary observation to a proportion. See [MV] *measure_option* for details on this subject and for the formulas for all the available (dis)similarity measures.

Also See

Complementary:	[MV] **cluster notes**, [MV] **cluster stop**, [MV] **cluster utility**
Related:	[MV] **cluster kmeans**
Background:	[MV] **cluster**

Title

> **cluster medianlinkage** — Median-linkage cluster analysis

Syntax

Cluster analysis of data

> cluster <u>med</u>ianlinkage [*varlist*] [*if*] [*in*] [, *cluster_options*]

Cluster analysis of a dissimilarity matrix

> clustermat <u>med</u>ianlinkage *matname* [*if*] [*in*] [, *clustermat_options*]

cluster_options	description
Main	
<u>mea</u>sure(*measure*)	similarity or dissimilarity measures; see *Options* for available measures; default is L2squared
<u>n</u>ame(*clname*)	name of resulting cluster analysis
Advanced	
<u>g</u>enerate(*stub*)	prefix for generated variables; default prefix is *clname*

clustermat_options	description
Main	
<u>sh</u>ape(*shape*)	shape (storage method) of *matname*
add	add cluster information to data currently in memory
clear	replace data in memory with cluster information
<u>labelv</u>ar(*varname*)	place dissimilarity matrix row names in *varname*
<u>n</u>ame(*clname*)	name of resulting cluster analysis
Advanced	
force	perform clustering after fixing *matname* problems
<u>g</u>enerate(*stub*)	prefix for generated variables; default prefix is *clname*

shape	*matname* is stored as a
<u>f</u>ull	square symmetric matrix; the default
<u>l</u>ower	vector of rowwise lower triangle (with diagonal)
<u>ll</u>ower	vector of rowwise strict lower triangle (no diagonal)
<u>u</u>pper	vector of rowwise upper triangle (with diagonal)
<u>uu</u>pper	vector of rowwise strict upper triangle (no diagonal)

144

Description

cluster medianlinkage performs hierarchical agglomerative median-linkage cluster analysis. See [MV] **cluster** for a general discussion of cluster analysis and a description of the other cluster commands.

clustermat medianlinkage performs hierarchical agglomerative median-linkage cluster analysis on the dissimilarity matrix *matname*. See [MV] **clustermat** for a general discussion of cluster analysis of dissimilarity matrices and a description of the other clustermat commands.

After cluster medianlinkage or clustermat medianlinkage, the cluster dendrogram command (see [MV] **cluster dendrogram**) displays the resulting dendrogram, the cluster stop or clustermat stop commands (see [MV] **cluster stop**) help determine the number of groups, and the cluster generate command (see [MV] **cluster generate**) produces grouping variables.

Options for cluster medianlinkage

Main

measure(*measure*) is one of the similarity or dissimilarity measures allowed by Stata. This option is not case sensitive. See [MV] *measure_option* for a discussion of these measures.

The available measures designed for continuous data are L2 (synonym Euclidean); L2squared, which is the default for cluster medianlinkage; L1 (synonyms absolute, cityblock, and manhattan); Linfinity (synonym maximum); L(#); Lpower(#); Canberra; correlation; and angular (synonym angle).

The available measures designed for binary data are matching, Jaccard, Russell, Hamann, Dice, antiDice, Sneath, Rogers, Ochiai, Yule, Anderberg, Kulczynski, Gower2, and Pearson.

Several authors advise using the L2squared *measure* exclusively with median linkage. See *(Dis)similarity transformations and the Lance and Williams formula* and *Warning concerning (dis)similarity choice* in [MV] **cluster** for details.

name(*clname*) specifies the name to attach to the resulting cluster analysis. If name() is not specified, Stata finds an available cluster name, displays it for your reference, and attaches the name to your cluster analysis.

Advanced

generate(*stub*) provides a prefix for the variable names created by cluster medianlinkage. By default, the variable-name prefix will be the name specified in name(). Three variables are created and attached to the cluster-analysis results with the suffixes _id, _ord, and _hgt. Users generally will not need to access these variables directly.

Median linkage can produce reversals or crossovers; see [MV] **cluster** for details. When reversals happen, cluster medianlinkage also creates a fourth variable with the suffix _pht. This is a pseudoheight variable that is used by some of the postclustering commands to properly interpret the _hgt variable.

Options for clustermat medianlinkage

Main

shape(*shape*) specifies the storage mode of *matname*, the matrix of dissimilarities. The following shapes are allowed:

full specifies that *matname* is an $n \times n$ symmetric matrix.

lower specifies that *matname* is a row or column vector of length $n(n + 1)/2$, with the rowwise lower triangle of the dissimilarity matrix including the diagonal of zeros.

$$D_{11}\ D_{21}\ D_{22}\ D_{31}\ D_{32}\ D_{33}\ \ldots\ D_{n1}\ D_{n2}\ \ldots\ D_{nn}$$

llower specifies that *matname* is a row or column vector of length $n(n - 1)/2$, with the rowwise lower triangle of the dissimilarity matrix excluding the diagonal.

$$D_{21}\ D_{31}\ D_{32}\ D_{41}\ D_{42}\ D_{43}\ \ldots\ D_{n1}\ D_{n2}\ \ldots\ D_{n,n-1}$$

upper specifies that *matname* is a row or column vector of length $n(n + 1)/2$, with the rowwise upper triangle of the dissimilarity matrix including the diagonal of zeros.

$$D_{11}\ D_{12}\ \ldots\ D_{1n}\ D_{22}\ D_{23}\ \ldots\ D_{2n}\ D_{33}\ D_{34}\ \ldots\ D_{3n}\ \ldots\ D_{nn}$$

uupper specifies that *matname* is a row or column vector of length $n(n - 1)/2$, with the rowwise upper triangle of the dissimilarity matrix excluding the diagonal.

$$D_{12}\ D_{13}\ \ldots\ D_{1n}\ D_{23}\ D_{24}\ \ldots\ D_{2n}\ D_{34}\ D_{35}\ \ldots\ D_{3n}\ \ldots\ D_{n-1,n}$$

add specifies that clustermat's results be added to the dataset currently in memory. The number of observations (selected observations based on the if and in qualifiers) must equal the number of rows and columns of *matname*. Either clear or add is required if a dataset is currently in memory.

clear drops all the variables and cluster solutions in the current dataset in memory (even if that dataset has changed since the data were last saved) before generating clustermat's results. Either clear or add is required if a dataset is currently in memory.

labelvar(*varname*) specifies the name of a new variable to be created containing the row names of matrix *matname*.

name(*clname*) specifies the name to attach to the resulting cluster analysis. If name() is not specified, Stata finds an available cluster name, displays it for your reference, and attaches the name to your cluster analysis.

Advanced

force allows computations to continue when *matname* is nonsymmetric or has nonzeros on the diagonal. By default, clustermat will complain and exit when it encounters these conditions. force specifies that clustermat operate on the symmetric matrix $(matname * matname')/2$, with any nonzero diagonal entries treated as if they were zero.

generate(*stub*) provides a prefix for the variable names created by clustermat. By default, the variable-name prefix is the name specified in name(). Three variables are created and attached to the cluster-analysis results with the suffixes _id, _ord, and _hgt. Users generally will not need to access these variables directly.

Median linkage can produce reversals or crossovers; see [MV] **cluster** for details. When reversals happen, cluster medianlinkage also creates a fourth variable with the suffix _pht. This is a pseudoheight variable that is used by some of the postclustering commands to properly interpret the _hgt variable.

Remarks

An example using the default L2squared (squared Euclidean) distance and L2 (Euclidean) distance on continuous data and an example using the matching coefficient on binary data illustrate the cluster medianlinkage command. A third example illustrates the use of clustermat medianlinkage in clustering variables instead of observations. These are the same datasets introduced in [MV] **cluster singlelinkage**, which are used as examples for all the hierarchical clustering methods so that you can compare the results from using different hierarchical clustering methods.

▷ Example 1

As explained in the first example of [MV] **cluster singlelinkage**, as the senior data analyst for a small biotechnology firm, you are given a dataset with 4 chemical laboratory measurements on 50 different samples of a particular plant gathered from the rain forest. The head of the expedition that gathered the samples thinks, based on information from the natives, that an extract from the plant might reduce the negative side effects associated with your company's best-selling nutritional supplement.

While the company chemists and botanists continue exploring the possible uses of the plant and plan future experiments, the head of product development asks you to look at the preliminary data and to report anything that might be helpful to the researchers.

While all 50 of the plants are supposed to be of the same type, you decide to perform a cluster analysis to see if there are subgroups or anomalies among them. Single-linkage clustering helped you discover an anomaly in the data. You now wish to see if you discover the same thing using median-linkage clustering with the default squared Euclidean distance and with Euclidean distance.

You first call cluster medianlinkage, letting the distance default to L2squared (squared Euclidean distance), and use the name() option to attach the name med to the resulting cluster analysis. You then use the cluster list command (see [MV] **cluster utility**) to list the components of your cluster analysis.

```
. use http://www.stata-press.com/data/r9/labtech
. cluster medianlinkage x1 x2 x3 x4, name(med)
. cluster list med
med  (type: hierarchical,  method: median,  dissimilarity: L2squared)
        vars: med_id (id variable)
              med_ord (order variable)
              med_hgt (real_height variable)
              med_pht (pseudo_height variable)
       other: range: 0 .
              cmd: cluster medianlinkage x1 x2 x3 x4, name(med)
              varlist: x1 x2 x3 x4
```

You repeat the same process, this time using L2 (Euclidean distance) and giving it the name L2med.

```
. cluster medianlinkage x1 x2 x3 x4, name(L2med) measure(L2)
. cluster list L2med
L2med  (type: hierarchical,  method: median,  dissimilarity: L2)
       vars: L2med_id (id variable)
             L2med_ord (order variable)
             L2med_hgt (real_height variable)
             L2med_pht (pseudo_height variable)
      other: range: 0 .
             cmd: cluster medianlinkage x1 x2 x3 x4, name(L2med) measure(L2)
             varlist: x1 x2 x3 x4
```

You wish to use the `cluster dendrogram` command to graph the dendrogram (see [MV] **cluster dendrogram**), but since this particular cluster analysis produces reversals, Stata refuses to draw the dendrogram.

You decide to use the `cluster generate` command (see [MV] **cluster generate**) to produce grouping variables for 2 to 10 groups for each of the two cluster analyses. You also wish to examine the cross-tabulation of each of these generated groups against a variable that identifies which laboratory technician produced the data. (For the sake of brevity, only one cross-tabulation is shown for each cluster analysis.)

```
. cluster gen gm = groups(2/10), name(med)
. cluster gen gL2m = groups(2/10), name(L2med)
. table labtech gm2
```

labtech	gm2 1	2
Al	10	
Bill	10	
Deb	10	
Jen	10	
Sam		10

```
. table labtech gL2m4
```

labtech	gL2m4 1	2	3	4
Al	10			
Bill	10			
Deb	8		1	1
Jen	10			
Sam		10		

The samples analyzed by Sam appear to be clustered together more strongly than the samples analyzed by the other technicians. The reason for this phenomenon is not as obvious from this analysis as it was when we viewed the dendrogram from the single-linkage clustering (see [MV] **cluster singlelinkage**).

◁

▷ Example 2

This example analyzes the same data that was introduced in the second example of [MV] **cluster singlelinkage**. The sociology professor of your graduate-level class gives, as homework, a dataset containing 30 observations on 60 binary variables, with the assignment to tell him something about the 30 subjects represented by the observations.

In addition to examining single-linkage clustering of these data, you decide to see what median-linkage clustering shows. As with the single-linkage clustering, you pick the simple matching binary coefficient to measure the similarity between groups. The name() option is used to attach the name medlink to the cluster analysis. cluster list displays the details; see [MV] **cluster utility**.

```
. use http://www.stata-press.com/data/r9/homework, clear
. cluster median a1-a60, measure(match) name(medlink)
. cluster list medlink
medlink  (type: hierarchical,  method: median,  similarity: matching)
      vars: medlink_id (id variable)
            medlink_ord (order variable)
            medlink_hgt (real_height variable)
            medlink_pht (pseudo_height variable)
     other: range: 1 0
            cmd: cluster medianlinkage a1-a60, measure(match) name(medlink)
            varlist: a1 a2 a3 a4 a5 a6 a7 a8 a9 a10 a11 a12 a13 a14 a15 a16 a17
                a18 a19 a20 a21 a22 a23 a24 a25 a26 a27 a28 a29 a30 a31 a32
                a33 a34 a35 a36 a37 a38 a39 a40 a41 a42 a43 a44 a45 a46 a47
                a48 a49 a50 a51 a52 a53 a54 a55 a56 a57 a58 a59 a60
```

You attempt to use the cluster dendrogram command to display the dendrogram, but since this particular cluster analysis produced reversals, cluster dendrogram refuses to produce the dendrogram. You realize that with reversals, the resulting dendrogram would not be easy to interpret anyway.

You decide to compare the three-group solution from this median-linkage clustering with the variable called truegrp provided by the teacher. You use the cluster generate command (see [MV] **cluster generate**) to create a grouping variable, based on your centroid clustering, to compare with truegrp.

```
. cluster gen medgrp3 = group(3)
. table medgrp3 truegrp
```

medgrp3	truegrp 1	2	3
1		10	
2	10		
3			10

Other than the numbers arbitrarily assigned to the three groups, your teacher's conclusions and the results from the three-group median-linkage clustering are in complete agreement.

◁

▷ Example 3

The wclub dataset contains answers from 30 women to 35 yes/no questions. The variables are described in example 3 of [MV] **clustermat**. We are interested in seeing how median-linkage clustering will cluster the 35 variables (instead of the observations).

We use the `matrix dissimilarity` command to produce a dissimilarity matrix equal to one minus the Jaccard similarity; see [MV] **matrix dissimilarity**.

```
. use http://www.stata-press.com/data/r9/wclub, clear
. matrix dissimilarity clubD = , variables Jaccard dissim(oneminus)
. clustermat medianlink clubD, name(clubmedian) clear labelvar(question)
obs was 0, now 35
```

We cannot draw the dendrogram due to reversals, but we can examine some of the groupings produced by the cluster analysis.

```
. cluster generate g = groups(2/4)
. table g2
```

g2	Freq.
1	34
2	1

```
. table g3
```

g3	Freq.
1	33
2	1
3	1

```
. table g4
```

g4	Freq.
1	32
2	1
3	1
4	1

```
. list g2 g3 g4 question if g4 > 1
```

	g2	g3	g4	question
29.	1	2	3	brdg
31.	2	3	4	hors
34.	1	1	2	bird

`bird` (have a bird), `brdg` (enjoy playing bridge), and `hors` (have a horse) seem to be the least related to the other variables. These three variables, in turn, merge last into the super group containing the remaining variables.

◁

❑ Technical Note

`cluster medianlinkage` requires more memory and more execution time than `cluster singlelinkage`. With a large number of observations, the execution time may be significant.

❑

Methods and Formulas

[MV] **cluster** discusses hierarchical clustering and places median-linkage clustering in this general framework. Conceptually, hierarchical agglomerative clustering proceeds as follows. The N observations start out as N separate groups, each of size one. The two closest observations are merged into one group, producing $N - 1$ total groups. The closest two groups are then merged so that there are $N - 2$ total groups. This process continues until all the observations are merged into one large group, producing a hierarchy of groupings from one group to N groups. The difference between the various hierarchical-linkage methods depends on how they define "closest" when comparing groups.

Median-linkage clustering is a variation on centroid-linkage clustering. The difference is in how groups of unequal size are treated. Centroid linkage gives each observation equal weight. Median linkage gives each group of observations equal weight, meaning that with unequal group sizes, the observations in the smaller group will have more weight than the observations in the larger group.

The median-linkage clustering algorithm produces two variables that together act as a pointer representation of a dendrogram. To this, Stata adds a third variable used to restore the sort order, as needed, so that the two variables of the pointer representation remain valid. The first variable of the pointer representation gives the order of the observations. The second variable has one less element and gives the height in the dendrogram at which the adjacent observations in the order-variable join. When reversals happen, which they often do, a fourth variable, called a pseudoheight, is produced. This is used by postclustering commands with the height variable to properly interpret the ordering of the hierarchy.

See [MV] *measure_option* for the details and formulas of the available *measure*s, which include (dis)similarity measures for continuous and binary data. See [MV] **cluster** for a warning concerning (dis)similarity measure choice.

Also See

Complementary:	[MV] **cluster dendrogram**, [MV] **cluster generate**, [MV] **cluster notes**, [MV] **cluster stop**, [MV] **cluster utility**
Related:	[MV] **cluster averagelinkage**, [MV] **cluster centroidlinkage**, [MV] **cluster completelinkage**, [MV] **cluster singlelinkage**, [MV] **cluster wardslinkage**, [MV] **cluster waveragelinkage**
Background:	[MV] **cluster**, [MV] **clustermat**

Title

> **cluster notes** — Place notes in cluster analysis

Syntax

Add a note to a cluster analysis

> cluster <u>notes</u> *clname* : *text*

List all cluster notes

> cluster <u>notes</u>

List cluster notes associated with specified cluster analyses

> cluster <u>notes</u> *clnamelist*

Drop cluster notes

> cluster <u>notes</u> drop *clname* $\left[\,\text{in } \textit{numlist}\,\right]$

Description

The `cluster notes` command attaches notes to a previously run cluster analysis. The notes become part of the data and are saved when the data are saved and retrieved when the data are used; see [D] **save**.

To add a note to a cluster analysis, type `cluster notes`, the cluster-analysis name, a colon, and the text.

Typing `cluster notes` by itself lists all cluster notes associated with all defined cluster analyses. `cluster notes` followed by one or more cluster names lists the notes for those cluster analyses.

`cluster notes drop` allows you to drop cluster notes.

Remarks

The cluster-analysis system in Stata has many features that allow you to manage the various cluster analyses that you perform. See [MV] **cluster** for information on all the available cluster-analysis commands; see [MV] **cluster utility** for other `cluster` commands, including `cluster list`, that help you manage your analyses. The `cluster notes` command is modeled after Stata's `notes` command (see [D] **notes**), but they are different systems and do not interact.

▷ Example 1

We illustrate the `cluster notes` command starting with three cluster analyses that have already been performed. The `cluster dir` command shows us the names of all the existing cluster analyses; see [MV] **cluster utility**.

```
. cluster dir
sngeuc
sngabs
kmn3abs

. cluster note sngabs : I used single linkage with absolute value distance

. cluster note sngeuc : Euclidean distance and single linkage

. cluster note kmn3abs : This has the kmeans cluster results for 3 groups

. cluster notes
sngeuc
      notes:   1. Euclidean distance and single linkage

sngabs
      notes:   1. I used single linkage with absolute value distance

kmn3abs
      notes:   1. This has the kmeans cluster results for 3 groups
```

After adding a note to each of the three cluster analyses, we used the `cluster notes` command without arguments to list all the notes for all the cluster analyses.

The * and ? characters may be used when referring to cluster names; see [U] **11.2 Abbreviation rules**.

```
. cluster note k* : Verify that observation 5 is correct.  I am suspicious that
> there was a typographical error or instrument failure in recording the
> information.

. cluster notes kmn3abs
kmn3abs
      notes:   1. This has the kmeans cluster results for 3 groups
               2. Verify that observation 5 is correct. I am suspicious that
                  there was a typographical error or instrument failure in
                  recording the information.
```

`cluster notes` expanded k* to kmn3abs, the only cluster name that begins with a k. Notes that extend to multiple lines are automatically wrapped when displayed. When entering long notes, you just continue to type until your note is finished. Pressing *Return* signals that you are done with that note.

After examining the dendrogram (see [MV] **cluster dendrogram**) for the sngeuc single-linkage cluster analysis and seeing one small group of data that split off from the main body of data at a very large distance, you investigate further and find data problems. You decide to add some notes to the sngeuc analysis.

```
. cluster note *euc : All of Sam's data look wrong to me.

. cluster note *euc : I think Sam should be fired.

. cluster notes sng?*
sngeuc
      notes:   1. Euclidean distance and single linkage
               2. All of Sam's data look wrong to me.
               3. I think Sam should be fired.

sngabs
      notes:   1. I used single linkage with absolute value distance
```

Sam, one of the lab technicians, who happens to be the owner's nephew and is paid more than you, really messed up. After adding these notes, you get second thoughts about keeping the notes attached to the cluster analysis (and the data). You decide you really want to delete those notes and to add a more politically correct note.

```
. cluster note sngeuc : Ask Jennifer to help Sam re-evaluate his data.
. cluster note sngeuc
sngeuc
    notes:    1. Euclidean distance and single linkage
              2. All of Sam's data looks wrong to me.
              3. I think Sam should be fired.
              4. Ask Jennifer to help Sam re-evaluate his data.
. cluster note drop sngeuc in 2/3
. cluster notes kmn3abs s*
kmn3abs
    notes:    1. This has the kmeans cluster results for 3 groups
              2. Verify that observation 5 is correct. I am suspicious that
                 there was a typographical error or instrument failure in
                 recording the information.
sngeuc
    notes:    1. Euclidean distance and single linkage
              2. Ask Jennifer to help Sam re-evaluate his data.
sngabs
    notes:    1. I used single linkage with absolute value distance
```

Just for illustration purposes, the new note was added before deleting the two offending notes. `cluster notes drop` can take an `in` argument followed by a list of note numbers. The numbers correspond to those shown in the listing provided by the `cluster notes` command. After the deletions, the note numbers are reassigned to remove gaps. So, `sngeuc` note 4 becomes note 2 after the deletion of notes 2 and 3 as shown above.

Without an `in` argument, the `cluster notes drop` command drops all notes associated with the named cluster.

◁

Remember that the cluster notes are stored with the data and, as with other updates you make to the data, the additions and deletions are not permanent until you save the data; see [D] **save**.

❏ Technical Note

Programmers can access the notes (and all the other cluster attributes) using the `cluster query` command; see [MV] **cluster programming utilities**.

❏

Also See

Complementary:	[MV] **cluster programming utilities**, [MV] **cluster utility**, [D] **save**
Related:	[D] **notes**
Background:	[MV] **cluster**

Title

cluster programming subroutines — Add cluster-analysis routines

Description

This entry describes how to extend Stata's cluster command; see [MV] **cluster**. Programmers can add subcommands to cluster, add functions to cluster generate (see [MV] **cluster generate**), add stopping rules to cluster stop (see [MV] **cluster stop**), and set up an alternative command to be executed when cluster dendrogram is called (see [MV] **cluster dendrogram**).

The cluster command also provides utilities for programmers; see [MV] **cluster programming utilities** to learn more.

Remarks

Remarks are presented under the headings

> *Adding a cluster subroutine*
> *Adding a cluster generate function*
> *Adding a cluster stopping rule*
> *Applying an alternate cluster dendrogram routine*

Adding a cluster subroutine

You add a cluster subroutine by creating a Stata program with the name cluster_*subcmdname*. For example, to add the subcommand xyz to cluster, create cluster_xyz.ado. Users could then execute the xyz subcommand with

    cluster xyz ...

Everything entered on the command line following cluster xyz is passed to the cluster_xyz command.

You can add new clustering methods, new cluster-management tools, and new postclustering programs. The cluster command has subcommands that can be helpful to cluster-analysis programmers; see [MV] **cluster programming utilities**.

▷ Example 1

We will add a cluster subroutine by writing a simple postcluster-analysis routine that provides a cross-tabulation of two cluster-analysis grouping variables. The syntax of the new command will be

$$\text{cluster mycrosstab } clname1 \; clname2 \; \big[\, , \; tabulate\_options \big]$$

Here is the program:

```
program cluster_mycrosstab
        version 9
        gettoken clname1 0 : 0 , parse(" ,")
        gettoken clname2 rest : 0 , parse(" ,")
        cluster query 'clname1'
        local groupvar1 'r(groupvar)'
        cluster query 'clname2'
        local groupvar2 'r(groupvar)'
        tabulate 'groupvar1' 'groupvar2' 'rest'
end
```

See [P] **gettoken** for information on the gettoken command, and see [R] **tabulate twoway** for information on the tabulate command. The cluster query command is one of the cluster programming utilities that is documented in [MV] **cluster programming utilities**.

We can demonstrate cluster mycrosstab in action. This example starts with two cluster analyses, cl1 and cl2, that have already been performed. The dissimilarity measure and the variables included in the two cluster analyses differ. We want to see how closely the two cluster analyses match.

```
. cluster list, type method dissim var
cl2 (type: partition, method: kmeans, dissimilarity: L(1.5))
      vars: gvar2 (group variable)

cl1 (type: partition, method: kmeans, dissimilarity: L1)
      vars: cl1gvar (group variable)

. cluster mycrosstab cl1 cl2, chi2
```

| | | | gvar2 | | | | |
cl1gvar	1	2	3	4	5	Total
1	0	0	10	0	0	10
2	1	4	0	5	6	16
3	8	0	0	4	1	13
4	9	0	8	4	0	21
5	0	8	0	0	6	14
Total	18	12	18	13	13	74

```
          Pearson chi2(16) =  98.4708    Pr = 0.000
```

The chi2 option was included to demonstrate that we were able to exploit the existing options of tabulate with very little programming effort. We just pass along to tabulate any of the extra arguments received by cluster_mycrosstab.

◁

Adding a cluster generate function

Programmers can add functions to the cluster generate command (see [MV] **cluster generate**) by creating a command called clusgen_*name*. For example, to add a function called abc() to cluster generate, you could create clusgen_abc.ado. Users could then execute

cluster generate *newvarname* = abc(...) ...

Everything entered on the command line following cluster generate is passed to clusgen_abc.

▷ Example 2

Here is the beginning of a clusgen_abc program that expects an integer argument and has one option called name(*clname*), which gives the name of the cluster. If name() is not specified, the name defaults to that of the most-recently performed cluster analysis. We will assume, for illustration purposes, that the cluster analysis must be hierarchical and will check for this in the clusgen_abc program.

```
program clusgen_abc
      version 9
      // we use gettoken to work our way through the parsing
      gettoken newvar 0 : 0 , parse(" =")
      gettoken temp 0 : 0 , parse(" =")
      if '"'temp'"' != "=" {
             error 198
      }
      gettoken temp 0 : 0 , parse(" (")
      if '"'temp'"' != "abc" {
             error 198
      }
      gettoken funcarg 0 : 0 , parse(" (") match(temp)
      if '"'temp'"' != "(" {
             error 198
      }

      // funcarg holds the integer argument to abc()
      confirm integer number 'funcarg'

      // we can now use syntax to parse the option
      syntax [, Name(str) ]

      // cluster query will give us the list of cluster names
      if '"'name'"' == "" {
             cluster query
             local clnames 'r(names)'
             if "'clnames'" == "" {
                    di as err "no cluster solutions defined"
                    exit 198
             }
             // first name in the list is the latest clustering
             local name : word 1 of 'clnames'
      }

      // cluster query followed by name will tell us the type
      cluster query 'name'
      if "'r(type)'" != "hierarchical" {
             di as err "only allowed with hierarchical clustering"
             exit 198
      }
      /*
         you would now pull more information from the call of
                   cluster query 'name'
         and do your computations and generate 'newvar'
      */
      ...

end
```

See [MV] **cluster programming utilities** for details on the cluster query command.

◁

Adding a cluster stopping rule

Programmers can add stopping rules to the rules() option of the cluster stop command (see [MV] **cluster stop**) by creating a Stata program with the name clstop_*name*. For example, to add a stopping rule named mystop so that cluster stop would now have a rules(mystop) option, you could create clstop_mystop.ado defining the clstop_mystop program. Users could then execute

cluster stop [*clname*], rule(mystop) ...

The `clstop_mystop` program is passed the cluster name (*clname*) provided by the user (or the name of the current cluster result if no name is specified), followed by a comma and all the options entered by the user except for the `rule(mystop)` option.

▷ Example 3

We will add a `rule(stepsize)` option to `cluster stop`. This option implements the simple step-size stopping rule (see Milligan and Cooper 1985), which computes the difference in fusion values between levels in a hierarchical cluster analysis. (A fusion value is the similarity or dissimilarity measure at which clusters are fused or split in the hierarchical cluster structure.) Large values of the step-size stopping rule indicate groupings with more distinct cluster structure.

Examining cluster dendrograms (see [MV] **cluster dendrogram**) to visually determine the number of clusters is equivalent to using a visual approximation to the step-size stopping rule.

Here is the `clstop_stepsize` program:

```
program clstop_stepsize, sortpreserve rclass
        version 9
        syntax anything(name=clname) [, Depth(integer -1) ]

        cluster query 'clname'
        if "'r(type)'" != "hierarchical" {
                di as error ///
                    "rule(stepsize) only allowed with hierarchical clustering"
                exit 198
        }
        if "'r(pseudo_heightvar)'" != "" {
                di as error "dendrogram reversals encountered"
                exit 198
        }

        local hgtvar 'r(heightvar)'
        if '""'r(similarity)'""' != "" {
                sort 'hgtvar'
                local negsign "-"
        }
        else if '""'r(dissimilarity)'""' != "" {
                gsort -'hgtvar'
        }
        else {
                di as error "dissimilarity or similarity not set"
                exit 198
        }

        quietly count if !missing('hgtvar')
        local depth = cond('depth'<=1, r(N), min('depth',r(N)))

        tempvar diff
        qui gen double 'diff'='negsign'('hgtvar'-'hgtvar'[_n+1]) if _n<'depth'

        di
        di as txt "Depth" _col(10) "Stepsize"
        di as txt "{hline 17}"
        forvalues i = 1/'= 'depth'-1' {
                local j = 'i' + 1
                di as res 'j' _col(10) %8.0g 'diff'['i']
                return scalar stepsize_'j' = 'diff'['i']
        }
        return local rule "stepsize"
end
```

See [P] **syntax** for information about the `syntax` command, [P] **forvalues** for information about the `forvalues` looping command, and [P] **macro** for information about the '`= ... `' macro function. The `cluster query` command is one of the cluster programming utilities that is documented in [MV] **cluster programming utilities**.

With this program, users can obtain the step-size stopping rule. We demonstrate this using the average-linkage hierarchical cluster analysis found in the second example of [MV] **cluster averagelinkage**. The dataset contains 30 observations on 60 binary variables. The simple matching coefficient is used as the similarity measure in the average-linkage clustering.

```
. cluster a a1-a60, measure(match) name(alink)
. cluster stop alink, rule(stepsize) depth(15)
```

Depth	Stepsize
2	.065167
3	.187333
4	.00625
5	.007639
6	.002778
7	.005952
8	.002381
9	.008333
10	.005556
11	.002778
12	0
13	0
14	.006667
15	.01

In the `clstop_stepsize` program, we included a `depth()` option. `cluster stop`, when called with the new `rule(stepsize)` option, can also have the `depth()` option. In this case, we specified that it stop at a depth of 15.

The largest step-size, .187, happens at the three-group level of the hierarchy. This number, .187, represents the difference between the matching coefficient created when two groups are formed and that created when three groups are formed in this hierarchical cluster analysis.

The `clstop_stepsize` program could be enhanced by using a better output table format. An option could also be added that saves the results to a matrix.

◁

Applying an alternate cluster dendrogram routine

Programmers can change the behavior of the `cluster dendrogram` command (alias `cluster tree`); see [MV] **cluster dendrogram**. This is accomplished by using the `other()` option of the `cluster set` command (see [MV] **cluster programming utilities**) with a *tag* of `treeprogram` and with *text* giving the name of the command to be used in place of the standard Stata program for `cluster dendrogram`. For example, if you had created a new hierarchical cluster-analysis method for Stata that needed a different algorithm for producing dendrograms, you would use the command

```
cluster set clname, other(treeprogram progname)
```

to set *progname* as the program to be executed when `cluster dendrogram` is called.

▷ Example 4

If we were creating a new hierarchical cluster-analysis method called `myclus`, we could create a program called `cluster_myclus` (see the discussion at the beginning of *Remarks*). If `myclus` needed a different dendrogram routine from the standard one used within Stata, we could include the following line inside `cluster_myclus.ado` at the point where we set the cluster attributes.

```
cluster set 'clname', other(treeprogram myclustree)
```

We could then create a program called `myclustree` in a file called `myclustree.ado` that implements the particular dendrogram program needed by `myclus`.

◁

Reference

Milligan, G. W. and M. C. Cooper. 1985. An examination of procedures for determining the number of clusters in a dataset. *Psychometrika* 50: 159–179.

Also See

Background: [MV] **cluster**, [MV] **cluster programming utilities**

Title

> **cluster programming utilities** — Cluster-analysis programming utilities

Syntax

Obtain various attributes of a cluster analysis

> cluster query [*clname*]

Set various attributes of a cluster analysis

> cluster set [*clname*] [, *set_options*]

Delete attributes from a cluster analysis

> cluster delete *clname* [, *delete_options*]

Check similarity and dissimilarity measure name

> cluster parsedistance *measure*

Compute similarity and dissimilarity measure

> cluster measures *varlist* [*if*] [*in*], compare(*numlist*) generate(*newvarlist*)
>
> [*measures_options*]

set_options	description
addname	add *clname* to the master list of cluster analyses
type(*type*)	set the cluster type for *clname*
method(*method*)	set the name of the clustering method for the cluster analysis
similarity(*measure*)	set the name of the similarity measure used for the cluster analysis
dissimilarity(*measure*)	set the name of the dissimilarity measure used for the cluster analysis
var(*tag varname*)	set *tag* that points to *varname*
char(*tag charname*)	set *tag* that points to *charname*
other(*tag text*)	set *tag* with *text* attached to the tag marker
note(*text*)	add a note to the *clname*

delete_options	description
zap	delete all possible settings for *clname*
delname	remove *clname* from the master list of current cluster analyses
type	delete the cluster type entry from *clname*
method	delete the cluster method entry from *clname*
dissimilarity	delete the dissimilarity entries from *clname*
similarity	delete the similarity entries from *clname*
notes(*numlist*)	delete the specified numbered notes from *clname*
allnotes	remove all notes from *clname*
var(*tag*)	remove *tag* from *clname*
allvars	remove all the entries pointing to variables for *clname*
varzap(*tag*)	same as var(), but also delete the referenced variable
allvarzap	same as allvars, but also delete the variables
char(*tag*)	remove *tag* that points to a Stata characteristic from *clname*
allchars	remove all entries pointing to Stata characteristics for *clname*
charzap(*tag*)	same as char(), but also delete the characteristic
allcharzap	same as allchars, but also delete the characteristics
other(*tag*)	delete *tag* and its associated text from *clname*
allothers	delete all entries from *clname* that have been set using other()

measures_options	description
* compare(*numlist*)	use *numlist* as the comparison observations
* generate(*newvarlist*)	create new variables *newvarlist*
measure	(dis)similarity measure; see *Options* for available measures; default is L2
propvars	interpret observations implied by if and in as proportions of binary observations
propcompares	interpret comparison observations as proportions of binary observations

* compare(*numlist*) and generate(*newvarlist*) are required.

Description

The cluster query, cluster set, cluster delete, cluster parsedistance, and cluster measures commands provide tools for programmers to add their own cluster-analysis subroutines to Stata's cluster command; see [MV] **cluster** and [MV] **cluster programming subroutines**. These commands make it possible for the new command to take advantage of Stata's cluster-management facilities.

cluster query provides a way to obtain the various attributes of a cluster analysis in Stata. If *clname* is omitted, cluster query returns in r(names) a list of the names of all currently defined cluster analyses. If *clname* is provided, the various attributes of the specified cluster analysis are returned in r(). These attributes include the type, method, (dis)similarity used, created variable names, notes, and any other information attached to the cluster analysis.

cluster set allows you to set the various attributes that define a cluster analysis in Stata, including naming your cluster results and adding the name to the master list of currently defined cluster results.

With `cluster set`, you can provide information on the type, method, and (dis)similarity measure of your cluster-analysis results. You can associate variables and Stata characteristics (see [P] **char**) with your cluster analysis. `cluster set` also allows you to add notes and other specified fields to your cluster-analysis result. These items become part of the dataset and are saved with the data.

`cluster delete` allows you to delete attributes from a cluster analysis in Stata. This command is the inverse of `cluster set`.

`cluster parsedistance` takes the similarity or dissimilarity *measure* name and checks it against the list of those provided by Stata, taking account of allowed minimal abbreviations and aliases. Aliases are resolved (for instance, `Euclidean` is changed into the equivalent L2).

`cluster measures` computes the similarity or dissimilarity *measure* between the observations listed in the `compare()` option and the observations included based on the `if` and `in` conditions and places the results in the variables specified by the `generate()` option.

Stata also provides a method for programmers to extend the `cluster` command by providing subcommands; see [MV] **cluster programming subroutines**.

Options for cluster set

`addname` adds *clname* to the master list of currently defined cluster analyses. When *clname* is not specified, the `addname` option is mandatory, and in this case, `cluster set` automatically finds a cluster name that is not currently in use and uses this as the cluster name. `cluster set` returns the name of the cluster in `r(name)`. If `addname` is not specified, the *clname* must have been added to the master list previously (for instance, through a previous call to `cluster set`).

`type(`*type*`)` sets the cluster type for *clname*. `type(hierarchical)` indicates that the cluster analysis is hierarchical-style clustering, and `type(partition)` indicates that it is a partition-style clustering. You are not restricted to these types. For instance, you might program some kind of fuzzy partition-clustering analysis, so you, in that case, use `type(fuzzy)`.

`method(`*method*`)` sets the name of the clustering method for the cluster analysis. For instance, Stata uses `method(kmeans)` to indicate a kmeans cluster analysis and uses `method(single)` to indicate single-linkage cluster analysis. You are not restricted to the names currently employed within Stata.

`dissimilarity(`*measure*`)` and `similarity(`*measure*`)` set the name of the dissimilarity or similarity measure used for the cluster analysis. For example, Stata uses `dissimilarity(L2)` to indicate the L2 or Euclidean distance. You are not restricted to the names currently employed within Stata. See [MV] *measure_option* and [MV] **cluster** for a listing and discussion of (dis)similarity measures.

`var(`*tag varname*`)` sets a marker called *tag* in the cluster analysis that points to the variable *varname*. For instance, Stata uses `var(group `*varname*`)` to set a grouping variable from a kmeans cluster analysis. With single-linkage clustering, Stata uses `var(id `*idvarname*`)`, `var(order `*ordervarname*`)`, and `var(height `*hgtvarname*`)` to set the id, order, and height variables that define the cluster-analysis result. You are not restricted to the names currently employed within Stata. Up to ten `var()` options may be specified with a single `cluster set` command.

`char(`*tag charname*`)` sets a marker called *tag* in the cluster analysis that points to the Stata characteristic named *charname*; see [P] **char**. This can be either an `_dta[]` dataset characteristic or a variable characteristic. Up to ten `char()` options may be specified with a single `cluster set` command.

`other(`*tag text*`)` sets a marker called *tag* in the cluster analysis with *text* attached to the *tag* marker. Stata uses `other(k #)` to indicate that k (the number of groups) was # in a kmeans cluster analysis. You are not restricted to the names currently employed within Stata. Up to ten `other()` options may be specified with a single `cluster set` command.

note(*text*) adds a note to the *clname* cluster analysis. The cluster notes command (see [MV] **cluster notes**) is the command to add, delete, or view cluster notes. The cluster notes command uses the note() option of cluster set to add a note to a cluster analysis. Up to ten note() options may be specified with a single cluster set command.

Options for cluster delete

zap deletes all possible settings for cluster analysis *clname*. It is the same as specifying the delname, type, method, dissimilarity, similarity, allnotes, allcharzap, allothers, and allvarzap options.

delname removes *clname* from the master list of current cluster analyses. This option does not affect the various settings that make up the cluster analysis. To remove them, use the other options of cluster delete.

type deletes the cluster type entry from *clname*.

method deletes the cluster method entry from *clname*.

dissimilarity and similarity delete the dissimilarity and similarity entries, respectively, from *clname*.

notes(*numlist*) deletes the specified numbered notes from *clname*. The numbering corresponds to the returned results from the cluster query *clname* command. The cluster notes drop command (see [MV] **cluster notes**) drops a cluster note. It, in turn, calls cluster delete, using the notes() option to drop the notes.

allnotes removes all notes from the *clname* cluster analysis.

var(*tag*) removes from *clname* the entry labeled *tag* that points to a variable. This option does not delete the variable.

allvars removes all the entries pointing to variables for *clname*. This option does not delete the corresponding variables.

varzap(*tag*) is the same as var() and actually deletes the referenced variable.

allvarzap is the same as allvars and actually deletes the variables.

char(*tag*) removes from *clname* the entry labeled *tag* that points to a Stata characteristic (see [P] **char**). This option does not delete the characteristic.

allchars removes all the entries pointing to Stata characteristics for *clname*. This option does not delete the characteristics.

charzap(*tag*) is the same as char() and actually deletes the characteristic.

allcharzap is the same as allchars and actually deletes the characteristics.

other(*tag*) deletes from *clname* the *tag* entry and its associated text, which were set using the other() option of the cluster set command.

allothers deletes all entries from *clname* that have been set using the other() option of the cluster set command.

Options for cluster measures

compare(*numlist*) is required and specifies the observations to use as the comparison observations. Each of these observations will be compared with the observations implied by the if and in conditions using the specified (dis)similarity *measure*. The results are stored in the corresponding new variable from the generate() option. There must be the same number of elements in *numlist* as there are variable names in the generate() option.

generate(*newvarlist*) is required and specifies the names of the variables to be created. There must be as many elements in *newvarlist* as there are numbers specified in the compare() option.

measure is one of the similarity or dissimilarity measures allowed by Stata. This option is not case sensitive. See [MV] ***measure_option*** for a discussion of these measures.

The available measures designed for continuous data are L2 (synonym Euclidean), which is the default; L2squared; L1 (synonyms absolute, cityblock, and manhattan); Linfinity (synonym maximum); L(*#*); Lpower(*#*); Canberra; correlation; and angular (synonym angle).

The available measures designed for binary data are matching, Jaccard, Russell, Hamann, Dice, antiDice, Sneath, Rogers, Ochiai, Yule, Anderberg, Kulczynski, Gower2, and Pearson.

propvars is for use with binary measures and specifies that the observations implied by the if and in conditions be interpreted as proportions of binary observations. The default action with binary measures treats all nonzero values as one (excluding missing values). With propvars, the values are confirmed to be between zero and one, inclusive. See [MV] ***measure_option*** for a discussion of the use of proportions with binary measures.

propcompares is for use with binary measures. It indicates that the comparison observations (those specified in the compare() option) are to be interpreted as proportions of binary observations. The default action with binary measures treats all nonzero values as one (excluding missing values). With propcompares, the values are confirmed to be between zero and one, inclusive. See [MV] ***measure_option*** for a discussion of the use of proportions with binary measures.

Remarks

▷ Example 1

Programmers can determine which cluster solutions currently exist by using the cluster query command without specifying a cluster name to return the names of all currently defined clusters.

```
. use http://www.stata-press.com/data/r9/auto
(1978 Automobile Data)
. cluster k gear turn trunk mpg displ, k(6) name(grpk6L2) measure(L2) gen(g612)
. cluster k gear turn trunk mpg displ, k(7) name(grpk7L2) measure(L2) gen(g712)
. cluster kmed gear turn trunk mpg displ, k(6) name(grpk6L1) measure(L1) gen(g611)
. cluster kmed gear turn trunk mpg displ, k(7) name(grpk7L1) measure(L1) gen(g711)
. cluster dir
grpk7L1
grpk6L1
grpk7L2
grpk6L2
```

```
. cluster query
. return list
macros:
             r(names) : "grpk7L1 grpk6L1 grpk7L2 grpk6L2"
```

In this case, there are four cluster solutions. A programmer can further process the `r(names)` returned macro. For example, to determine which current cluster solutions used kmeans clustering, we would loop through these four cluster solution names and, for each one, call `cluster query` to determine its properties.

```
. local clusnames 'r(names)'
. foreach cname of local clusnames {
  2.          cluster query 'cname'
  3.          if "'r(method)'" == "kmeans" {
  4.                  local kmeancls 'kmeancls' 'cname'
  5.          }
  6. }
. di "{tab}Cluster analyses using kmeans: 'kmeancls'"
        Cluster analyses using kmeans: grpk7L2 grpk6L2
```

In this case, we examined `r(method)`, which records the name of the cluster-analysis method. Two of the four cluster solutions used kmeans.

◁

▷ Example 2

We interactively demonstrate `cluster set`, `cluster delete`, and `cluster query`, though in practice these would be used within a program.

First, we add the name `myclus` to the master list of cluster analyses and, at the same time, set the type, method, and similarity.

```
. cluster set myclus, addname type(madeup) method(fake) similarity(who knows)
. cluster query
. return list
macros:
             r(names) : "myclus grpk7L1 grpk6L1 grpk7L2 grpk6L2"
. cluster query myclus
. return list
macros:
              r(name) : "myclus"
        r(similarity) : "who knows"
            r(method) : "fake"
              r(type) : "madeup"
```

`cluster query` shows that `myclus` was successfully added to the master list of cluster analyses and that the attributes that were `cluster set` can also be obtained.

Now we add a reference to a variable. We will use the word `group` as the *tag* for a variable `mygrpvar`. We also add another item called `xyz` and associate some text with the `xyz` item.

```
. cluster set myclus, var(group mygrpvar) other(xyz some important info)
. cluster query myclus
```

```
. return list

macros:
                r(name) : "myclus"
              r(o1_val) : "some important info"
              r(o1_tag) : "xyz"
            r(groupvar) : "mygrpvar"
             r(v1_name) : "mygrpvar"
              r(v1_tag) : "group"
          r(similarity) : "who knows"
              r(method) : "fake"
                r(type) : "madeup"
```

The `cluster query` command returned the `mygrpvar` information in two ways. The first way is with `r(v#_tag)` and `r(v#_name)`. In this case, there is only one variable associated with `myclus`, so we have `r(v1_tag)` and `r(v1_name)`. This allows the programmer to loop over all the saved variable names without knowing beforehand what the *tag*s might be or how many there are. You could loop as follows:

```
local i 1
while "`r(v`i'_tag)'" != "" {
        ...
        local ++i
}
```

The second way the variable information is returned is in an `r()` result with the *tag* name appended by var, `r(`*tag*`var)`. In our example, this is `r(groupvar)`. This second method is convenient when, as the programmer, you know exactly which *varname* information you are seeking.

The same logic applies to characteristic attributes that are `cluster set`.

Now we continue with our interactive example:

```
. cluster delete myclus, method var(group)

. cluster set myclus, note(a note) note(another note) note(a third note)

. cluster query myclus

. return list

macros:
                r(name) : "myclus"
               r(note3) : "a third note"
               r(note2) : "another note"
               r(note1) : "a note"
              r(o1_val) : "some important info"
              r(o1_tag) : "xyz"
          r(similarity) : "who knows"
                r(type) : "madeup"
```

We used `cluster delete` to remove the method and the `group` variable we had associated with `myclus`. Three notes were then added simultaneously using the `note()` option of `cluster set`. In practice, users will use the `cluster notes` command (see [MV] **cluster notes**) to add and delete cluster notes. The `cluster notes` command is implemented using the `cluster set` and `cluster delete` programming commands.

We finish our interactive demonstration of these commands by deleting more attributes from `myclus` and then completely eliminating `myclus`. In practice, users would remove a cluster analysis with the `cluster drop` command (see [MV] **cluster utility**), which is implemented using the `zap` option of the `cluster delete` command.

```
. cluster delete myclus, allnotes similarity

. cluster query myclus
```

```
. return list

macros:
                r(name) : "myclus"
             r(o1_val) : "some important info"
             r(o1_tag) : "xyz"
                r(type) : "madeup"

. cluster delete myclus, zap

. cluster query

. return list

macros:
             r(names) : "grpk7L1 grpk6L1 grpk7L2 grpk6L2"
```

The cluster attributes that are `cluster set` become a part of the dataset. They are saved with the dataset when it is saved and are available again when the dataset is used; see [D] **save**.

◁

❏ Technical Note

You may wonder how Stata's cluster-analysis data structures are implemented. Stata data characteristics (see [P] **char**) hold the information. The details of the implementation are not important, and in fact, we encourage you to use the `set`, `delete`, and `query` subcommands to access the cluster attributes. This way, if we ever decide to change the underlying implementation, you will be protected through Stata's version-control feature.

❏

▷ Example 3

The `cluster parsedistance` programming command takes as an argument the name of a similarity or dissimilarity measure. Stata then checks this name against those that are implemented within Stata (and available to you through the `cluster measures` command). Uppercase or lowercase letters are allowed, and minimal abbreviations are checked. Some of the measures have aliases, which are resolved so that a standard measure name is returned. We demonstrate the `cluster parsedistance` command interactively:

```
. cluster parsedistance max

. sreturn list

macros:
             s(drange) : "0 ."
             s(dtype) : "dissimilarity"
               s(dist) : "Linfinity"

. cluster parsedistance Eucl

. sreturn list

macros:
             s(drange) : "0 ."
             s(dtype) : "dissimilarity"
               s(dist) : "L2"

. cluster parsedistance correl

. sreturn list

macros:
             s(drange) : "1 -1"
             s(dtype) : "similarity"
               s(dist) : "correlation"
```

```
. cluster parsedistance jacc

. sreturn list

macros:
            s(drange) : "1 0"
            s(binary) : "binary"
             s(dtype) : "similarity"
              s(dist) : "Jaccard"
```

`cluster parsedistance` returns `s(dtype)` as either `similarity` or `dissimilarity` and returns `s(dist)` as the standard Stata name for the (dis)similarity. `s(drange)` gives the range of the measure (most similar to most dissimilar). If the measure is designed for binary variables, `s(binary)` is returned with the word `binary`, as seen above.

See [MV] ***measure_option*** for a listing of the similarity and dissimilarity measures and their properties.

◁

▷ Example 4

`cluster measures` computes the similarity or dissimilarity measure between each comparison observation and the observations implied by the `if` and `in` conditions (or all the data if no `if` or `in` conditions are specified).

We demonstrate using the auto dataset:

```
. use http://www.stata-press.com/data/r9/auto, clear

. cluster measures turn trunk gear_ratio in 1/10, compare(3 11) gen(z3 z11) L1

. format z* %8.2f

. list turn trunk gear_ratio z3 z11 in 1/11
```

	turn	trunk	gear_r~o	z3	z11
1.	40	11	3.58	6.50	14.30
2.	40	11	2.53	6.55	13.25
3.	35	12	3.08	0.00	17.80
4.	40	16	2.93	9.15	8.65
5.	43	20	2.41	16.67	1.13
6.	43	21	2.73	17.35	2.45
7.	34	10	2.87	3.21	20.59
8.	42	16	2.93	11.15	6.65
9.	43	17	2.93	13.15	4.65
10.	42	13	3.08	8.00	9.80
11.	44	20	2.28	.	.

Using the three variables `turn`, `trunk`, and `gear_ratio`, we computed the L1 (or absolute value) distance between the third observation and the first ten observations and placed the results in the variable `z3`. The distance between the eleventh observation and the first ten was placed in variable `z11`.

There are many measures designed for binary data. Below we illustrate `cluster measures` with the matching coefficient binary similarity measure. We have eight observations on ten binary variables, and we will compute the matching similarity measure between the last three observations and all eight observations.

```
. cluster measures x1-x10, compare(6/8) gen(z6 z7 z8) matching
. format z* %4.2f
. list
```

	x1	x2	x3	x4	x5	x6	x7	x8	x9	x10	z6	z7	z8
1.	1	0	0	0	1	1	0	0	1	1	0.60	0.80	0.40
2.	1	1	1	0	0	1	0	1	1	0	0.70	0.30	0.70
3.	0	0	1	0	0	0	1	0	0	1	0.60	0.40	0.20
4.	1	1	1	1	0	0	0	1	1	1	0.40	0.40	0.60
5.	0	1	0	1	1	0	1	0	0	1	0.20	0.60	0.40
6.	1	0	1	0	0	1	0	0	0	0	1.00	0.40	0.60
7.	0	0	0	1	1	1	0	0	1	1	0.40	1.00	0.40
8.	1	1	0	1	0	1	0	1	0	0	0.60	0.40	1.00

Stata treats all nonzero observations as one (except missing values, which are treated as missing values) when computing these binary measures.

When the similarity measure between binary observations and the means of groups of binary observations is needed, the propvars and propcompares options of cluster measures provide the solution. The mean of binary observations is a proportion. The value 0.2 would indicate that 20 percent of the values were one and 80 percent were zero for the group. See [MV] *measure_option* for a discussion of binary measures. The propvars option indicates that the main body of observations should be interpreted as proportions. The propcompares option specifies that the comparison observations be treated as proportions.

We compare ten binary observations on five variables to two observations holding proportions by using the propcompares option:

```
. cluster measures a* in 1/10, compare(11 12) gen(c1 c2) matching propcompare
. list
```

	a1	a2	a3	a4	a5	c1	c2
1.	1	1	1	0	1	.6	.56
2.	0	0	1	1	1	.36	.8
3.	1	0	1	0	0	.76	.56
4.	1	1	0	1	1	.36	.44
5.	1	0	0	0	0	.68	.4
6.	0	0	1	1	1	.36	.8
7.	1	0	1	0	1	.64	.76
8.	1	0	0	0	1	.56	.6
9.	0	1	1	1	1	.32	.6
10.	1	1	1	1	1	.44	.6
11.	.8	.4	.7	.1	.2	.	.
12.	.5	0	.9	.6	1	.	.

◁

Saved Results

cluster query with no arguments saves in r():

Macros

r(names)   cluster solution names

cluster query with an argument saves in r():

Macros

r(name)	cluster name
r(type)	type of cluster analysis
r(method)	cluster-analysis method
r(similarity)	similarity measure name
r(dissimilarity)	dissimilarity measure name
r(note#)	cluster note number #
r(v#_tag)	variable tag number #
r(v#_name)	varname associated with r(v#_tag)
r(*tag*var)	varname associated with *tag*
r(c#_tag)	characteristic tag number #
r(c#_name)	characteristic name associated with r(c#_tag)
r(c#_val)	characteristic value associated with r(c#_tag)
r(*tag*char)	characteristic name associated with *tag*
r(o#_tag)	other tag number #
r(o#_val)	other value associated with r(o#_tag)

cluster set saves in r():

Macros

r(name)   cluster name

cluster parsedistance saves in s():

Macros

s(dist)	(dis)similarity measure name
s(darg)	argument of (dis)similarities that take them, such as L(#)
s(dtype)	the word similarity or dissimilarity
s(drange)	range of measure (most similar to most dissimilar)
s(binary)	the word binary if the measure is for binary observations

cluster measures saves in r():

Macros

r(generate)	variable names from the generate() option
r(compare)	observation numbers from the compare() option
r(dtype)	the word similarity or dissimilarity
r(distance)	the name of the (dis)similarity measure
r(binary)	the word binary if the measure is for binary observations

Also See

Background: [MV] **cluster**, [MV] **cluster programming subroutines**

Title

cluster singlelinkage — Single-linkage cluster analysis

Syntax

Cluster analysis of data

> cluster singlelinkage [*varlist*] [*if*] [*in*] [, *cluster_options*]

Cluster analysis of a dissimilarity matrix

> clustermat singlelinkage *matname* [*if*] [*in*] [, *clustermat_options*]

cluster_options	description
Main	
measure(*measure*)	similarity or dissimilarity measures; see *Options* for available measures; default is L2
name(*clname*)	name of resulting cluster analysis
Advanced	
generate(*stub*)	prefix for generated variables; default prefix is *clname*

clustermat_options	description
Main	
shape(*shape*)	shape (storage method) of *matname*
add	add cluster information to data currently in memory
clear	replace data in memory with cluster information
labelvar(*varname*)	place dissimilarity matrix row names in *varname*
name(*clname*)	name of resulting cluster analysis
Advanced	
force	perform clustering after fixing *matname* problems
generate(*stub*)	prefix for generated variables; default prefix is *clname*

shape	*matname* is stored as a
full	square symmetric matrix; the default
lower	vector of rowwise lower triangle (with diagonal)
llower	vector of rowwise strict lower triangle (no diagonal)
upper	vector of rowwise upper triangle (with diagonal)
uupper	vector of rowwise strict upper triangle (no diagonal)

Description

cluster singlelinkage performs hierarchical agglomerative single-linkage cluster analysis on observations. Single-linkage cluster analysis is also known as the nearest-neighbor technique. See [MV] **cluster** for a general discussion of cluster analysis and a description of the other cluster commands.

clustermat singlelinkage performs hierarchical agglomerative single-linkage cluster analysis on the dissimilarity matrix *matname*. See [MV] **clustermat** for a general discussion of cluster analysis of dissimilarity matrices and a description of the other clustermat commands.

After cluster singlelinkage or clustermat singlelinkage, the cluster dendrogram command (see [MV] **cluster dendrogram**) displays the resulting dendrogram, the cluster stop or clustermat stop commands (see [MV] **cluster stop**) help determine the number of groups, and the cluster generate command (see [MV] **cluster generate**) produces grouping variables.

Options for cluster singlelinkage

⌐ Main ⌐

measure(*measure*) is one of the similarity or dissimilarity measures allowed by Stata. This option is not case sensitive. See [MV] *measure_option* for a discussion of these measures.

The available measures designed for continuous data are L2 (synonym Euclidean), which is the default; L2squared; L1 (synonyms absolute, cityblock, and manhattan); Linfinity (synonym maximum); L(#); Lpower(#); Canberra; correlation; and angular (synonym angle).

The available measures designed for binary data are matching, Jaccard, Russell, Hamann, Dice, antiDice, Sneath, Rogers, Ochiai, Yule, Anderberg, Kulczynski, Gower2, and Pearson.

name(*clname*) specifies the name to attach to the resulting cluster analysis. If name() is not specified, Stata finds an available cluster name, displays it for your reference, and attaches the name to your cluster analysis.

⌐ Advanced ⌐

generate(*stub*) provides a prefix for the variable names created by cluster singlelinkage. By default, the variable-name prefix will be the name specified in name(). Three variables are created and attached to the cluster-analysis results with the suffixes _id, _ord, and _hgt. Users generally will not need to access these variables directly.

Options for clustermat singlelinkage

⌐ Main ⌐

shape(*shape*) specifies the storage mode of *matname*, the matrix of dissimilarities. The following shapes are allowed:

full specifies that *matname* is an $n \times n$ symmetric matrix.

lower specifies that *matname* is a row or column vector of length $n(n+1)/2$, with the rowwise lower triangle of the dissimilarity matrix including the diagonal of zeros.

$$D_{11}\ D_{21}\ D_{22}\ D_{31}\ D_{32}\ D_{33}\ \ldots\ D_{n1}\ D_{n2}\ \ldots\ D_{nn}$$

`llower` specifies that *matname* is a row or column vector of length $n(n-1)/2$, with the rowwise lower triangle of the dissimilarity matrix excluding the diagonal.

$$D_{21}\ D_{31}\ D_{32}\ D_{41}\ D_{42}\ D_{43}\ \ldots\ D_{n1}\ D_{n2}\ \ldots\ D_{n,n-1}$$

`upper` specifies that *matname* is a row or column vector of length $n(n+1)/2$, with the rowwise upper triangle of the dissimilarity matrix including the diagonal of zeros.

$$D_{11}\ D_{12}\ \ldots\ D_{1n}\ D_{22}\ D_{23}\ \ldots\ D_{2n}\ D_{33}\ D_{34}\ \ldots\ D_{3n}\ \ldots\ D_{nn}$$

`uupper` specifies that *matname* is a row or column vector of length $n(n-1)/2$, with the rowwise upper triangle of the dissimilarity matrix excluding the diagonal.

$$D_{12}\ D_{13}\ \ldots\ D_{1n}\ D_{23}\ D_{24}\ \ldots\ D_{2n}\ D_{34}\ D_{35}\ \ldots\ D_{3n}\ \ldots\ D_{n-1,n}$$

`add` specifies that `clustermat`'s results be added to the dataset currently in memory. The number of observations (selected observations based on the `if` and `in` qualifiers) must equal the number of rows and columns of *matname*. Either `clear` or `add` is required if a dataset is currently in memory.

`clear` drops all the variables and cluster solutions in the current dataset in memory (even if that dataset has changed since the data were last saved) before generating `clustermat`'s results. Either `clear` or `add` is required if a dataset is currently in memory.

`labelvar(`*varname*`)` specifies the name of a new variable to be created containing the row names of matrix *matname*.

`name(`*clname*`)` specifies the name to attach to the resulting cluster analysis. If `name()` is not specified, Stata finds an available cluster name, displays it for your reference, and attaches the name to your cluster analysis.

⌐ Advanced ⌐

`force` allows computations to continue when *matname* is nonsymmetric or has nonzeros on the diagonal. By default, `clustermat` will complain and exit when it encounters these conditions. `force` specifies that `clustermat` operate on the symmetric matrix $(matname * matname')/2$, with any nonzero diagonal entries treated as if they were zero.

`generate(`*stub*`)` provides a prefix for the variable names created by `clustermat`. By default, the variable-name prefix is the name specified in `name()`. Three variables are created and attached to the cluster-analysis results with the suffixes `_id`, `_ord`, and `_hgt`. Users generally will not need to access these variables directly.

Remarks

An example using the default L2 (Euclidean) distance on continuous data and an example using the `matching` coefficient on binary data illustrate the `cluster singlelinkage` command. A third example illustrates the use of `clustermat singlelinkage` in clustering variables instead of observations. The data from these three examples are also used in [MV] **cluster averagelinkage**, [MV] **cluster centroidlinkage**, [MV] **cluster completelinkage**, [MV] **cluster medianlinkage**, [MV] **cluster wardslinkage**, and [MV] **cluster waveragelinkage**, so that you can compare different hierarchical clustering methods.

▷ Example 1

As the senior data analyst for a small biotechnology firm, you are given a dataset with 4 chemical laboratory measurements on 50 different samples of a particular plant gathered from the rain forest. The head of the expedition that gathered the samples thinks, based on information from the natives, that an extract from the plant might reduce the negative side effects associated with your company's best-selling nutritional supplement.

While the company chemists and botanists continue exploring the possible uses of the plant and plan future experiments, the head of product development asks you to look at the preliminary data and to report anything that might be helpful to the researchers.

While all 50 of the plants are supposed to be of the same type, you decide to perform a cluster analysis to see if there are subgroups or anomalies among them. You arbitrarily decide to use single-linkage clustering with the default Euclidean distance.

```
. use http://www.stata-press.com/data/r9/labtech
. cluster singlelinkage x1 x2 x3 x4, name(sngeuc)
. cluster list sngeuc
sngeuc  (type: hierarchical,  method: single,  dissimilarity: L2)
      vars: sngeuc_id (id variable)
            sngeuc_ord (order variable)
            sngeuc_hgt (height variable)
     other: range: 0 .
            cmd: cluster singlelinkage x1 x2 x3 x4, name(sngeuc)
            varlist: x1 x2 x3 x4
```

The `cluster singlelinkage` command generated some variables and created a cluster object with the name `sngeuc`, which you supplied as an argument. `cluster list` provides details about the cluster object; see [MV] **cluster utility**.

What you really want to see is the dendrogram for this cluster analysis; see [MV] **cluster dendrogram**.

```
. cluster dendrogram sngeuc, xlabel(, angle(90) labsize(*.75))
```

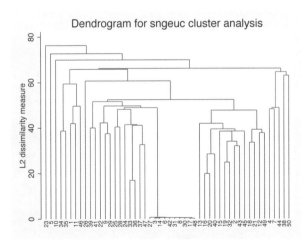

Dendrogram for sngeuc cluster analysis

From your experience looking at dendrograms, two things jump out at you about this cluster analysis. The first is the observations showing up in the middle of the dendrogram that are all very close to each other (very short vertical bars) and are far from any other observations (the long vertical bar connecting them to the rest of the dendrogram). Next you notice that if you ignore those ten observations, the rest of the dendrogram does not indicate strong clustering, as evidenced by the relatively short vertical bars in the upper portion of the dendrogram.

You start to look for clues as to why these ten observations are so peculiar. Looking at scatterplots is usually helpful, and so you examine the matrix of scatterplots.

```
. graph matrix x1 x2 x3 x4
```

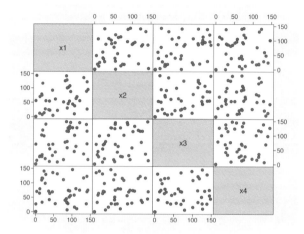

Unfortunately, these scatterplots do not indicate what might be going on.

Suddenly, based on your past experience with the laboratory technicians, you have an idea of what to check next. Because of past data mishaps, the company started the policy of placing within each dataset a variable giving the name of the technician who produced the measurement. You decide to view the dendrogram, using the technician's name as the label instead of the default observation number.

```
. cluster dendrogram sngeuc, labels(labtech) xlabel(, angle(90) labsize(*.75))
```

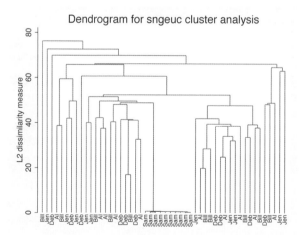

Dendrogram for sngeuc cluster analysis

Your suspicions are confirmed. Sam, one of the laboratory technicians, has messed up once again. You list the data and see that all of his observations are between zero and one, while the other four technicians' data range up to about 150, as expected. It looks like Sam forgot, once again, to calibrate his sensor before analyzing his samples. You decide to save a note of your findings with this cluster analysis (see [MV] **cluster notes** for the details) and to send the data back to the laboratory to be fixed.

◁

▷ Example 2

The sociology professor of your graduate-level class gives, as homework, a dataset containing 30 observations on 60 binary variables, with the assignment to tell him something about the 30 subjects represented by the observations. You feel that this assignment is too vague, but, since your grade depends on it, you get to work trying to figure something out.

Among the analyses you try is the following cluster analysis. You decide to use single-linkage clustering with the simple matching binary coefficient since it is easy to understand. Just for fun, though it makes no difference to you, you specify the **generate()** option to force the generated variables to have **zstub** as a prefix. You let Stata pick a name for your cluster analysis by not specifying the **name()** option.

```
. use http://www.stata-press.com/data/r9/homework, clear
. cluster s a1-a60, measure(matching) gen(zstub)
cluster name: _cl_1
. cluster list
_cl_1  (type: hierarchical,  method: single,  similarity: matching)
      vars: zstub_id (id variable)
            zstub_ord (order variable)
            zstub_hgt (height variable)
     other: range: 1 0
            cmd: cluster singlelinkage a1-a60, measure(matching) gen(zstub)
            varlist: a1 a2 a3 a4 a5 a6 a7 a8 a9 a10 a11 a12 a13 a14 a15 a16 a17
                a18 a19 a20 a21 a22 a23 a24 a25 a26 a27 a28 a29 a30 a31 a32
                a33 a34 a35 a36 a37 a38 a39 a40 a41 a42 a43 a44 a45 a46 a47
                a48 a49 a50 a51 a52 a53 a54 a55 a56 a57 a58 a59 a60
```

Stata selected _cl_1 as the cluster name and created the variables zstub_id, zstub_ord, and zstub_hgt.

You display the dendrogram using the **cluster tree** command, which is a synonym for **cluster dendrogram**. Since Stata uses the most recently performed cluster analysis by default, you do not need to type the name.

(Continued on next page)

. cluster tree

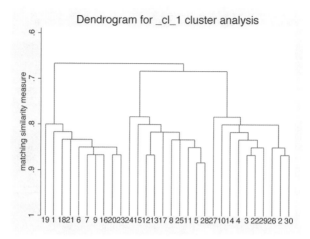

Dendrogram for _cl_1 cluster analysis

The dendrogram seems to indicate the presence of 3 groups among the 30 observations. You decide that this is probably the structure your teacher wanted you to find, and you begin to write up your report. You want to examine the three groups further, so you use the `cluster generate` command (see [MV] **cluster generate**) to create a grouping variable to make the task easier. You examine various summary statistics and tables for the three groups and finish your report.

After the assignment is turned in, your professor gives you the identical dataset with the addition of one more variable, `truegrp`, which indicates the groupings he feels are in the data. You do a cross-tabulation of the `truegrp` and `grp3`, your grouping variable, to see if you are going to get a good grade on the assignment.

. cluster gen grp3 = group(3)

. table grp3 truegrp

	truegrp		
grp3	1	2	3
1		10	
2			10
3	10		

Other than the numbers arbitrarily assigned to the three groups, both you and your professor are in complete agreement. You rest easier that night knowing that you may survive one more semester.

◁

▷ Example 3

The `wclub` dataset contains answers from 30 women to 35 yes/no questions. The variables are described in example 3 of [MV] **clustermat**. We are interested in seeing how single-linkage clustering will cluster the 35 variables (instead of the observations).

We use the `matrix dissimilarity` command to produce a dissimilarity matrix equal to one minus the Jaccard similarity; see [MV] **matrix dissimilarity**.

```
. use http://www.stata-press.com/data/r9/wclub, clear

. matrix dissimilarity clubD = , variables Jaccard dissim(oneminus)

. clustermat singlelink clubD, name(clubsing) clear labelvar(question)
obs was 0, now 35

. cluster dendrogram clubsing, labels(question)
                xlabel(, angle(90) labsize(*.75))
                title(Single-linkage clustering)
                ytitle(1 - Jaccard similarity, suffix)
```

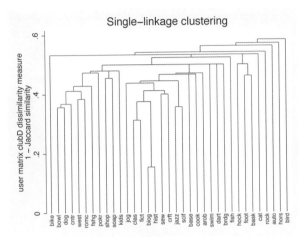

From these 30 women, we see that the `biog` (enjoy reading biographies) and `hist` (enjoy reading history) questions were most closely related. `auto` (enjoy automobile mechanics), `hors` (have a horse), and `bird` (have a bird) seem to be the least related to the other variables. These three variables, in turn, merge last into the super group containing the remaining variables.

◁

Methods and Formulas

[MV] **cluster** discusses hierarchical clustering and places single-linkage clustering in this general framework. It compares single linkage with complete and average linkage.

Conceptually, hierarchical agglomerative single-linkage clustering proceeds as follows. The N observations start out as N separate groups, each of size one. The two closest observations are merged into one group, producing $N - 1$ total groups. The closest two groups are then merged so that there are $N - 2$ total groups. This process continues until all the observations are merged into one large group, producing a hierarchy of groupings from one group to N groups. For single-linkage clustering, the "closest two groups" are determined by the closest observations between the two groups.

Stata's implementation of single-linkage clustering is modeled after the algorithm presented in Sibson (1973) and mentioned in Rohlf (1982). This algorithm produces two variables that act as a pointer representation of a dendrogram. To this, Stata adds a third variable used to restore the sort order, as needed, so that the two variables of the pointer representation remain valid. The first variable gives the order of the observations. The second variable has one less element and gives the height in the dendrogram at which the adjacent observations in the order-variable join.

See [MV] *measure_option* for the details and formulas of the available *measure*s, which include (dis)similarity measures for continuous and binary data.

References

See [MV] **cluster** for references related to cluster analysis, including single-linkage clustering. The following references are especially important in terms of the implementation of single-linkage clustering:

Day, W. H. E. and H. Edelsbrunner. 1984. Efficient algorithms for agglomerative hierarchical clustering methods. *Journal of Classification* 1: 7–24.

Rohlf, F. J. 1982. Single-link clustering algorithms. In *Handbook of Statistics*, Vol. 2, ed. P. R. Krishnaiah and L. N. Kanal, 267–284. Amsterdam: North-Holland.

Sibson, R. 1973. SLINK: An optimally efficient algorithm for the single-link cluster method. *Computer Journal* 16: 30–34.

Also See

Complementary:	[MV] **cluster dendrogram**, [MV] **cluster generate**, [MV] **cluster notes**, [MV] **cluster stop**, [MV] **cluster utility**
Related:	[MV] **cluster averagelinkage**, [MV] **cluster centroidlinkage**, [MV] **cluster completelinkage**, [MV] **cluster medianlinkage**, [MV] **cluster wardslinkage**, [MV] **cluster waveragelinkage**
Background:	[MV] **cluster**, [MV] **clustermat**

Title

> **cluster stop** — Cluster-analysis stopping rules

Syntax

Cluster analysis of data

> cluster stop [*clname*] [, *options*]

Cluster analysis of a dissimilarity matrix

> clustermat stop [*clname*] , <u>var</u>iables(*varlist*) [*options*]

options	description
<u>rule</u>(<u>cal</u>inski)	use Caliński–Harabasz pseudo-F index stopping rule; the default
<u>rule</u>(duda)	use Duda–Hart Je(2)/Je(1) index stopping rule
† <u>rule</u>(*rule_name*)	use *rule_name* stopping rule; see *Options* for details
<u>groups</u>(*numlist*)	compute stopping rule for specified groups
<u>matrix</u>(*matname*)	save results in matrix *matname*
* <u>var</u>iables(*varlist*)	compute the stopping rule using *varlist*

† rule(*rule_name*) is not shown in the dialog box. See [MV] **cluster programming subroutines** for information on how to add stopping rules to the cluster stop command.

* variables(*varlist*) is required with a clustermat solution and optional with a cluster solution.

Description

Cluster-analysis stopping rules are used to determine the number of clusters. A stopping-rule value (also called an index) is computed for each cluster solution (e.g., at each level of the hierarchy in a hierarchical cluster analysis). Larger values (or smaller, depending on the particular stopping rule) indicate more distinct clustering. See [MV] **cluster** for background information on cluster analysis and on the cluster and clustermat commands.

The cluster stop and clustermat stop commands currently provide two stopping rules, the Caliński and Harabasz (1974) pseudo-F index and the Duda and Hart (1973) Je(2)/Je(1) index. For both of these rules, larger values indicate more distinct clustering. Presented with the Duda and Hart Je(2)/Je(1) values are pseudo-T-squared values. Smaller pseudo-T-squared values indicate more distinct clustering.

clname specifies the name of the cluster analysis. The default is the most-recently performed cluster analysis, which can be reset using the cluster use command; see [MV] **cluster utility**.

Additional stop rules may be added; see [MV] **cluster programming subroutines**, which illustrates this by showing a program that adds the step-size stopping rule.

Options

rule(calinski | duda | *rule_name*) indicates the stopping rule. rule(calinski), the default, specifies the Caliński and Harabasz pseudo-F index. rule(duda) specifies the Duda and Hart Je(2)/Je(1) index.

rule(calinski) is allowed for both hierarchical and nonhierarchical cluster analyses. rule(duda) is only allowed for hierarchical cluster analyses.

You can add stopping rules to the cluster stop command (see [MV] **cluster programming subroutines**) using the rule(*rule_name*) option. [MV] **cluster programming subroutines** illustrates how to add stopping rules by showing a program that adds a rule(stepsize) option, which implements the simple step-size stopping rule mentioned in Milligan and Cooper (1985).

groups(*numlist*) specifies the cluster groupings for which the stopping rule is to be computed. groups(3/20) specifies that the measure be computed for the 3-group solution, the 4-group solution, ..., and the 20-group solution.

With rule(duda), the default is groups(1/15). With rule(calinski) for a hierarchical cluster analysis, the default is groups(2/15). groups(1) is not allowed with rule(calinski) since the measure is not defined for the degenerate 1-group cluster solution. The groups() option is unnecessary (and not allowed) for a nonhierarchical cluster analysis.

If there are ties in the hierarchical cluster-analysis structure, some (or possibly all) of the requested stopping-rule solutions may not be computable. cluster stop passes over, without comment, the groups() for which ties in the hierarchy cause the stopping rule to be undefined.

matrix(*matname*) saves the results in a matrix named *matname*.

With rule(calinski), the matrix has two columns, the first giving the number of clusters, and the second giving the corresponding Caliński and Harabasz pseudo-F stopping-rule index.

With rule(duda), the matrix has three columns: the first column gives the number of clusters, the second column gives the corresponding Duda and Hart Je(2)/Je(1) stopping-rule index, and the third column provides the corresponding pseudo-T-squared values.

variables(*varlist*) specifies the variables to be used in the computation of the stopping rule. By default, the variables used for the cluster analysis are used. variables() is required for cluster solutions produced by clustermat.

Remarks

Everitt, Landau, and Leese (2001) and Gordon (1999) discuss the problem of determining the number of clusters and describe several stopping rules, including the Caliński and Harabasz (1974) pseudo-F index and the Duda and Hart (1973) Je(2)/Je(1) index. There are a large number of cluster stopping rules. Milligan and Cooper (1985) provide an evaluation of 30 stopping rules, singling out the Caliński and Harabasz index and the Duda and Hart index as two of the best rules.

Large values of the Caliński and Harabasz pseudo-F index indicate distinct clustering. The Duda and Hart Je(2)/Je(1) index has an associated pseudo-T-squared value. A large Je(2)/Je(1) index value and a small pseudo-T-squared value indicate distinct clustering. See *Methods and Formulas* at the end of this entry for details.

Example 2 of [MV] **clustermat** shows the use of the clustermat stop command.

Some stopping rules such as the Duda and Hart index only work with a hierarchical cluster analysis. The Caliński and Harabasz index, however, may be applied to both nonhierarchical and hierarchical cluster analyses.

▷ Example 1

Previously, you ran kmeans and kmedians cluster analyses on data where you measured the flexibility, speed, and strength of the 80 students in your physical education class; see the first examples of [MV] **cluster kmeans** and [MV] **cluster kmedians**. Your original goal was to split the class into four groups, though you also examined the three- and five-group kmeans and kmedian cluster solutions as possible alternatives.

As described in [MV] **cluster kmedians**, you finally decided that while the four-group solution seemed the best from a clustering standpoint, the five-group solution given by kmedian clustering would work best for your situation.

Now out of curiosity, you wonder what the Caliński and Harabasz stopping rule shows for the three-, four-, and five-group solutions from your kmedian clustering.

```
. use http://www.stata-press.com/data/r9/physed
. cluster kmed flex speed strength, k(3) name(kmed3) measure(abs) start(lastk)
. cluster kmed flex speed strength, k(4) name(kmed4) measure(abs) start(kr(11736))
. cluster kmed flex speed strength, k(5) name(kmed5) measure(abs) start(prand(8723))
. cluster stop kmed3
```

Number of clusters	Calinski/ Harabasz pseudo-F
3	132.75

```
. cluster stop kmed4
```

Number of clusters	Calinski/ Harabasz pseudo-F
4	337.10

```
. cluster stop kmed5
```

Number of clusters	Calinski/ Harabasz pseudo-F
5	300.45

The four-group solution with a Caliński and Harabasz pseudo-F value of 337.10 is largest, indicating that the four-group solution is the most distinct compared with the three-group and five-group solutions.

The three-group solution has a much lower stopping-rule value of 132.75. The five-group solution, with a value of 300.45, is reasonably close to the four-group solution.

Though you do not think it will change your decision on how to split your class into groups, you are curious to see what a hierarchical cluster analysis might produce. You decide to try an average-linkage cluster analysis using the default Euclidean distance; see [MV] **cluster averagelinkage**. You examine the resulting cluster analysis with the `cluster tree` command, which is an easier-to-type alias for the `cluster dendrogram` command; see [MV] **cluster dendrogram**.

```
. cluster averagelink flex speed strength, name(avglnk)
. cluster tree avglnk, xlabel(, angle(90) labsize(*.75))
```

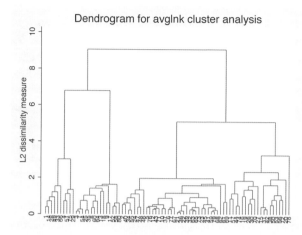

Dendrogram for avglnk cluster analysis

You are curious to see how the four- and five-group solutions from this hierarchical cluster analysis compare with the four- and five-group solutions from the kmedian clustering.

```
. cluster gen avgg = groups(4/5), name(avglnk)
. table kmed4 avgg4
```

kmed4	avgg4			
	1	2	3	4
1			35	
2		15		
3				20
4	10			

```
. table kmed5 avgg5
```

kmed5	avgg5				
	1	2	3	4	5
1		15			
2				19	1
3			20		
4	10				
5			15		

The four-group solutions are identical, except for the numbers used to label the groups. The five-group solutions are different. The kmedian clustering split the 35-member group into subgroups having 20 and 15 members. The average-linkage clustering instead split one member off from the 20-member group.

Now you examine the Caliński and Harabasz pseudo-F stopping-rule values associated with the kmedian hierarchical cluster analysis.

```
. cluster stop avglnk, rule(calinski)
```

Number of clusters	Calinski/ Harabasz pseudo-F
2	131.86
3	126.62
4	337.10
5	269.07
6	258.40
7	259.37
8	290.78
9	262.86
10	258.53
11	249.93
12	247.85
13	247.53
14	236.98
15	226.51

Since `rule(calinski)` is the default, you could have obtained this same table by typing

```
. cluster stop avglnk
```

or, since `avglnk` was the most-recent cluster analysis performed, by typing

```
. cluster stop
```

You did not specify the number of groups to examine from the hierarchical cluster analysis, so it defaulted to examining up to 15 groups. The highest Caliński and Harabasz pseudo-F value is 337.10 for the four-group solution.

What does the Duda and Hart stopping rule produce for this hierarchical cluster analysis?

```
. cluster stop avglnk, rule(duda) groups(1/10)
```

Number of clusters	Duda/Hart	
	Je(2)/Je(1)	pseudo T-squared
1	0.3717	131.86
2	0.1349	147.44
3	0.2283	179.19
4	0.8152	4.08
5	0.2232	27.85
6	0.5530	13.74
7	0.5287	29.42
8	0.6887	3.16
9	0.4888	8.37
10	0.7621	7.80

This time, we asked to see the results for one to ten groups. The largest Duda and Hart Je(2)/Je(1) stopping-rule value is 0.8152 corresponding to four groups. The smallest pseudo-T-squared value is 3.16 for the eight-group solution, but notice that the pseudo-T-squared value for the four-group solution is also low, with a value of 4.08.

Distinct clustering is characterized by large Caliński and Harabasz pseudo-F values, large Duda and Hart Je(2)/Je(1) values, and small Duda and Hart pseudo-T-squared values.

The conventional wisdom for deciding the number of groups based on the Duda and Hart stopping-rule table is to find one of the largest Je(2)/Je(1) values that corresponds to a low pseudo-T-squared value that has much larger T-squared values next to it. This strategy, combined with the results from the Caliński and Harabasz results, indicates that the four-group solution is the most distinct from this hierarchical cluster analysis.

Despite all this, you decide to stick with the five-group solution found using kmedian clustering.

◁

❑ Technical Note

There is a good reason that the word "pseudo" appears in "pseudo-F" and "pseudo-T-squared". While these index values are based on well-known statistics, any p-values computed from these statistics would not be valid. Remember that cluster analysis searches for structure.

If you were to generate random observations, perform a cluster analysis, compute these stopping-rule statistics, and then follow that by computing what would normally be the p-values associated with the statistics, you would almost always end up with significant p-values.

Remember that you would expect, on average, 5 out of every 100 groupings of your random data to show up as significant when you use .05 as your threshold for declaring significance. Cluster-analysis methods search for the best groupings, so there is no surprise that p-values show high significance, even when none exists.

Examining the stopping-rule index values relative to one another is useful, however, in finding relatively reasonable groupings that may exist in the data.

❑

❑ Technical Note

As mentioned in *Methods and Formulas*, ties in the hierarchical cluster structure cause some of the stopping-rule index values to be undefined. Discrete (as opposed to continuous) data tend to cause ties in a hierarchical clustering. The more discrete the data, the more likely it is that ties will occur (and the more of them you will encounter) within a hierarchy.

Even with so-called continuous data, ties in the hierarchical clustering can occur. We say "so-called" because most continuous data are truncated or rounded. For instance, miles per gallon, length, weight, etc., which may really be continuous, may be observed and recorded only to the tens, ones, tenths, or hundredths of a unit.

You can have data with no ties in the observations and still have lots of ties in the hierarchy. Ties in distances (or similarities) between observations and groups of observations cause the ties in the hierarchy.

Because of this, do not be surprised when some (many) of the stopping-rule values that you request are not presented. Stata has decided not to break the ties arbitrarily, since the stopping-rule values may be very different, depending on which split is made.

❑

❑ Technical Note

The stopping rules also become less informative as the number of elements in the groups becomes small, that is, having many groups, each with few observations. We recommend that if you need to examine the stopping-rule values deep within your hierarchical cluster analysis, you do so skeptically.

❑

Saved Results

cluster stop and clustermat stop with rule(calinski) saves in r():

Scalars

    r(calinski_#)   Caliński and Harabasz pseudo-F for # groups

Macros

    r(rule)   calinski

    r(label)   C-H pseudo-F

    r(longlabel)   Caliński and Harabasz pseudo-F

cluster stop and clustermat stop with rule(duda) saves in r():

Scalars

    r(duda_#)   Duda and Hart Je(2)/Je(1) value for # groups

    r(dudat2_#)   Duda and Hart pseudo-T-squared value for # groups

Macros

    r(rule)   duda

    r(label)   D-H Je(2)/Je(1)

    r(longlabel)   Duda and Hart Je(2)/Je(1)

    r(label2)   D-H pseudo-T-squared

    r(longlabel2)   Duda and Hart pseudo-T-squared

Methods and Formulas

cluster stop and clustermat stop are implemented as ado-files.

The Caliński and Harabasz pseudo-F stopping-rule index for g groups and N observations is

$$\frac{\text{trace}(\mathbf{B})/(g-1)}{\text{trace}(\mathbf{W})/(N-g)}$$

where $\mathbf{B}$ is the between-cluster sum of squares and cross-products matrix, and $\mathbf{W}$ is the within-cluster sum of squares and cross-products matrix.

Large values of the Caliński and Harabasz pseudo-F stopping-rule index indicate distinct cluster structure. Small values indicate less clearly defined cluster structure.

The Duda and Hart Je(2)/Je(1) stopping-rule index value is literally Je(2) divided by Je(1). Je(1) is the sum of squared errors within the group that is to be divided. Je(2) is the sum of squared errors in the two resulting subgroups.

Large values of the Duda and Hart pseudo-T-squared stopping-rule index indicate distinct cluster structure. Small values indicate less clearly defined cluster structure.

The Duda and Hart Je(2)/Je(1) index requires hierarchical clustering information. It needs to know at each level of the hierarchy which group is to be split and how. The Duda and Hart index is also local since the only information used comes from the group being split. The information in the rest of the groups does not enter the computation.

In comparison, the Caliński and Harabasz rule does not require hierarchical information and is global since the information from each group is used in the computation.

A pseudo-T-squared value is also presented with the Duda and Hart Je(2)/Je(1) index. The relationship is

$$\frac{1}{\text{Je}(2)/\text{Je}(1)} = 1 + \frac{T^2}{N_1 + N_2 - 2}$$

where N_1 and N_2 are the numbers of observations in the two subgroups.

Notice that Je(2)/Je(1) will be zero when Je(2) is zero, i.e., when the two subgroups each have no variability. An example of this is when the cluster being split has two distinct values that are being split into singleton subgroups. Je(1) will never be zero because we do not split groups that have no variability. When Je(2)/Je(1) is zero, the pseudo-T-squared value is undefined.

Ties in splitting a hierarchical cluster analysis create an ambiguity for the Je(2)/Je(1) measure. For example, to compute the measure for the case of going from five clusters to six, you need to identify the one cluster that will be split. With a tie in the hierarchy, you would instead go from five clusters directly to seven (just as an example). Stata refuses to produce an answer in this situation.

References

Caliński, T. and J. Harabasz. 1974. A dendrite method for cluster analysis. *Communications in Statistics* 3: 1–27.

Duda, R. O. and P. E. Hart. 1973. *Pattern Classification and Scene Analysis.* New York: Wiley.

Everitt, B. S., S. Landau, and M. Leese. 2001. *Cluster Analysis.* 4th ed. London: Arnold.

Gordon, A. D. 1999. *Classification.* 2nd ed. Boca Raton, FL: CRC.

Milligan, G. W. and M. C. Cooper. 1985. An examination of procedures for determining the number of clusters in a dataset. *Psychometrika* 50: 159–179.

Also See

Complementary: [MV] **cluster averagelinkage**, [MV] **cluster centroidlinkage**,

[MV] **cluster completelinkage**, [MV] **cluster kmeans**,

[MV] **cluster kmedians**, [MV] **cluster medianlinkage**,

[MV] **cluster programming subroutines**, [MV] **cluster singlelinkage**,

[MV] **cluster wardslinkage**, [MV] **cluster waveragelinkage**

Related: [MV] **cluster dendrogram**, [MV] **cluster generate**

Background: [MV] **cluster**, [MV] **clustermat**

Title

cluster utility — List, rename, use, and drop cluster analyses

Syntax

Directory style listing of currently defined clusters

    cluster dir

Detailed listing of clusters

    cluster list [clnamelist] [, list_options]

Completely drop the named clusters

    cluster drop { clnamelist | _all }

Mark a cluster analysis as the most current one

    cluster use clname

Rename a cluster

    cluster rename oldclname newclname

Rename variables attached to a cluster

    cluster renamevar oldvarname newvarname [, name(clname)]

    cluster renamevar oldstub newstub , prefix [name(clname)]

list_options	description
Options	
<u>note</u>s	list cluster notes
<u>t</u>ype	list cluster analysis type
<u>m</u>ethod	list cluster analysis method
<u>di</u>ssimilarity	list cluster analysis dissimilarity measure
<u>s</u>imilarity	list cluster analysis similarity measure
<u>v</u>ars	list variable names attached to the cluster analysis
<u>c</u>hars	list any characteristics attached to the cluster analysis
<u>o</u>ther	list any "other" information
† <u>a</u>ll	list all items and information attached to the cluster; the default

† all is not shown in the dialog box.

Description

These `cluster` utility commands allow you to view and manipulate the cluster objects that you have created. See [MV] **cluster** for an overview of cluster analysis and for the available `cluster` commands. If you desire even more control over your cluster objects, or if you are programming new cluster subprograms, additional `cluster` programmer utilities are available; see [MV] **cluster programming utilities** for details.

The `cluster dir` command provides a directory-style listing of all the currently defined clusters. `cluster list` provides a detailed listing of the specified clusters or of all current clusters if no cluster names are specified. The default action is to list all the information attached to the clusters. You may limit the type of information listed by specifying particular options.

The `cluster drop` command removes the named clusters. The keyword `_all` specifies that all current cluster analyses be dropped.

Stata cluster analyses are referenced by name. Many `cluster` commands default to using the most-recently defined cluster analysis if no cluster name is provided. The `cluster use` command sets the specified cluster analysis as the most-recently executed cluster analysis, so that, by default, this cluster analysis will be used if the cluster name is omitted from many of the `cluster` commands. You may use the * and ? name-matching characters to shorten the typing of cluster names; see [U] **11.2 Abbreviation rules**.

`cluster rename` allows you to rename a cluster analysis without changing any of the variable names attached to the cluster analysis. The `cluster renamevar` command, on the other hand, allows you to rename the variables attached to a cluster analysis and to update the cluster object with the new variable names. Do not use the **rename** command (see [D] **rename**) to rename variables attached to a cluster analysis since this would invalidate the cluster object. Use the `cluster renamevar` command instead.

Options for cluster list

> Options

`notes` specifies that cluster notes be listed.

`type` specifies that the type of cluster analysis be listed.

`method` specifies that the cluster analysis method be listed.

`dissimilarity` specifies that the dissimilarity measure be listed.

`similarity` specifies that the similarity measure be listed.

`vars` specifies that the variables attached to the clusters be listed.

`chars` specifies that any Stata characteristics attached to the clusters be listed.

`other` specifies that information attached to the clusters under the heading "other" be listed.

The following option is available with `cluster list` but is not shown in the dialog box:

`all`, the default, specifies that all items and information attached to the cluster(s) be listed. You may instead pick among the `notes`, `type`, `method`, `dissimilarity`, `similarity`, `vars`, `chars`, and `other` options to limit what is presented.

Options for cluster renamevar

name(*clname*) indicates the cluster analysis within which the variable renaming is to take place. If
name() is not specified, the most-recently performed cluster analysis (or the one specified by
cluster use) will be used.

prefix specifies that all variables attached to the cluster analysis that have *oldstub* as the beginning
of their name be renamed, with *newstub* replacing *oldstub*.

Remarks

▷ Example 1

We demonstrate these cluster utility commands by beginning with four already-defined cluster
analyses. The dir and list subcommands provide listings of the cluster analyses.

```
. cluster dir
bcx3kmed
ayz5kmeans
abc_clink
xyz_slink
. cluster list xyz_slink
xyz_slink  (type: hierarchical,  method: single,  dissimilarity: L2)
       vars: xyz_slink_id (id variable)
             xyz_slink_ord (order variable)
             xyz_slink_hgt (height variable)
      other: range: 0 .
             cmd: cluster singlelinkage x y z, name(xyz_slink)
             varlist: x y z

. cluster list
bcx3kmed  (type: partition,  method: kmedians,  dissimilarity: L2)
       vars: bcx3kmed (group variable)
      other: k: 3
             start: krandom
             range: 0 .
             cmd: cluster kmedians b c x, k(3) name(bcx3kmed)
             varlist: b c x
ayz5kmeans  (type: partition,  method: kmeans,  dissimilarity: L2)
       vars: ayz5kmeans (group variable)
      other: k: 5
             start: krandom
             range: 0 .
             cmd: cluster kmeans a y z, k(5) name(ayz5kmeans)
             varlist: a y z
abc_clink  (type: hierarchical,  method: complete,  dissimilarity: L2)
       vars: abc_clink_id (id variable)
             abc_clink_ord (order variable)
             abc_clink_hgt (height variable)
      other: range: 0 .
             cmd: cluster completelinkage a b c, name(abc_clink)
             varlist: a b c
xyz_slink  (type: hierarchical,  method: single,  dissimilarity: L2)
       vars: xyz_slink_id (id variable)
             xyz_slink_ord (order variable)
             xyz_slink_hgt (height variable)
      other: range: 0 .
             cmd: cluster singlelinkage x y z, name(xyz_slink)
             varlist: x y z
```

```
. cluster list a*, vars
ayz5kmeans
      vars: ayz5kmeans (group variable)

abc_clink
      vars: abc_clink_id (id variable)
            abc_clink_ord (order variable)
            abc_clink_hgt (height variable)
```

cluster dir listed the names of the four currently defined cluster analyses. cluster list followed by the name of one of the cluster analyses listed the information attached to that cluster analysis. The cluster list command, without an argument, listed the information for all currently defined cluster analyses. We demonstrated the vars option of cluster list to show that we can restrict the information that is listed. Notice also the use of a* as the cluster name. The *, in this case, indicates that any ending is allowed. For these four cluster analyses, it matches the names ayz5kmeans and abc_clink.

We now demonstrate the use of the renamevar subcommand.

```
. cluster renamevar ayz5kmeans g5km
variable ayz5kmeans not found in bcx3kmed
r(198);
. cluster renamevar ayz5kmeans g5km, name(ayz5kmeans)
. cluster list ayz5kmeans
ayz5kmeans (type: partition, method: kmeans, dissimilarity: L2)
      vars: g5km (group variable)
     other: k: 5
            start: krandom
            range: 0 .
            cmd: cluster kmeans a y z, k(5) name(ayz5kmeans)
            varlist: a y z
```

The first use of cluster renamevar failed because we did not specify which cluster object to use (with the name() option), and the most-recent cluster object, bcx3kmed, was not the appropriate one. After specifying the name() option with the appropriate cluster name, the renamevar subcommand changed the name as shown in the cluster list command that followed.

The cluster use command sets a particular cluster object as the default. We show this in conjunction with the prefix option of the renamevar subcommand.

```
. cluster use ayz5kmeans
. cluster renamevar g grp, prefix
. cluster renamevar xyz_slink_ wrk, prefix name(xyz*)
. cluster list ayz* xyz*
ayz5kmeans (type: partition, method: kmeans, dissimilarity: L2)
      vars: grp5km (group variable)
     other: k: 5
            start: krandom
            range: 0 .
            cmd: cluster kmeans a y z, k(5) name(ayz5kmeans)
            varlist: a y z
xyz_slink (type: hierarchical, method: single, dissimilarity: L2)
      vars: wrkid (id variable)
            wrkord (order variable)
            wrkhgt (height variable)
     other: range: 0 .
            cmd: cluster singlelinkage x y z, name(xyz_slink)
            varlist: x y z
```

The cluster use command placed ayz5kmeans as the current cluster object. The cluster renamevar command that followed capitalized on this by leaving off the name() option. The prefix option allowed us to change the variable names, as demonstrated in the `cluster list` of the two changed cluster objects.

`cluster rename` changes the name of cluster objects. `cluster drop` allows us to drop some or all of the cluster objects.

```
. cluster rename xyz_slink bob
. cluster rename ayz* sam
. cluster list, type method vars
sam  (type: partition,  method: kmeans)
      vars: grp5km (group variable)
bob  (type: hierarchical,  method: single)
      vars: wrkid (id variable)
            wrkord (order variable)
            wrkhgt (height variable)
bcx3kmed  (type: partition,  method: kmedians)
      vars: bcx3kmed (group variable)
abc_clink  (type: hierarchical,  method: complete)
      vars: abc_clink_id (id variable)
            abc_clink_ord (order variable)
            abc_clink_hgt (height variable)
. cluster drop bcx3kmed abc_clink
. cluster dir
sam
bob
. cluster drop _all
. cluster dir
```

We used options with `cluster list` to limit what was presented. The `_all` keyword with `cluster drop` removed all currently defined cluster objects.

◁

Also See

Related: [MV] **cluster notes**, [MV] **cluster programming utilities**,

                 [D] **notes**, [P] **char**

Background: [MV] **cluster**, [MV] **clustermat**

Title

cluster wardslinkage — Ward's linkage cluster analysis

Syntax

Cluster analysis of data

> cluster <u>wards</u>linkage [*varlist*] [*if*] [*in*] [, *cluster_options*]

Cluster analysis of a dissimilarity matrix

> clustermat <u>wards</u>linkage *matname* [*if*] [*in*] [, *clustermat_options*]

cluster_options	description
Main	
<u>measure</u>(*measure*)	similarity or dissimilarity measures; see *Options* for available measures; default is L2squared
<u>name</u>(*clname*)	name of resulting cluster analysis
Advanced	
<u>generate</u>(*stub*)	prefix for generated variables; default prefix is *clname*

clustermat_options	description
Main	
<u>shape</u>(*shape*)	shape (storage method) of *matname*
add	add cluster information to data currently in memory
clear	replace data in memory with cluster information
<u>labelvar</u>(*varname*)	place dissimilarity matrix row names in *varname*
<u>name</u>(*clname*)	name of resulting cluster analysis
Advanced	
force	perform clustering after fixing *matname* problems
<u>generate</u>(*stub*)	prefix for generated variables; default prefix is *clname*

shape	*matname* is stored as a
<u>full</u>	square symmetric matrix; the default
<u>lower</u>	vector of rowwise lower triangle (with diagonal)
<u>llower</u>	vector of rowwise strict lower triangle (no diagonal)
<u>upper</u>	vector of rowwise upper triangle (with diagonal)
<u>uupper</u>	vector of rowwise strict upper triangle (no diagonal)

Description

cluster wardslinkage performs hierarchical agglomerative Ward's linkage cluster analysis. See [MV] **cluster** for a general discussion of cluster analysis and a description of the other cluster commands.

clustermat wardslinkage performs hierarchical agglomerative Ward's linkage cluster analysis on the dissimilarity matrix *matname*. See [MV] **clustermat** for a general discussion of cluster analysis of dissimilarity matrices and a description of the other clustermat commands.

After cluster wardslinkage or clustermat wardslinkage, the cluster dendrogram command (see [MV] **cluster dendrogram**) displays the resulting dendrogram, the cluster stop or clustermat stop commands (see [MV] **cluster stop**) help determine the number of groups, and the cluster generate command (see [MV] **cluster generate**) produces grouping variables.

Options for cluster wardslinkage

$\boxed{\text{Main}}$

measure(*measure*) is one of the similarity or dissimilarity measures allowed by Stata. This option is not case sensitive. See [MV] ***measure_option*** for a discussion of these measures.

The available measures designed for continuous data are L2 (synonym Euclidean); L2squared, which is the default for cluster wardslinkage; L1 (synonyms absolute, cityblock, and manhattan); Linfinity (synonym maximum); L(#); Lpower(#); Canberra; correlation; and angular (synonym angle).

The available measures designed for binary data are matching, Jaccard, Russell, Hamann, Dice, antiDice, Sneath, Rogers, Ochiai, Yule, Anderberg, Kulczynski, Gower2, and Pearson.

Several authors advise using the L2squared *measure* exclusively with Ward's linkage. See *(Dis)similarity transformations and the Lance and Williams formula* and *Warning concerning (dis)similarity choice* in [MV] **cluster** for details.

name(*clname*) specifies the name to attach to the resulting cluster analysis. If name() is not specified, Stata finds an available cluster name, displays it for your reference, and attaches the name to your cluster analysis.

$\boxed{\text{Advanced}}$

generate(*stub*) provides a prefix for the variable names created by cluster wardslinkage. By default, the variable-name prefix is the name specified in name(). Three variables are created and attached to the cluster-analysis results with the suffixes _id, _ord, and _hgt. Users generally will not need to access these variables directly.

Options for clustermat wardslinkage

$\boxed{\text{Main}}$

shape(*shape*) specifies the storage mode of *matname*, the matrix of dissimilarities. The following shapes are allowed:

full specifies that *matname* is an $n \times n$ symmetric matrix.

lower specifies that *matname* is a row or column vector of length $n(n+1)/2$, with the rowwise lower triangle of the dissimilarity matrix including the diagonal of zeros.

$$D_{11}\ D_{21}\ D_{22}\ D_{31}\ D_{32}\ D_{33}\ \ldots\ D_{n1}\ D_{n2}\ \ldots\ D_{nn}$$

`llower` specifies that *matname* is a row or column vector of length $n(n-1)/2$, with the rowwise lower triangle of the dissimilarity matrix excluding the diagonal.

$$D_{21}\ D_{31}\ D_{32}\ D_{41}\ D_{42}\ D_{43}\ \ldots\ D_{n1}\ D_{n2}\ \ldots\ D_{n,n-1}$$

`upper` specifies that *matname* is a row or column vector of length $n(n+1)/2$, with the rowwise upper triangle of the dissimilarity matrix including the diagonal of zeros.

$$D_{11}\ D_{12}\ \ldots\ D_{1n}\ D_{22}\ D_{23}\ \ldots\ D_{2n}\ D_{33}\ D_{34}\ \ldots\ D_{3n}\ \ldots\ D_{nn}$$

`uupper` specifies that *matname* is a row or column vector of length $n(n-1)/2$, with the rowwise upper triangle of the dissimilarity matrix excluding the diagonal.

$$D_{12}\ D_{13}\ \ldots\ D_{1n}\ D_{23}\ D_{24}\ \ldots\ D_{2n}\ D_{34}\ D_{35}\ \ldots\ D_{3n}\ \ldots\ D_{n-1,n}$$

`add` specifies that `clustermat`'s results be added to the dataset currently in memory. The number of observations (selected observations based on the `if` and `in` qualifiers) must equal the number of rows and columns of *matname*. Either `clear` or `add` is required if a dataset is currently in memory.

`clear` drops all the variables and cluster solutions in the current dataset in memory (even if that dataset has changed since the data were last saved) before generating `clustermat`'s results. Either `clear` or `add` is required if a dataset is currently in memory.

`labelvar`(*varname*) specifies the name of a new variable to be created containing the row names of matrix *matname*.

`name`(*clname*) specifies the name to attach to the resulting cluster analysis. If `name()` is not specified, Stata finds an available cluster name, displays it for your reference, and attaches the name to your cluster analysis.

—————— Advanced ——————————————————————————————————————

`force` allows computations to continue when *matname* is nonsymmetric or has nonzeros on the diagonal. By default, `clustermat` will complain and exit when it encounters these conditions. `force` specifies that `clustermat` operate on the symmetric matrix $(matname * matname')/2$, with any nonzero diagonal entries treated as if they were zero.

`generate`(*stub*) provides a prefix for the variable names created by `clustermat`. By default, the variable-name prefix is the name specified in `name()`. Three variables are created and attached to the cluster-analysis results with the suffixes `_id`, `_ord`, and `_hgt`. Users generally will not need to access these variables directly.

Remarks

An example using the default `L2squared` (squared Euclidean) distance and `L2` (Euclidean) distance on continuous data and an example using the `matching` coefficient on binary data illustrate the `cluster wardslinkage` command. A third example illustrates the use of `clustermat wardslinkage` in clustering variables instead of observations. These are the same datasets introduced in [MV] **cluster singlelinkage**, which are used as examples for all the hierarchical clustering methods so that you can compare the results from using different hierarchical clustering methods.

▷ Example 1

As explained in the first example of [MV] **cluster singlelinkage**, as the senior data analyst for a small biotechnology firm, you are given a dataset with 4 chemical laboratory measurements on 50 different samples of a particular plant gathered from the rain forest. The head of the expedition that gathered the samples thinks, based on information from the natives, that an extract from the plant might reduce the negative side effects associated with your company's best-selling nutritional supplement.

While the company chemists and botanists continue exploring the possible uses of the plant and plan future experiments, the head of product development asks you to look at the preliminary data and to report anything that might be helpful to the researchers.

While all 50 of the plants are supposed to be of the same type, you decide to perform a cluster analysis to see if there are subgroups or anomalies among them. Single-linkage clustering helped you discover an anomaly in the data. You now wish to see if you discover the same thing using Ward's linkage clustering with the default squared Euclidean distance and with Euclidean distance.

You first call `cluster wardslinkage`, letting the distance default to L2squared (squared Euclidean distance), and use the `name()` option to attach the name `ward` to the resulting cluster analysis. You use the `cluster list` command (see [MV] **cluster utility**) to list the components of your cluster analysis. The `cluster dendrogram` command then graphs the dendrogram; see [MV] **cluster dendrogram**. As described in the [MV] **cluster singlelinkage** example, the `labels()` option is used, instead of the default action of showing the observation number, to identify which laboratory technician produced the data.

```
. use http://www.stata-press.com/data/r9/labtech

. cluster wardslinkage x1 x2 x3 x4, name(ward)

. cluster list ward
ward  (type: hierarchical,  method: wards,  dissimilarity: L2squared)
      vars: ward_id (id variable)
            ward_ord (order variable)
            ward_hgt (height variable)
     other: range: 0 .
            cmd: cluster wardslinkage x1 x2 x3 x4, name(ward)
            varlist: x1 x2 x3 x4

. cluster dendrogram ward, labels(labtech) xlabel(, angle(90) labsize(*.75))
```

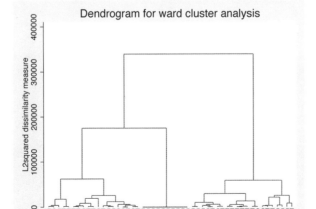

You now repeat the same analysis using the L2 option to obtain Euclidean distance instead of squared Euclidean distance. This time, you name the cluster L2ward.

```
. cluster wardslinkage x1 x2 x3 x4, name(L2ward) measure(L2)
. cluster list L2ward
L2ward  (type: hierarchical,  method: wards,  dissimilarity: L2)
      vars: L2ward_id (id variable)
            L2ward_ord (order variable)
            L2ward_hgt (height variable)
     other: range: 0 .
            cmd: cluster wardslinkage x1 x2 x3 x4, name(L2ward) measure(L2)
            varlist: x1 x2 x3 x4
. cluster dendrogram L2ward, labels(labtech) xlabel(, angle(90) labsize(*.75))
```

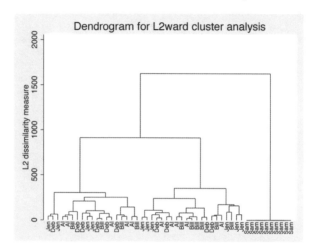

As with single-linkage clustering, you see that the samples analyzed by Sam, the lab technician, cluster together closely (dissimilarity measures near zero) and are separated from the rest of the data by a large dissimilarity gap (the long vertical line going up from Sam's cluster to eventually combine with other observations). When you examined the data, you discovered that Sam's data are all between zero and one, while the other four technicians have data that range from zero up to near 150. It appears that Sam has made a mistake.

◁

▷ Example 2

This example analyzes the same data that was introduced in the second example of [MV] **cluster singlelinkage**. The sociology professor of your graduate-level class gives, as homework, a dataset containing 30 observations on 60 binary variables, with the assignment to tell him something about the 30 subjects represented by the observations.

In addition to examining single-linkage clustering of these data, you decide to see what Ward's linkage clustering shows. As with the single-linkage clustering, you pick the simple matching binary coefficient to measure the similarity between groups. You use the name() option to attach the name wardlink to the cluster analysis. cluster list displays the details; see [MV] **cluster utility**. cluster tree, which is a synonym for cluster dendrogram, then displays the cluster tree (dendrogram); see [MV] **cluster dendrogram**.

```
. use http://www.stata-press.com/data/r9/homework

. cluster ward a1-a60, measure(match) name(wardlink)

. cluster list wardlink
wardlink  (type: hierarchical,  method: wards,  similarity: matching)
     vars: wardlink_id (id variable)
           wardlink_ord (order variable)
           wardlink_hgt (height variable)
    other: range: 1 0
           cmd: cluster wardslinkage a1-a60, measure(match) name(wardlink)
           varlist: a1 a2 a3 a4 a5 a6 a7 a8 a9 a10 a11 a12 a13 a14 a15 a16 a17
                    a18 a19 a20 a21 a22 a23 a24 a25 a26 a27 a28 a29 a30 a31 a32
                    a33 a34 a35 a36 a37 a38 a39 a40 a41 a42 a43 a44 a45 a46 a47
                    a48 a49 a50 a51 a52 a53 a54 a55 a56 a57 a58 a59 a60

. cluster tree wardlink
```

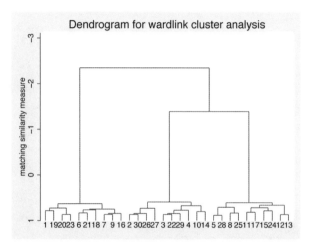

As with single-linkage clustering, the dendrogram from Ward's linkage clustering seems to indicate the presence of 3 groups among the 30 observations. However, notice the y-axis range for the resulting dendrogram. How can the matching similarity coefficient range from 1 to less than -2? By definition, the matching coefficient is bounded between 1 and 0. This is an artifact of the way Ward's linkage clustering is defined, and it underscores the warning mentioned in the discussion of the choice of *measure*. Also see *(Dis)similarity transformations and the Lance and Williams formula* and *Warning concerning (dis)similarity choice* in [MV] **cluster** for further details.

Later you receive another variable called `truegrp` that identifies the groups the teacher believes are in the data. You use the `cluster generate` command (see [MV] **cluster generate**) to create a grouping variable, based on your Ward's linkage clustering, to compare with `truegrp`. You do a cross-tabulation of `truegrp` and `wardgrp3`, your grouping variable, to see if your conclusions match those of the teacher.

(Continued on next page)

```
. cluster generate wardgrp3 = group(3)
. table wardgrp3 truegrp
```

	truegrp		
wardgrp3	1	2	3
1		10	
2	10		
3			10

Other than the numbers arbitrarily assigned to the three groups, your teacher's conclusions and the results from the Ward's linkage clustering are in complete agreement. So, despite the warning against using something other than squared Euclidean distance with Ward's linkage, you were still able to obtain a reasonable cluster-analysis solution with the matching similarity coefficient.

◁

▷ Example 3

The wclub dataset contains answers from 30 women to 35 yes/no questions. The variables are described in example 3 of [MV] **clustermat**. We are interested in seeing how Ward's linkage clustering will cluster the 35 variables (instead of the observations).

We use the matrix dissimilarity command to produce a dissimilarity matrix equal to one minus the Jaccard similarity; see [MV] **matrix dissimilarity**.

```
. use http://www.stata-press.com/data/r9/wclub
. matrix dissimilarity clubD = , variables Jaccard dissim(oneminus)
. clustermat ward clubD, name(clubward) clear labelvar(question)
obs was 0, now 35
. cluster dendrogram clubward, labels(question)
                xlabel(, angle(90) labsize(*.75))
                title(Ward's linkage clustering)
                ytitle(1 - Jaccard similarity, suffix)
```

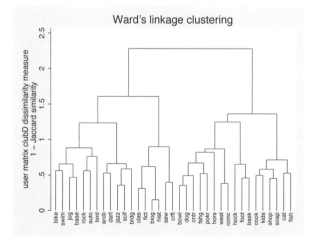

From these 30 women, we see that the biog (enjoy reading biographies) and hist (enjoy reading history) questions were most closely related. Other interesting relationships between the variables can be seen in the dendrogram.

◁

❑ Technical Note

> cluster wardslinkage requires more memory and more execution time than cluster singlelinkage. With a large number of observations, the execution time may be significant.

❑

Methods and Formulas

[MV] **cluster** discusses hierarchical clustering and places Ward's linkage clustering in this general framework. The Lance and Williams formula provides the basis for extending the well-known Ward's method of clustering into the general hierarchical-linkage framework that allows a choice of (dis)similarity measures.

Conceptually, hierarchical agglomerative Ward's linkage clustering proceeds as follows. The N observations start out as N separate groups each of size one. The two closest observations are merged into one group, producing $N - 1$ total groups. The closest two groups are then merged so that there are $N - 2$ total groups. This process continues until all the observations are merged into one large group, producing a hierarchy of groupings from one group to N groups. For Ward's method, the "closest two groups" are determined by minimizing the sum of squared errors.

The Ward's linkage clustering algorithm produces two variables that together act as a pointer representation of a dendrogram. To this, Stata adds a third variable used to restore the sort order, as needed, so that the two variables of the pointer representation remain valid. The first variable gives the order of the observations. The second variable has one less element and gives the height in the dendrogram at which the adjacent observations in the order-variable join.

See [MV] *measure_option* for the details and formulas of the available *measure*s, which include (dis)similarity measures for continuous and binary data. See [MV] **cluster** for a warning concerning (dis)similarity measure choice.

> Joe H. Ward, Jr. (1926–) obtained degrees in mathematics and educational psychology from the University of Texas. He worked as a personnel research psychologist for the U.S. Air Force Human Resources Laboratory applying educational psychology, statistics, and computers to a wide variety of problems.

Also See

Complementary: [MV] **cluster dendrogram**, [MV] **cluster generate**, [MV] **cluster notes**, [MV] **cluster stop**, [MV] **cluster utility**

Related: [MV] **cluster averagelinkage**, [MV] **cluster centroidlinkage**, [MV] **cluster completelinkage**, [MV] **cluster medianlinkage**, [MV] **cluster singlelinkage**, [MV] **cluster waveragelinkage**

Background: [MV] **cluster**, [MV] **clustermat**

Title

> **cluster waveragelinkage** — Weighted-average linkage cluster analysis

Syntax

Cluster analysis of data

> cluster <u>wa</u>veragelinkage [*varlist*] [*if*] [*in*] [, *cluster_options*]

Cluster analysis of a dissimilarity matrix

> clustermat <u>wa</u>veragelinkage *matname* [*if*] [*in*] [, *clustermat_options*]

cluster_options	description
Main	
<u>measure</u>(*measure*)	similarity or dissimilarity measures; see *Options* for available measures; default is **L2**
<u>name</u>(*clname*)	name of resulting cluster analysis
Advanced	
<u>generate</u>(*stub*)	prefix for generated variables; default prefix is *clname*

clustermat_options	description
Main	
<u>shape</u>(*shape*)	shape (storage method) of *matname*
add	add cluster information to data currently in memory
clear	replace data in memory with cluster information
<u>labelvar</u>(*varname*)	place dissimilarity matrix row names in *varname*
<u>name</u>(*clname*)	name of resulting cluster analysis
Advanced	
force	perform clustering after fixing *matname* problems
<u>generate</u>(*stub*)	prefix for generated variables; default prefix is *clname*

shape	*matname* is stored as a
<u>full</u>	square symmetric matrix; the default
<u>lower</u>	vector of rowwise lower triangle (with diagonal)
<u>ll</u>ower	vector of rowwise strict lower triangle (no diagonal)
<u>upper</u>	vector of rowwise upper triangle (with diagonal)
<u>uu</u>pper	vector of rowwise strict upper triangle (no diagonal)

Description

cluster waveragelinkage command performs hierarchical agglomerative weighted-average linkage cluster analysis. See [MV] **cluster** for a general discussion of cluster analysis and a description of the other cluster commands.

clustermat waveragelinkage performs hierarchical agglomerative weighted-average linkage cluster analysis on the dissimilarity matrix *matname*. See [MV] **clustermat** for a general discussion of cluster analysis of dissimilarity matrices and a description of the other clustermat commands.

After cluster waveragelinkage or clustermat waveragelinkage, the cluster dendrogram command (see [MV] **cluster dendrogram**) displays the resulting dendrogram, the cluster stop or clustermat stop commands (see [MV] **cluster stop**) help determine the number of groups, and the cluster generate command (see [MV] **cluster generate**) produces grouping variables.

Options for cluster waveragelinkage

Main

measure(*measure*) is one of the similarity or dissimilarity measures allowed by Stata. This option is not case sensitive. See [MV] *measure_option* for a discussion of these measures.

The available measures designed for continuous data are L2 (synonym Euclidean), which is the default; L2squared; L1 (synonyms absolute, cityblock, and manhattan); Linfinity (synonym maximum); L(#); Lpower(#); Canberra; correlation; and angular (synonym angle).

The available measures designed for binary data are matching, Jaccard, Russell, Hamann, Dice, antiDice, Sneath, Rogers, Ochiai, Yule, Anderberg, Kulczynski, Gower2, and Pearson.

name(*clname*) specifies the name to attach to the resulting cluster analysis. If name() is not specified, Stata finds an available cluster name, displays it for your reference, and attaches the name to your cluster analysis.

Advanced

generate(*stub*) provides a prefix for the variable names created by cluster waveragelinkage. By default, the variable-name prefix will be the name specified in name(). Three variables are created and attached to the cluster-analysis results with the suffixes _id, _ord, and _hgt. Users generally will not need to access these variables directly.

Options for clustermat waveragelinkage

Main

shape(*shape*) specifies the storage mode of *matname*, the matrix of dissimilarities. The following shapes are allowed:

full specifies that *matname* is an $n \times n$ symmetric matrix.

lower specifies that *matname* is a row or column vector of length $n(n+1)/2$, with the rowwise lower triangle of the dissimilarity matrix including the diagonal of zeros.

$$D_{11} \; D_{21} \; D_{22} \; D_{31} \; D_{32} \; D_{33} \; \ldots \; D_{n1} \; D_{n2} \; \ldots \; D_{nn}$$

llower specifies that *matname* is a row or column vector of length $n(n-1)/2$, with the rowwise lower triangle of the dissimilarity matrix excluding the diagonal.

$$D_{21} \; D_{31} \; D_{32} \; D_{41} \; D_{42} \; D_{43} \; \ldots \; D_{n1} \; D_{n2} \; \ldots \; D_{n,n-1}$$

upper specifies that *matname* is a row or column vector of length $n(n+1)/2$, with the rowwise upper triangle of the dissimilarity matrix including the diagonal of zeros.

$$D_{11} \; D_{12} \; \ldots \; D_{1n} \; D_{22} \; D_{23} \; \ldots \; D_{2n} \; D_{33} \; D_{34} \; \ldots \; D_{3n} \; \ldots \; D_{nn}$$

uupper specifies that *matname* is a row or column vector of length $n(n-1)/2$, with the rowwise upper triangle of the dissimilarity matrix excluding the diagonal.

$$D_{12} \; D_{13} \; \ldots \; D_{1n} \; D_{23} \; D_{24} \; \ldots \; D_{2n} \; D_{34} \; D_{35} \; \ldots \; D_{3n} \; \ldots \; D_{n-1,n}$$

add specifies that clustermat's results be added to the dataset currently in memory. The number of observations (selected observations based on the if and in qualifiers) must equal the number of rows and columns of *matname*. Either clear or add is required if a dataset is currently in memory.

clear drops all the variables and cluster solutions in the current dataset in memory (even if that dataset has changed since the data were last saved) before generating clustermat's results. Either clear or add is required if a dataset is currently in memory.

labelvar(*varname*) specifies the name of a new variable to be created containing the row names of matrix *matname*.

name(*clname*) specifies the name to attach to the resulting cluster analysis. If name() is not specified, Stata finds an available cluster name, displays it for your reference, and attaches the name to your cluster analysis.

___Advanced___

force allows computations to continue when *matname* is nonsymmetric or has nonzeros on the diagonal. By default, clustermat will complain and exit when it encounters these conditions. force specifies that clustermat operate on the symmetric matrix $(matname * matname')/2$, with any nonzero diagonal entries treated as if they were zero.

generate(*stub*) provides a prefix for the variable names created by clustermat. By default, the variable-name prefix is the name specified in name(). Three variables are created and attached to the cluster-analysis results with the suffixes _id, _ord, and _hgt. Users generally will not need to access these variables directly.

Remarks

An example using the default L2 (Euclidean) distance on continuous data and an example using the matching coefficient on binary data illustrate the cluster waveragelinkage command. A third example illustrates the use of clustermat waveragelinkage in clustering variables instead of observations. These are the same datasets that were introduced in [MV] **cluster singlelinkage**, which are used as examples for all the hierarchical clustering methods so that you can compare the results from using different hierarchical clustering methods.

▷ Example 1

As explained in the first example of [MV] **cluster singlelinkage**, as the senior data analyst for a small biotechnology firm, you are given a dataset with 4 chemical laboratory measurements on 50 different samples of a particular plant gathered from the rain forest. The head of the expedition that gathered the samples thinks, based on information from the natives, that an extract from the plant might reduce the negative side effects associated with your company's best-selling nutritional supplement.

While the company chemists and botanists continue exploring the possible uses of the plant and plan future experiments, the head of product development asks you to look at the preliminary data and to report anything that might be helpful to the researchers.

While all 50 of the plants are supposed to be of the same type, you decide to perform a cluster analysis to see if there are subgroups or anomalies among them. Single-linkage clustering helped you discover an anomaly in the data. You now wish to see if you discover the same thing using weighted-average linkage clustering with the default Euclidean distance.

You first call `cluster waveragelinkage` and use the `name()` option to attach the name L2wav to the resulting cluster analysis. You use the `cluster list` command (see [MV] **cluster utility**) to list the components of your cluster analysis. The `cluster dendrogram` command then graphs the dendrogram; see [MV] **cluster dendrogram**. As described in the [MV] **cluster singlelinkage** example, the `labels()` option is used, instead of the default action of showing the observation number, to identify which laboratory technician produced the data.

```
. use http://www.stata-press.com/data/r9/labtech
. cluster waveragelinkage x1 x2 x3 x4, name(L2wav)
. cluster list L2wav
L2wav (type: hierarchical,  method: waverage,  dissimilarity: L2)
      vars: L2wav_id (id variable)
            L2wav_ord (order variable)
            L2wav_hgt (height variable)
     other: range: 0 .
            cmd: cluster waveragelinkage x1 x2 x3 x4, name(L2wav)
            varlist: x1 x2 x3 x4
. cluster dendrogram L2wav, labels(labtech) xlabel(, angle(90) labsize(*.75))
```

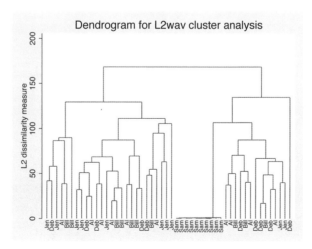

As with single-linkage clustering, you see that the samples analyzed by Sam, the lab technician, cluster together closely (dissimilarity measures near zero) and are separated from the rest of the data by a large dissimilarity gap (the long vertical line going up from Sam's cluster to eventually combine with other observations). When you examined the data, you discovered that Sam's data are all between zero and one, while the other four technicians have data that range from zero up to near 150. It appears that Sam has made a mistake.

◁

▷ Example 2

This example analyzes the same data that was introduced in the second example of [MV] **cluster singlelinkage**. The sociology professor of your graduate-level class gives, as homework, a dataset containing 30 observations on 60 binary variables, with the assignment to tell him something about the 30 subjects represented by the observations.

In addition to examining single-linkage clustering of these data, you decide to see what weighted-average linkage clustering shows. As with the single-linkage clustering, you pick the simple matching binary coefficient to measure the similarity between groups. You use the `name()` option to attach the name `wavlink` to the cluster analysis. `cluster list` displays the details; see [MV] **cluster utility**. `cluster tree`, which is a synonym for `cluster dendrogram`, then displays the cluster tree (dendrogram); see [MV] **cluster dendrogram**.

```
. use http://www.stata-press.com/data/r9/homework

. cluster waver a1-a60, measure(match) name(wavlink)

. cluster list wavlink
wavlink (type: hierarchical,  method: waverage,  similarity: matching)
      vars: wavlink_id (id variable)
            wavlink_ord (order variable)
            wavlink_hgt (height variable)
     other: range: 1 0
            cmd: cluster waveragelinkage a1-a60, measure(match) name(wavlink)
            varlist: a1 a2 a3 a4 a5 a6 a7 a8 a9 a10 a11 a12 a13 a14 a15 a16 a17
                a18 a19 a20 a21 a22 a23 a24 a25 a26 a27 a28 a29 a30 a31 a32
                a33 a34 a35 a36 a37 a38 a39 a40 a41 a42 a43 a44 a45 a46 a47
                a48 a49 a50 a51 a52 a53 a54 a55 a56 a57 a58 a59 a60

. cluster tree
```

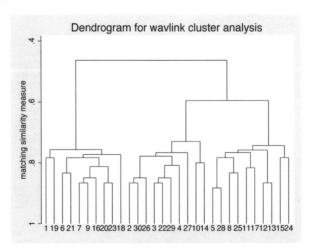

Since, by default, Stata uses the most-recently performed cluster analysis, you do not need to type the cluster name when calling `cluster tree`.

As with single-linkage clustering, the dendrogram from weighted-average linkage clustering seems to indicate the presence of 3 groups among the 30 observations. Later you receive another variable called `truegrp` that identifies the groups that the teacher believes are in the data. You use the `cluster generate` command (see [MV] **cluster generate**) to create a grouping variable, based on your weighted-average linkage clustering, to compare with `truegrp`. You do a cross-tabulation of `truegrp` and `wavgrp3`, your grouping variable, to see if your conclusions match those of the teacher.

```
. cluster gen wavgrp3 = group(3)
. table wavgrp3 truegrp
```

wavgrp3	truegrp 1	2	3
1		10	
2	10		
3			10

Other than the numbers arbitrarily assigned to the three groups, your teacher's conclusions and the results from the weighted-average linkage clustering are in complete agreement.

◁

▷ Example 3

The `wclub` dataset contains answers from 30 women to 35 yes/no questions. The variables are described in example 3 of [MV] **clustermat**. We are interested in seeing how weighted-average linkage clustering will cluster the 35 variables (instead of the observations).

We use the `matrix dissimilarity` command to produce a dissimilarity matrix equal to one minus the Jaccard similarity; see [MV] **matrix dissimilarity**.

```
. use http://www.stata-press.com/data/r9/wclub, clear
. matrix dissimilarity clubD = , variables Jaccard dissim(oneminus)
. clustermat waverage clubD, name(clubwav) clear labelvar(question)
obs was 0, now 35
```

(Continued on next page)

```
. cluster dendrogram clubwav, labels(question)
             xlabel(, angle(90) labsize(*.75))
             title(Weighted-average linkage clustering)
             ytitle(1 - Jaccard similarity, suffix)
```

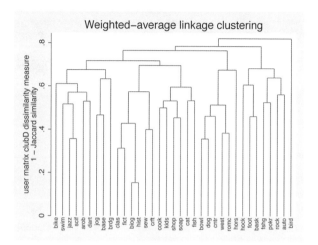

From these 30 women, we see that the `biog` (enjoy reading biographies) and `hist` (enjoy reading history) questions were most closely related. `bird` (have a bird) seems to be the least related to the other variables. It merges last into the super group containing the remaining variables.

◁

❏ Technical Note

cluster waveragelinkage requires more memory and execution time than cluster singlelinkage. With a large number of observations, the execution time may be significant.

❏

Methods and Formulas

[MV] **cluster** discusses hierarchical clustering and places weighted-average linkage clustering in this general framework. Conceptually, hierarchical agglomerative clustering proceeds as follows. The N observations start out as N separate groups each of size one. The two closest observations are merged into one group, producing $N-1$ total groups. The closest two groups are then merged so that there are $N-2$ total groups. This process continues until all the observations are merged into one large group, producing a hierarchy of groupings from one group to N groups. The difference between the various hierarchical-linkage methods depends on how "closest" is defined when comparing groups.

Weighted-average linkage clustering differs from on average-linkage clustering in how groups of unequal size are treated. Average linkage gives each observation equal weight. Weighted-average linkage gives each group of observations equal weight, meaning that with unequal group sizes, the observations in the smaller group will have more weight than the observations in the larger group.

The weighted-average linkage clustering algorithm produces two variables that together act as a pointer representation of a dendrogram. To this, Stata adds a third variable used to restore the sort order, as needed, so that the two variables of the pointer representation remain valid. The first variable gives the order of the observations. The second variable has one less element and gives the height in the dendrogram at which the adjacent observations in the order-variable join.

See [MV] *measure_option* for the details and formulas of the available *measures*, which include (dis)similarity measures for continuous and binary data. See [MV] **cluster** for a warning concerning (dis)similarity measure choice.

Also See

Complementary:	[MV] **cluster dendrogram**, [MV] **cluster generate**, [MV] **cluster notes**, [MV] **cluster stop**, [MV] **cluster utility**
Related:	[MV] **cluster averagelinkage**, [MV] **cluster centroidlinkage**, [MV] **cluster completelinkage**, [MV] **cluster medianlinkage**, [MV] **cluster singlelinkage**, [MV] **cluster wardslinkage**
Background:	[MV] **cluster**, [MV] **clustermat**

Title

> **factor** — Factor analysis

Syntax

Factor analysis of data

> <u>fac</u>tor [*varlist*] [*if*] [*in*] [*weight*] [, *method options*]

Factor analysis of a correlation matrix

> factormat *matname*, n(*#*) [*method options factormat_options*]

method	description
Model 2	
pf	principle factor; the default
pcf	principal-components factor
<u>ip</u>f	iterated principal factor
ml	maximum likelihood factor

options	description
Model 2	
<u>factor</u>s(*#*)	maximum number of factors to be retained
<u>minei</u>gen(*#*)	minimum value of eigenvalues to be retained
<u>cit</u>erate(*#*)	communality re-estimation iterations (ipf only)
Reporting	
<u>bl</u>anks(*#*)	display loadings as blank when |loadings| < #
<u>norot</u>ated	display unrotated solution, even if rotated results are available (replay only; not available in the dialog)
<u>altdivisor</u>	use trace of correlation matrix as the divisor for reported proportions
Max options	
<u>prot</u>ect(*#*)	perform # optimizations and report the best solution (ml only)
<u>random</u>	use random starting values (ml only); seldom used
seed(*seed*)	random-number seed (ml with protect() or random only)
maximize_options	control the maximization process; seldom used (ml only)

factormat_options	description
Model	
shape(**f**ull)	*matname* is a square symmetric matrix; the default
shape(**l**ower)	*matname* is a vector with the rowwise lower triangle (with diagonal)
shape(**u**pper)	*matname* is a vector with the rowwise upper triangle (with diagonal)
names(*namelist*)	variable names; required if *matname* is triangular
*n(#)	number of observations
sds(*matname$_2$*)	vector with standard deviations of variables
means(*matname$_3$*)	vector with means of variables

---

* n(#) is required for factormat.

bootstrap, by, jackknife, rolling, and statsby may be used with factor; see [U] **11.1.10 Prefix commands**.
aweights and fweights are allowed with factor; see [U] **11.1.6 weight**.
See [MV] **factor postestimation** for features available after estimation.

Description

factor and factormat perform a factor analysis of a correlation matrix. factor and factormat can produce principal factor, iterated principal factor, principal-components factor, and maximum-likelihood factor analyses. factor and factormat display the eigenvalues of the correlation matrix, the factor loadings, and the uniqueness ($= 1 -$ communality) of the variables.

factor expects data in the form of variables, allows weights, and can be run for subgroups (see [D] **by**). factormat is for use with a correlation or covariance matrix in the form of a square Stata matrix or a vector containing the rowwise upper or lower triangle of the correlation or covariance matrix. This is explained in more detail below; see option shape(). If a covariance matrix is provided to factormat, it is transformed into a correlation matrix for the factor analysis. To replay estimation results, you may type either factor or factormat.

Options for factor and factormat

> Model 2

pcf, pf, ipf, and ml indicate the type of estimation to be performed. The default is pf.

pcf specifies that the principal-components factor method be used to analyze the correlation matrix. The communalities are assumed to be 1.

pf specifies that the principal-factor method be used to analyze the correlation matrix. The factor loadings, sometimes called the factor patterns, are computed using the squared multiple correlations as estimates of the communality. pf is the default.

ipf specifies that the iterated principal-factor method be used to analyze the correlation matrix. This re-estimates the communalities iteratively.

ml specifies the maximum-likelihood factor method, assuming multivariate normal observations. This estimation method is equivalent to Rao's canonical-factor method, and maximizes the determinant of the partial correlation matrix. Hence, this solution is also meaningful as a descriptive method for non-normal data. ml is not available for singular correlation matrices. At least 3 variables must be specified with method ml.

factors(#) and mineigen(#) specify the maximum number of factors to be retained. factors() specifies the number directly, and mineigen() specifies it indirectly, keeping all factors with eigenvalues greater than the indicated value. The options can be specified individually, together, or not at all.

factors(#) sets the maximum number of factors to be retained for subsequent use by the postestimation commands. factor always prints the full set of eigenvalues but prints the corresponding eigenvectors only for retained factors. Specifying a number larger than the number of variables in the *varlist* is equivalent to specifying the number of variables in the *varlist*, and is the default.

mineigen(#) sets the minimum value of eigenvalues to be retained. The default for all methods except pcf is 5×10^{-6} (effectively zero), meaning that factors associated with negative eigenvalues will not be printed or retained. The default for pcf is 1. Many sources recommend mineigen(1), although the justification is complex and uncertain. If # is less than 5×10^{-6}, it is reset to 5×10^{-6}.

citerate(#) is used only with ipf and sets the number of iterations for re-estimating the communalities. If citerate() is not specified, iterations continue until the change in the communalities is small. ipf with citerate(0) produces the same results that pf does.

___

⌐ Reporting ⌐

blanks(#) specifies that factor loadings smaller than # (in absolute value) be displayed as blanks.

norotated specifies that the unrotated factor solution be displayed, even if a rotated factor solution is available. norotated is for use only with replaying results.

altdivisor specifies that reported proportions and cumulative proportions be computed using the trace of the correlation matrix, trace(e(C)), as the divisor. The default is to use the sum of all eigenvalues (even those that are negative) as the divisor.

___

⌐ Max options ⌐

protect(#) is used only with ml and requests that # optimizations with random starting values be performed along with squared multiple correlation coefficient starting values and that the best of the solutions be reported. The output also indicates whether all starting values converged to the same solution. When specified with a large number, such as protect(50), this provides reasonable assurance that the solution found is global and not just a local maximum. If trace is also specified (see [R] **maximize**), the parameters and likelihoods of each maximization will be printed.

random is used only with ml and requests that random starting values be used. This option is rarely used and should only be used after protect() has shown the presence of multiple maxima.

seed(*seed*) is used only with ml when the random or protect() options are also specified. seed() specifies the random-number seed; see [D] **generate**.

maximize_options: iterate(#), [no]log, trace, tolerance(#), ltolerance(#), see [R] **maximize**. These options are seldom used.

Options unique to factormat

___

⌐ Model ⌐

shape(*shape*) specifies the shape (storage method) for the covariance or correlation matrix *matname*. The following shapes are supported:

full specifies that the correlation or covariance structure of k variables is stored as a symmetric $k \times k$ matrix. This is the default.

lower specifies that the correlation or covariance structure of k variables is stored as a vector with $k(k+1)/2$ elements in rowwise lower-triangular order,

$$C_{11}\ C_{21}\ C_{22}\ C_{31}\ C_{32}\ C_{33}\ \ldots\ C_{k1}\ C_{k2}\ \ldots\ C_{kk}$$

upper specifies that the correlation or covariance structure of k variables is stored as a vector with $k(k+1)/2$ elements in rowwise upper-triangular order,

$$C_{11}\ C_{12}\ C_{13}\ \ldots\ C_{1k}\ C_{22}\ C_{23}\ \ldots C_{2k}\ \ldots\ C_{(k-1k-1)}\ C_{(k-1k)}\ C_{kk}$$

names(*namelist*) specifies a list of k different names to be used to document output and label estimation results and as variable names by predict. names() is required if the correlation or covariance matrix is in vectorized storage mode (i.e., shape(lower) or shape(upper) are specified). By default, factormat verifies that the row and column names of *matname* and the column or row names of *matname*$_2$ and *matname*$_3$ from the sds() and means() options are in agreement. Using the names() option turns off this check.

n(#), a required option, specifies the number of observations on which *matname* is based.

sds(*matname*$_2$) specifies a $k \times 1$ or $1 \times k$ matrix with the standard deviations of the variables. The row or column names should match the variable names, unless the names() option is specified. sds() may only be specified if *matname* is a correlation matrix. Specify sds() if you have variables in your dataset and want to use predict after factormat. sds() does not affect the computations of factormat, but provides information so that predict does not assume the standard deviations are one.

means(*matname*$_3$) specifies a $k \times 1$ or $1 \times k$ matrix with the means of the variables. The row or column names should match the variable names, unless the names() option is specified. Specify means() if you have variables in your dataset and want to use predict after factormat. means() does not affect the computations of factormat, but provides information so that predict does not assume the means are zero.

Remarks

Remarks are presented under the headings

> *Introduction*
> *Factor analysis*
> *Factor analysis from a correlation matrix*

Introduction

Factor analysis, in the sense of exploratory factor analysis, is a statistical technique for data reduction. It reduces the number of variables in an analysis by describing linear combinations of the variables that contain most of the information and that, hopefully, admit meaningful interpretations.

Factor analysis originated with the work of Spearman (1904), and has since witnessed an explosive growth, especially in the social sciences and, interestingly, in chemometrics. For an introduction, we refer to Kim and Mueller (1978a, 1978b), van Bell et al. (2004, chapter 14), and Hamilton (2004, chapter 12). Intermediate-level treatments include Gorsuch (1983) and Harman (1976). For mathematically more advanced discussions, see Mulaik (1972), Mardia, Kent, and Bibby (1979, chapter 9), and Fuller (1987).

Factor analysis

Factor analysis finds a small number of common factors (say q of them) that linearly reconstruct the p original variables

$$y_{ij} = z_{i1}b_{1j} + z_{i2}b_{2j} + \cdots + z_{iq}b_{qj} + e_{ij}$$

where y_{ij} is the value of the ith observation on the jth variable, z_{ik} is the ith observation on the kth common factor, b_{kj} is the set of linear coefficients called the factor loadings, and e_{ij} is similar to a residual but is known as the jth variable's unique factor. Note that everything except the left-hand-side variable is to be estimated, so the model has an infinite number of solutions. Various constraints are introduced to make the model determinate.

"Reconstruction" is typically defined in terms of prediction of the correlation matrix of the original variables, unlike principal components (see [MV] **pca**), where reconstruction means minimum residual variance summed across all equations (variables).

Once the factors and their loadings have been estimated, they are interpreted—an admittedly subjective process. Interpretation typically means examining the b_{kj}s and assigning names to each factor. Due to the indeterminacy of the factor solution, we are not limited to examining solely the b_{kj}s. The loadings could be rotated—that is, we could look at another set of b_{kj}s that, while appearing different, are every bit as good as (and no better than) the original loadings. Such "rotations" come in two flavors—orthogonal and oblique. Since there are an infinite number of potential rotations, different rotations could lead to different interpretations of the same data. These are not to be viewed as conflicting, but instead as two different ways of looking at the same thing. See [MV] **factor postestimation** and [MV] **rotate** for more information on rotation.

▷ Example 1

We wish to analyze physicians' attitudes toward cost and have a dataset of responses to six questions about cost asked of 568 physicians in the Medical Outcomes Study from Tarlov et al. (1989). Factor analysis is often used to validate a combination of questions that looks meaningful at first glance. In this case, we wish to create a variable that summarizes the information on each physician's attitude toward cost.

Each of the responses is coded on a 5-point scale, where 1 means "agree" and 5 means "disagree":

```
. use http://www.stata-press.com/data/r9/bg2
(Physician-cost data)

. describe

Contains data from http://www.stata-press.com/data/r9/bg2.dta
  obs:           568                          Physician-cost data
  vars:            7                          11 Feb 2005 21:54
  size:        17,040 (98.4% of memory free)  (_dta has notes)
```

variable name	storage type	display format	value label	variable label
clinid	int	%9.0g		Physician identifier
bg2cost1	float	%9.0g		Best health care is expensive
bg2cost2	float	%9.0g		Cost is a major consideration
bg2cost3	float	%9.0g		Determine cost of tests first
bg2cost4	float	%9.0g		Monitor likely complications only
bg2cost5	float	%9.0g		Use all means regardless of cost
bg2cost6	float	%9.0g		Prefer unnecessary tests to missing tests

```
Sorted by:  clinid
```

We perform the factorization on bg2cost1, bg2cost2, ..., bg2cost6.

```
. factor bg2cost1-bg2cost6
(obs=568)
Factor analysis/correlation                Number of obs    =     568
    Method: principal factors              Retained factors =       3
    Rotation: (unrotated)                  Number of params =      15
```

Factor	Eigenvalue	Difference	Proportion	Cumulative
Factor1	0.85389	0.31282	1.0310	1.0310
Factor2	0.54107	0.51786	0.6533	1.6844
Factor3	0.02321	0.17288	0.0280	1.7124
Factor4	-0.14967	0.03951	-0.1807	1.5317
Factor5	-0.18918	0.06197	-0.2284	1.3033
Factor6	-0.25115	.	-0.3033	1.0000

```
    LR test: independent vs. saturated:  chi2(15) =  269.07 Prob>chi2 = 0.0000
Factor loadings (pattern matrix) and unique variances
```

Variable	Factor1	Factor2	Factor3	Uniqueness
bg2cost1	0.2470	0.3670	-0.0446	0.8023
bg2cost2	-0.3374	0.3321	-0.0772	0.7699
bg2cost3	-0.3764	0.3756	0.0204	0.7169
bg2cost4	-0.3221	0.1942	0.1034	0.8479
bg2cost5	0.4550	0.2479	0.0641	0.7274
bg2cost6	0.4760	0.2364	-0.0068	0.7175

factor retained only the first three factors because the eigenvalues associated with the remaining factors are negative. According to the default mineigen(0) criterion, a factor must have an eigenvalue greater than zero to be retained. You can set this threshold higher by specifying mineigen(#). Although factor elected to retain three factors, only the first two appear to be meaningful.

The first factor seems to describe the physician's average position on cost since it affects the responses to all the questions "positively", as shown by the signs in the first column of the factor-loading table. We say "positively" because, obviously, the signs on three of the loadings are negative. When we look back at the results of describe, however, we find that the direction of the responses on bg2cost2, bg2cost3, and bg2cost4 are reversed. If the physician feels that cost should not be a major influence on medical treatment, he or she is likely to disagree with these three items and to agree with the other three.

The second factor loads positively (absolutely, not logically) on all six items and could be interpreted as describing the physician's tendency to agree with any good-sounding idea put forth. Psychologists refer to this as the "positive response set". On statistical grounds, we would probably keep this second factor, although on substantive grounds, we would be tempted to drop it.

We finally point to the column with the header "uniqueness". Uniqueness is the percentage of variance for the variable that is not explained by the common factors. The quantity "1 – uniqueness" is called communality. Uniqueness could be pure measurement error, or it could represent something that is measured reliably in that particular variable, but not by any of the others. The greater the uniqueness, the more likely that it is more than just measurement error. Values over 0.6 are usually considered high; all the variables in this problem are even higher—over 0.71. If the uniqueness is high, then the variable is not well explained by the factors.

◁

▷ Example 2

Did you notice that the cumulative proportions of the eigenvalues exceeded 1.0 in our factor analysis? This is due to the negative eigenvalues. By default, the proportion and cumulative proportion columns are computed using the sum of all eigenvalues as the divisor. The altdivisor option allows you to display the proportions and cumulative proportions using the trace of the correlation matrix as the divisor. This option is allowed at estimation time or when replaying results. We demonstrate by replaying the results with this option.

```
. factor, altdivisor
Factor analysis/correlation                    Number of obs    =      568
     Method: principal factors                 Retained factors =        3
     Rotation: (unrotated)                      Number of params =       15
```

Factor	Eigenvalue	Difference	Proportion	Cumulative
Factor1	0.85389	0.31282	0.1423	0.1423
Factor2	0.54107	0.51786	0.0902	0.2325
Factor3	0.02321	0.17288	0.0039	0.2364
Factor4	-0.14967	0.03951	-0.0249	0.2114
Factor5	-0.18918	0.06197	-0.0315	0.1799
Factor6	-0.25115	.	-0.0419	0.1380

```
LR test: independent vs. saturated:  chi2(15) =  269.07 Prob>chi2 = 0.0000
Factor loadings (pattern matrix) and unique variances
```

Variable	Factor1	Factor2	Factor3	Uniqueness
bg2cost1	0.2470	0.3670	-0.0446	0.8023
bg2cost2	-0.3374	0.3321	-0.0772	0.7699
bg2cost3	-0.3764	0.3756	0.0204	0.7169
bg2cost4	-0.3221	0.1942	0.1034	0.8479
bg2cost5	0.4550	0.2479	0.0641	0.7274
bg2cost6	0.4760	0.2364	-0.0068	0.7175

Among the sources we examined, there was not a consensus on which divisor is most appropriate. As such, both are available.

◁

▷ Example 3

factor provides a number of alternative estimation strategies for the factor model. We specified no options on the factor command when we fitted our first model, so we obtained the principal-factor solution. The *communalities* (defined as $1 - uniqueness$) were estimated using the squared multiple correlation coefficients.

We could have instead obtained the estimates from "principal-component factors", treating the communalities as all 1—meaning that there are no unique factors—by specifying the pcf option:

```
. factor bg2cost1-bg2cost6, pcf
(obs=568)
```

Factor analysis/correlation   Number of obs   =   568
    Method: principal-component factors   Retained factors =   2
    Rotation: (unrotated)   Number of params =   11

Factor	Eigenvalue	Difference	Proportion	Cumulative
Factor1	1.70622	0.30334	0.2844	0.2844
Factor2	1.40288	0.49422	0.2338	0.5182
Factor3	0.90865	0.18567	0.1514	0.6696
Factor4	0.72298	0.05606	0.1205	0.7901
Factor5	0.66692	0.07456	0.1112	0.9013
Factor6	0.59236	.	0.0987	1.0000

LR test: independent vs. saturated: chi2(15) = 269.07 Prob>chi2 = 0.0000

Factor loadings (pattern matrix) and unique variances

Variable	Factor1	Factor2	Uniqueness
bg2cost1	0.3581	0.6279	0.4775
bg2cost2	-0.4850	0.5244	0.4898
bg2cost3	-0.5326	0.5725	0.3886
bg2cost4	-0.4919	0.3254	0.6521
bg2cost5	0.6238	0.3962	0.4539
bg2cost6	0.6543	0.3780	0.4290

In this case, we find that the principal-component factor model is inappropriate. It is based on the assumption that the uniquenesses are 0, but we find that there is considerable uniqueness—there is considerable variability left over after our two factors. We should use some other method.

◁

▷ Example 4

We could have fitted our model using iterated principal factors by specifying the ipf option. In this case, the initial estimates of the communalities would be the squared multiple correlation coefficients, but the solution would then be iterated to obtain different (better) estimates:

```
. factor bg2cost1-bg2cost6, ipf
(obs=568)
```

Factor analysis/correlation   Number of obs   =   568
    Method: iterated principal factors   Retained factors =   5
    Rotation: (unrotated)   Number of params =   15

Factor	Eigenvalue	Difference	Proportion	Cumulative
Factor1	1.08361	0.31752	0.5104	0.5104
Factor2	0.76609	0.53816	0.3608	0.8712
Factor3	0.22793	0.19469	0.1074	0.9786
Factor4	0.03324	0.02085	0.0157	0.9942
Factor5	0.01239	0.01256	0.0058	1.0001
Factor6	-0.00017	.	-0.0001	1.0000

LR test: independent vs. saturated: chi2(15) = 269.07 Prob>chi2 = 0.0000

Factor loadings (pattern matrix) and unique variances

Variable	Factor1	Factor2	Factor3	Factor4	Factor5	Uniqueness
bg2cost1	0.2471	0.4059	-0.1349	-0.1303	0.0288	0.7381
bg2cost2	-0.4040	0.3959	-0.2636	0.0349	0.0040	0.6093
bg2cost3	-0.4479	0.4570	0.1290	0.0137	-0.0564	0.5705
bg2cost4	-0.3327	0.1943	0.2655	0.0091	0.0810	0.7744
bg2cost5	0.5294	0.3338	0.2161	-0.0134	-0.0331	0.5604
bg2cost6	0.5174	0.2943	-0.0801	0.1208	0.0265	0.6240

In this case, we retained too many factors. Unlike in principal factors or principal-component factors, we cannot simply ignore the unnecessary factors because the uniquenesses are re-estimated from the data and therefore depend on the number of retained factors. We need to re-estimate. We use the opportunity to demonstrate the option blanks(#) for displaying "small loadings" as blanks for easier reading:

```
. factor bg2cost1-bg2cost6, ipf factors(2) blanks(.30)
(obs=568)
```

Factor analysis/correlation   Number of obs   =   568
    Method: iterated principal factors   Retained factors =   2
    Rotation: (unrotated)   Number of params =   11

Factor	Eigenvalue	Difference	Proportion	Cumulative
Factor1	1.03954	0.30810	0.5870	0.5870
Factor2	0.73144	0.60785	0.4130	1.0000
Factor3	0.12359	0.11571	0.0698	1.0698
Factor4	0.00788	0.03656	0.0045	1.0743
Factor5	-0.02867	0.07418	-0.0162	1.0581
Factor6	-0.10285	.	-0.0581	1.0000

LR test: independent vs. saturated: chi2(15) = 269.07 Prob>chi2 = 0.0000

Factor loadings (pattern matrix) and unique variances

Variable	Factor1	Factor2	Uniqueness
bg2cost1		0.3941	0.7937
bg2cost2	-0.3590		0.7827
bg2cost3	-0.5189	0.4935	0.4872
bg2cost4	-0.3230		0.8699
bg2cost5	0.4667	0.3286	0.6742
bg2cost6	0.5179	0.3325	0.6212

(blanks represent abs(loading)<.3)

It is instructive to compare the reported uniquenesses for this model and the previous one, where five factors were retained. Also note that compared with the results we obtained from principal factors, these results do not differ much.

◁

▷ Example 5

Finally, we could have fitted our model using the maximum likelihood method by specifying the ml option. The maximum likelihood method assumes that the data are multivariate normal distributed. If the factor model provides an adequate approximation to the data, maximum likelihood estimates have favorable properties compared to the other estimation methods. Rao (1955) has shown that his canonical factor method is equivalent to the maximum likelihood method. This method seeks to maximize canonical correlations between the manifest variables and the common factors. Thus `ml` may be used descriptively, even if we are unwilling to assume multivariate normality.

As with `ipf`, if we do not specify the number of factors, Stata retains more than two factors (it retained three), and, as with `ipf`, we will need to re-estimate with the number of factors that we really want. To save paper, we will start by retaining two factors:

```
. factor bg2cost1-bg2cost6, ml factors(2)
(obs=568)
Iteration 0:   log likelihood = -28.702162
Iteration 1:   log likelihood = -7.0065234
Iteration 2:   log likelihood = -6.8513798
Iteration 3:   log likelihood = -6.8429502
Iteration 4:   log likelihood = -6.8424747
Iteration 5:   log likelihood = -6.8424491
Iteration 6:   log likelihood = -6.8424477
```

```
Factor analysis/correlation                    Number of obs     =      568
    Method: maximum likelihood                 Retained factors  =        2
    Rotation: (unrotated)                      Number of params  =       11
                                               Schwarz's BIC     =  83.4482
Log likelihood = -6.842448                     (Akaike's) AIC    =  35.6849
```

Factor	Eigenvalue	Difference	Proportion	Cumulative
Factor1	1.02766	0.28115	0.5792	0.5792
Factor2	0.74651	.	0.4208	1.0000

```
LR test: independent vs. saturated:  chi2(15) =  269.07 Prob>chi2 = 0.0000
LR test:    2 factors vs. saturated:  chi2(4)  =   13.58 Prob>chi2 = 0.0087
```

Factor loadings (pattern matrix) and unique variances

Variable	Factor1	Factor2	Uniqueness
bg2cost1	-0.1371	0.4235	0.8018
bg2cost2	0.4140	0.1994	0.7888
bg2cost3	0.6199	0.3692	0.4794
bg2cost4	0.3577	0.0909	0.8638
bg2cost5	-0.3752	0.4355	0.6695
bg2cost6	-0.4295	0.4395	0.6224

`factor` displays a likelihood ratio test of independence against the saturated model with each estimation method. Since we are factor analyzing a correlation matrix, independence implies sphericity. Passing this test is necessary for a factor analysis to be meaningful.

In addition to the "standard" output, when you use the `ml` option, Stata reports a likelihood-ratio test of the number of factors in the model against the saturated model. This test is only approximately chi-squared, and we have used the correction recommended by Bartlett (1951). Be aware that there are many variations on this test in use by different statistical packages.

The following comments were made by the analyst looking at these results: "There is, in my opinion, weak evidence of more than two factors. The χ^2 test for more than two factors is really a test of how well you are fitting the correlation matrix. It is not surprising that the model does not fit it perfectly. The significance of 1%, however, suggests to me that there might be a third factor. As for the loadings, they yield a similar interpretation to other factor models we fitted, although there are some noteworthy differences." When we challenged the analyst on this last statement, he added that he would want to rotate the resulting factors before committing himself further.

◁

❑ Technical Note

Going back to the tests, Stata will sometimes comment, "Note: test formally not valid because a Heywood case was encountered". The approximations used in computing the χ^2 value and degrees of freedom are mathematically justified on the assumption that an *interior* solution to the factor maximum likelihood was found. This is the case in our example above, but that will not always be so.

Boundary solutions, called Heywood solutions, often produce uniquenesses of 0, and in that case, at least at a formal level, the test cannot be justified. Nevertheless, we believe that the reported tests are useful, even in such circumstances, provided that they are interpreted cautiously. The maximum likelihood method seems to be particularly prone to producing Heywood solutions.

This message is also printed when, in principle, there are enough free parameters to completely fit the correlation matrix, another sort of boundary solution. We say "in principle" because the correlation matrix frequently cannot be fitted perfectly, so you will see a positive χ^2 with zero degrees of freedom. This warning note is printed because the geometric assumptions underlying the likelihood-ratio test are violated.

❑

❑ Technical Note

In a factor analysis with factors estimated with the maximum likelihood method, there may possibly be more than one local maximum, and you may want assurances that the maximum reported is the global maximum. Multiple maxima are especially likely when there is more than one group of variables, the groups are reasonably uncorrelated, and you attempt to fit a model with too few factors.

When you specify the `protect(#)` option, Stata performs # optimizations of the likelihood function, beginning each with random starting values, before continuing with the squared multiple correlations initialized solution. Stata then selects the maximum of the maxima and reports it, along with a note informing you if other local maxima were found. `protect(50)` provides considerable assurance.

If you then wish to explore any of the nonglobal maxima, include the `random` option. This option, which is never specified with `protect()`, uses random starting values and reports the solution to which those random values converge. In the case of multiple maxima, giving the command repeatedly will eventually report all local maxima. You are advised to set the random-number seed to ensure that your results are reproducible; see [D] **generate**.

❑

Factor analysis from a correlation matrix

You may want to perform a factor analysis directly from a correlation matrix rather than from variables in a dataset. You may not have access to the dataset, or you may have used another method of estimating a correlation matrix—e.g., as a matrix of tetrachoric correlations; see [R] **tetrachoric**. You can provide either a correlation or a covariance matrix—`factormat` will translate a covariance matrix into a correlation matrix.

▷ Example 6

We illustrate with a small example with three variables on respondent's senses (visual, hearing, and taste), with a correlation matrix.

```
. matrix C = ( 1.000, 0.943,  0.771 \
               0.943, 1.000,  0.605 \
               0.771, 0.605,  1.000 )
```

Note that elements within a row are separated by a comma, while rows are separated by a backslash \. We now use `factormat` to analyze C. There are two required options in this case. First, the option `n(979)` specifies that the sample size is 979. Second, `factormat` has to have labels for the variables. It is possible to define row and column names for C. We did not explicitly set the names of C, so Stata has generated default row and columns names—r1 r2 r3 for the rows, and c1 c2 c3 for the columns. This will confuse `factormat`: Why does a symmetric correlation matrix have different names for the rows and for the columns? `factormat` would complain about the problem and stop. We could set the row and column names of C to be the same and invoke `factormat` again. Alternatively, we can specify the `names()` option with the variable names to be used.

```
. factormat C, n(979) names(visual hearing taste) fac(1) ipf
(obs=979)

Factor analysis/correlation                 Number of obs    =      979
    Method: iterated principal factors      Retained factors =        1
    Rotation: (unrotated)                   Number of params =        3

    Beware: solution is a Heywood case
            (i.e., invalid or boundary values of uniqueness)
```

Factor	Eigenvalue	Difference	Proportion	Cumulative
Factor1	2.43622	2.43609	1.0000	1.0000
Factor2	0.00013	0.00028	0.0001	1.0001
Factor3	-0.00015	.	-0.0001	1.0000

```
    LR test: independent vs. saturated:  chi2(3)  = 3425.87 Prob>chi2 = 0.0000
Factor loadings (pattern matrix) and unique variances
```

Variable	Factor1	Uniqueness
visual	1.0961	-0.2014
hearing	0.8603	0.2599
taste	0.7034	0.5053

If we have the correlation matrix already in electronic form, this is a fine method. But if we have to enter a correlation matrix by hand, we may rather want to exploit its symmetry to enter just the upper triangle or lower triangle. This is not an issue with our small 3 variable example, but what about a correlation matrix of 25 variables? However, there is an advantage to entering the correlation matrix in full symmetric form: redundancy offers some protection against making data-entry errors; `factormat` will complain if the matrix is not symmetric.

`factormat` allows us to enter just one of the triangles of the correlation matrix as a vector, i.e., a matrix with one row or column. We enter the upper triangle, including the diagonal,

```
. matrix Cup = ( 1.000, 0.943, 0.771,
                        1.000, 0.605,
                               1.000 )
```

Note that all elements are separated by a comma; indentation and the use of three lines are done for readability. We could have typed, all the numbers "in a row".

```
. matrix Cup = ( 1.000, 0.943, 0.771, 1.000, 0.605, 1.000)
```

We have to specify the option `shape(upper)` to inform `factormat` that the elements in the vector Cup are the upper triangle in rowwise order.

```
. factormat Cup, n(979) shape(upper) fac(2) names(visual hearing taste)
  (output omitted )
```

If we had entered the lower triangle of C, a vector Clow, it would have been defined as

    . matrix Clow = (1.000, 0.943, 1.000, 0.771, 0.605, 1.000)

The features of factormat and factor are the same for estimation. Postestimation facilities are also the same—except that predict will not work after factormat, unless variables corresponding to the names() option exist in the dataset; see [MV] **factor postestimation**.

◁

Saved Results

factor and factormat save in e():

Scalars

e(N)	number of observations
e(f)	number of retained factors
e(evsum)	sum of all eigenvalues
e(df_m)	model degrees of freedom
e(df_r)	residual degrees of freedom
e(chi2_i)	likelihood-ratio test of "independence vs. saturated"
e(df_i)	degrees of freedom of test of "independence vs. saturated"
e(p_i)	p-value of "independence vs. saturated"
e(ll_0)	log likelihood of null model (ml only)
e(ll)	log likelihood (ml only)
e(aic)	Akaike's AIC (ml only)
e(bic)	Schwarz's BIC (ml only)
e(chi2_1)	likelihood-ratio test of "# factors vs. saturated" (ml only)
e(df_1)	degrees of freedom of test of "# factors vs. saturated" (ml only)

Macros

e(cmd)	factor
e(title)	Factor analysis
e(method)	pf, pcf, ipf, or ml
e(mtitle)	description of method (e.g., principal factors)
e(heywood)	Heywood case (when encountered)
e(predict)	factor_p
e(estat_cmd)	factor_estat
e(rotate_cmd)	factor_rotate
e(matrixname)	input matrix (factormat only)
e(mineigen)	specified mineigen() option
e(factors)	specified factors() option
e(seed)	starting random-number seed (seed() option only)
e(wtype)	weight type (factor only)
e(wexp)	weight expression (factor only)
e(properties)	nob noV eigen

Matrices

e(sds)	standard deviations of analyzed variables
e(means)	means of analyzed variables
e(C)	analyzed correlation matrix
e(Phi)	variance matrix common factors
e(L)	factor loadings
e(Psi)	uniqueness (variance of specific factors)
e(Ev)	eigenvalues

Functions

e(sample)	marks estimation sample (factor only)

rotate after factor and factormat stores items in e() along with the estimation command. See *Results* of [MV] **factor postestimation** and [MV] **rotate** for details.

Before Stata version 9, factor returned results in r(). This behavior is retained under version control.

Methods and Formulas

factor and factormat are implemented as ado-files.

This section describes the statistical factor model. Suppose that there are p variables and q factors. Let Ψ represent the $p \times p$ diagonal matrix of uniquenesses, and let Λ represent the $p \times q$ factor loading matrix. Let f be a $1 \times q$ matrix of factors. The standardized (mean 0, variance 1) vector of observed variables x $(1 \times p)$ is given by the system of regression equations

$$x = f\Lambda' + e$$

where e is a $1 \times p$ vector of errors with diagonal covariance equal to the uniqueness matrix Ψ. In addition it is assumed that the common factors f and the specific factors e are uncorrelated.

Under the factor model, the correlation matrix of x, called Σ, is decomposed by factor analysis as

$$\Sigma = \Lambda\Phi\Lambda' + \Psi$$

There is an obvious freedom in re-expressing a given decomposition of Σ. The default and unrotated form assumes uncorrelated common factors, $\Phi = I$. Stata performs this decomposition by an eigenvector calculation. First, an estimate is found for the uniqueness Ψ, and then the columns of Λ are computed as the q leading eigenvectors, scaled by the square root of the appropriate eigenvalue.

See Harman (1976); Mardia, Kent, and Bibby (1979); Rencher (1998, chapter 10); and Rencher (2002, chapter 13) for discussions of estimation methods in factor analysis. For details about maximum likelihood estimation, see also Lawley and Maxwell (1971) and Clarke (1970).

References

Bartlett, M. S. 1951. The effect of standardization on a χ^2 approximation in factor analysis. *Biometrika* 38: 337–344.

Clarke, M. R. B. 1970. A rapidly convergent method for maximum-likelihood factor analysis. *British Journal of Mathematical and Statistical Psychology* 23: 43–52.

Fuller, W. A. 1987. *Measurement Error Models.* New York: Wiley.

Gorsuch, R. L. 1983. *Factor Analysis.* 2nd ed. Hillsdale, NJ: Lawrence Erlbaum Associates.

Hamilton, L. C. 2004. *Statistics with Stata.* Belmont, CA: Brooks/Cole.

Harman, H. H. 1976. *Modern Factor Analysis.* 3rd ed. Chicago: University of Chicago Press.

Kim, J. O. and C. W. Mueller. 1978a. *Introduction to factor analysis. What it is and how to do it.* Sage University Paper Series on Quantitative Applications in the Social Sciences, 07–013. Thousand Oaks, CA: Sage.

———. 1978b. *Factor analysis: Statistical methods and practical issues.* Sage University Paper Series on Quantitative Applications in the Social Sciences, 07–014. Thousand Oaks, CA: Sage.

Lawley, D. N. and A. E. Maxwell. 1971. *Factor Analysis as a Statistical Method.* London: Butterworths.

Mardia, K. V., J. T. Kent, and J. M. Bibby. 1979. *Multivariate Analysis.* New York: Academic Press.

Mulaik, S. A. 1972. *The Foundations of Factor Analysis.* New York: McGraw–Hill.

Rao, C. R. 1955. Estimation and tests of significance in factor analysis. *Psychometrika* 20: 93–111.

Rencher, A. C. 1998. *Multivariate Statistical Inference and Applications.* New York: Wiley.

———. 2002. *Methods of Multivariate Analysis.* 2nd ed. New York: Wiley.

Spearman, C. 1904. General intelligence objectively determined and measured. *American Journal of Psychology* 15: 201–293.

Tarlov, A. R., J. E. Ware, Jr., S. Greenfield, E. C. Nelson, E. Perrin, and M. Zubkoff. 1989. The medical outcomes study. *Journal of the American Medical Association* 262: 925–930.

van Belle, G., L. D. Fisher, P. J. Heagerty, and T. Lumley. 2004. *Biostatistics: A Methodology for the Health Sciences.* 2nd ed. New York: Wiley.

Also See

Complementary:	[MV] **factor postestimation**,
	[D] **impute**
Related:	[MV] **canon**, [MV] **pca**,
	[R] **alpha**
Background:	[U] **11.1.10 Prefix commands**,
	[U] **20 Estimation and postestimation commands**,
	[R] **maximize**

Title

factor postestimation — Postestimation tools for factor and factormat

Description

The following postestimation commands are of special interest after `factor` and `factormat`:

command	description
estat anti	anti-image correlation and covariance matrices
estat common	correlation matrix of the common factors
estat factors	AIC and BIC model-selection criteria for different numbers of factors
estat kmo	Kaiser–Meyer–Olkin measure of sampling adequacy
estat residuals	matrix of correlation residuals
estat rotatecompare	compare rotated and unrotated loadings
estat smc	squared multiple correlations between each variable and the rest
estat structure	correlations between variables and common factors
* estat summarize	estimation sample summary
loadingplot	plot factor loadings
rotate	rotate factor loadings
scoreplot	plot score variables
screeplot	plot eigenvalues

* `estat summarize` is not available after `factormat`.

For information about `loadingplot` and `scoreplot`, see [MV] **scoreplot**; for information about `screeplot`, see [MV] **screeplot**; and for all other commands, see below.

In addition, the following standard postestimation commands are available:

command	description
* estimates	cataloging estimation results; see [R] **estimates**
† predict	predict regression or Bartlett scores

* `estimates table` is not allowed, and `estimates stats` is only allowed with the `ml` factor method.

† `predict` after `factormat` works only if you have variables in memory that match the names specified in `factormat`. `predict` assumes mean zero and standard deviation one unless the `means()` and `sds()` options of `factormat` were provided.

See the corresponding entries in the *Stata Base Reference Manual* for details.

Special-interest postestimation commands

`estat anti` displays the anti-image correlation and anti-image covariance matrices. These are minus the partial covariance and minus the partial correlation matrices of all pairs of variables, holding all other variables constant.

225

estat common displays the correlation matrix of the common factors. For orthogonal factor loadings, the common factors are uncorrelated, and hence an identity matrix is shown. estat common is of more interest after oblique rotations.

estat factors displays model-selection criteria (AIC and BIC) for models with 1, 2, ..., # factors. Each model is estimated using maximum likelihood (i.e., using the ml option of factor).

estat kmo specifies that the Kaiser–Meyer–Olkin measure of sampling adequacy (KMO) be displayed. KMO takes values between 0 and 1, with small values meaning that overall the variables have too little in common to warrant a factor analysis. Historically, the following labels are given to values of KMO (Kaiser 1974):

0.00 to 0.49	unacceptable
0.50 to 0.59	miserable
0.60 to 0.69	mediocre
0.70 to 0.79	middling
0.80 to 0.89	meritorious
0.90 to 1.00	marvelous

estat residuals displays the raw or standardized residuals of the observed correlations with respect to the fitted (reproduced) correlation matrix.

estat rotatecompare displays the unrotated factor loadings and the most-recent rotated factor loadings.

estat smc displays the squared multiple correlations between each variable and all other variables. SMC is a theoretical lower bound for communality, so it is an upper bound for uniqueness. Note that the pf factor method estimates the communalities by smc.

estat structure displays the factor structure, i.e., the correlations between the variables and the common factors.

estat summarize displays summary statistics of the variables in the factor analysis over the estimation sample. This subcommand is, of course, not available after factormat.

rotate modifies the results of the last factor or factormat command to create a set of loadings that are more interpretable than those originally produced. A variety of orthogonal and oblique rotations are available, including varimax, orthomax, promax, and oblimin. See [MV] **rotate** for additional details. rotate stores results along with the original estimation results so that replaying factor or factormat and other postestimation commands may refer to the unrotated results, as well as the rotated results.

Syntax for predict

predict *newvarlist* [*if*] [*in*] [, *statistic options*]

statistic	description
Main	
regression	regression scoring method
bartlett	Bartlett scoring method

options	description
Main	
<u>noro</u>tated	use unrotated results, even when rotated results are available
<u>not</u>able	suppress table of scoring coefficients
<u>f</u>ormat(*%fmt*)	format for displaying the scoring coefficients

Options for predict

⌐ Main ⌐──

regression produces factors scored by the regression method.

bartlett produces factors scored by the method suggested by Bartlett (1937, 1938). This method produces unbiased factors, but they may be less accurate than those produced by the default regression method suggested by Thomson (1951). Regression-scored factors have the smallest mean squared error from the true factors but may be biased.

norotated specifies that unrotated factors be scored even when you have previously issued a rotate command. The default is to use rotated factors if they are available and unrotated factors otherwise.

notable suppresses the table of scoring coefficients.

format(*%fmt*) specifies the display format for scoring coefficients.

Syntax for estat

Anti-image correlation/covariance matrices

     estat anti [, nocorr nocov <u>for</u>mat(*%fmt*)]

Correlation of common factors

     estat <u>com</u>mon [, <u>noro</u>tated <u>for</u>mat(*%fmt*)]

Model-selection criteria

     estat <u>fac</u>tors [, <u>fac</u>tors(*#*) <u>det</u>ail]

Sample adequacy measures

     estat kmo [, <u>nov</u>ar <u>for</u>mat(*%fmt*)]

Residuals of correlation matrix

     estat <u>res</u>iduals [, <u>f</u>itted <u>o</u>bs <u>sr</u>esiduals <u>for</u>mat(*%fmt*)]

Comparison of rotated and unrotated loadings

     estat <u>rot</u>atecompare [, <u>for</u>mat(*%fmt*)]

Squared multiple correlations

     estat smc [, <u>for</u>mat(*%fmt*)]

Correlations between variables and common factors

> estat <u>structure</u> [, <u>noro</u>tated <u>form</u>at(% *fmt*)]

Summarize variables for estimation sample

> estat <u>summarize</u> [, <u>label</u> <u>nohea</u>der <u>noweig</u>hts]

Options for estat

nocorr, an option used with estat anti, suppresses the display of the anti-image correlation matrix.

nocov, an option used with estat anti, suppresses the display of the anti-image covariance matrix.

format(%*fmt*) specifies the display format. The defaults differ between the subcommands.

norotated, an option used with estat common and estat structure, requests that the displayed and returned results be based on the unrotated original factor solution rather than on the last rotation (orthogonal or oblique).

factors(#), an option used with estat factors, specifies the maximum number of factors to include in the summary table.

detail, an option used with estat factors, presents the output from each run of factor (or factormat) used in the computations of the AIC and BIC values.

novar, an option used with estat kmo, suppresses the KMO measures of sampling adequacy for the variables in the factor analysis, displaying the overall KMO measure only.

fitted, an option used with estat residuals, displays the fitted (reconstructed) correlation matrix based on the retained factors.

obs, an option used with estat residuals, displays the observed correlation matrix.

sresiduals, an option used with estat residuals, displays the matrix of standardized residuals of the correlations. Be careful when interpreting these residuals; see Jöreskog and Sörbom (1988).

label, noheader, and noweights are the same as for the generic estat summarize command; see [R] **estat**.

Remarks

Remarks are presented under the headings

> *Postestimation statistics*
> *Plots of eigenvalues, factor loadings, and scores*
> *Rotating the factor loadings*
> *Factor scores*

Postestimation statistics

Many postestimation statistics are available after factor and factormat.

▷ Example 1

   After `factor` and `factormat` there are a number of "classical" methods for assessing whether the variables have enough in common to have warranted the use of a factor model. One method is to examine the squared multiple-correlations of each variable with all other variables—this is usually an upper bound to *communality* and thus a lower bound to $1 - uniqueness(= communality)$ of the variables.

```
. use http://www.stata-press.com/data/r9/bg2
(Physician-cost data)

. quietly factor bg2cost1-bg2cost6, factors(2) ml

. estat smc
Squared multiple correlations of variables with all other variables
```

Variable	smc
bg2cost1	0.1054
bg2cost2	0.1370
bg2cost3	0.1637
bg2cost4	0.0866
bg2cost5	0.1671
bg2cost6	0.1683

   Additional diagnostic tools, such as examining the anti-image correlation and anti-image covariance matrices (`estat anti`) and the Kaiser–Meyer–Olkin measure of sampling adequacy (`estat kmo`), are also available. See [MV] **pca postestimation** for an illustration of their use.

◁

▷ Example 2

   Another set of postestimation tools help in determining the number of factors that should be retained. Later we will show the use of `screeplot` for producing a scree plot—a plot of the explained variance by the common factors. This is often used as a visual guide for selecting the number of factors to retain.

   Some authors advocate the standard model information criteria AIC and BIC for determining the number of factors (Schwarz 1978; Akaike 1987). This presupposes that the factors are extracted by maximum likelihood. `estat factors` provides these measures.

```
. estat factors
Factor analysis with different numbers of factors (maximum likelihood)
```

#factors	loglik	df_m	df_r	AIC	BIC
1	-60.53727	6	9	133.0745	159.1273
2	-6.842448	11	4	35.6849	83.44823
3	-3.37e-12	15	0	30	95.13182

```
no Heywood cases encountered
```

   The table shows the AIC and BIC statistics for the models with 1, 2, and 3 factors. Note that the 3-factor model is saturated, with 0 degrees of freedom. In this trivial case, and excluding the saturated case, both criteria select the 2-factor model.

◁

▷ Example 3

Two `estat` subcommands display statistics that help in interpreting the model and the results—in particular after an oblique rotation. `estat structure` displays the *structure* matrix containing the correlations between the (manifest) variables and the common factors.

```
. estat structure
```
Structure matrix: correlations between variables and common factors

Variable	Factor1	Factor2
bg2cost1	-0.1371	0.4235
bg2cost2	0.4140	0.1994
bg2cost3	0.6199	0.3692
bg2cost4	0.3577	0.0909
bg2cost5	-0.3752	0.4355
bg2cost6	-0.4295	0.4395

Notice that this matrix of correlations coincides with the pattern matrix, i.e., the matrix with factor loadings. This holds true for the unrotated factor solution as well as after an orthogonal rotation, such as a varimax rotation. It does not hold true after an oblique rotation. After an oblique rotation, the common factors are correlated. This correlation between the common factors also influences the correlation between the common factors and the manifest variables. The correlation matrix of the common factors is displayed by the `common` subcommand of `estat`. Since we have not yet rotated, we would only see an identity matrix. Later we show `estat common` output after an oblique rotation.

To assess the quality of a factor model, we may compare the observed correlation matrix $\mathbf{C}$ with the fitted ("reconstructed") matrix $\widehat{\mathbf{\Sigma}} = \widehat{\mathbf{\Lambda}}\widehat{\mathbf{\Phi}}\widehat{\mathbf{\Lambda}}' + \widehat{\mathbf{\Psi}}$ by examining the raw residuals $\mathbf{C} - \widehat{\mathbf{\Sigma}}$.

```
. estat residuals, obs fit
```
Observed correlations

Variable	bg2co~1	bg2co~2	bg2co~3	bg2co~4	bg2co~5	bg2co~6
bg2cost1	1.0000					
bg2cost2	0.0920	1.0000				
bg2cost3	0.0540	0.3282	1.0000			
bg2cost4	-0.0380	0.1420	0.2676	1.0000		
bg2cost5	0.2380	-0.1394	-0.0550	-0.0567	1.0000	
bg2cost6	0.2431	-0.0671	-0.1075	-0.1329	0.3524	1.0000

Fitted ("reconstructed") values for correlations

Variable	bg2co~1	bg2co~2	bg2co~3	bg2co~4	bg2co~5	bg2co~6
bg2cost1	1.0000					
bg2cost2	0.0277	1.0000				
bg2cost3	0.0714	0.3303	0.9999			
bg2cost4	-0.0106	0.1662	0.2553	1.0000		
bg2cost5	0.2359	-0.0685	-0.0718	-0.0946	1.0000	
bg2cost6	0.2450	-0.0902	-0.1040	-0.1137	0.3525	1.0000

```
Raw residuals of correlations (observed-fitted)
```

Variable	bg2co~1	bg2co~2	bg2co~3	bg2co~4	bg2co~5	bg2co~6
bg2cost1	-0.0000					
bg2cost2	0.0643	-0.0000				
bg2cost3	-0.0174	-0.0021	0.0001			
bg2cost4	-0.0274	-0.0242	0.0124	-0.0000		
bg2cost5	0.0021	-0.0709	0.0168	0.0379	0.0000	
bg2cost6	-0.0019	0.0231	-0.0035	-0.0193	-0.0002	-0.0000

To gauge the size of the residuals, `estat residuals` can also display the standardized residuals.

```
. estat residuals, sres
Standardized residuals of correlations
```

Variable	bg2co~1	bg2co~2	bg2co~3	bg2co~4	bg2co~5	bg2co~6
bg2cost1	-0.0001					
bg2cost2	1.5324	-0.0003				
bg2cost3	-0.4140	-0.0480	0.0011			
bg2cost4	-0.6538	-0.5693	0.2859	-0.0000		
bg2cost5	0.0484	-1.6848	0.3993	0.9003	0.0001	
bg2cost6	-0.0434	0.5480	-0.0836	-0.4560	-0.0037	-0.0000

Be careful when interpreting these standardized residuals, as they tend to be smaller than normalized residuals; i.e., these residuals tend to have a smaller variance than 1 if the model is true (see Bollen 1989).

◁

Plots of eigenvalues, factor loadings, and scores

Scree plots, factor loading plots, and score plots are easily obtained after `factor` and `factormat`.

▷ Example 4

The scree plot is a popular tool for determining the number of factors to be retained. A scree plot is a plot of the eigenvalues shown in decreasing order (Cattell 1966). We fit a factor model, extracting factors with the principal factor method.

```
. use http://www.stata-press.com/data/r9/sp2
. factor ghp31-ghp05, pcf
  (output omitted)
```

How many factors should we retain? We issue the `screeplot` command with the `mean` option, specifying that a horizontal line be plotted at the mean of the eigenvalues (a height of 1 since we are dealing with the eigenvalues of a correlation matrix).

. screeplot, mean

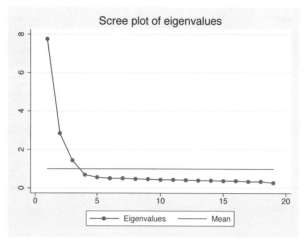

The plot suggests that we retain three factors, both because of the shape of the scree plot and because of Kaiser's well-known criterion suggesting that we retain factors with eigenvalue larger than 1. We may specify the option `mineigen(1)` during estimation to enforce this criterion. In this case, there is no need—`mineigen(1)` is the default with `pcf`.

◁

▷ Example 5

A second plot that is sometimes useful is the factor loadings plot. We display the plot with the loadings of the leading two factors.

. loadingplot, xline(0) yline(0) aspect(1) note(unrotated principal factors)

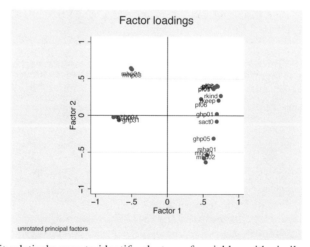

The plot makes it relatively easy to identify clusters of variables with similar loadings. With more than two factors, we can choose to see the multiple plots in a matrix style or a combined-graph style. The default is matrix style, but the combined style allows better control over various graph options—for instance, the addition of `xline(0)` and `yline(0)`. Here is a combined style graph.

```
. loadingplot, factors(3) combined xline(0) yline(0) aspect(1)
              xlabel(-0.8(0.4)0.8) ylabel(-0.8(0.4)0.8)
```

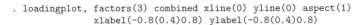

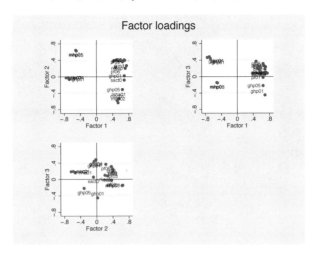

◁

▷ Example 6

Common factor scores can also be plotted for the observations using the `scoreplot` command. (See the discussion of `predict` to see how you can produce score variables.)

```
. scoreplot, msymbol(smcircle) msize(tiny)
```

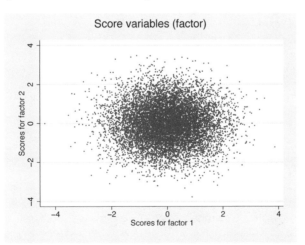

With so many observations, the plot's main purpose is to identify extreme cases. With smaller datasets with meaningful descriptions of the observations (e.g., country names, brands, etc.), the score plot is good for visually clustering observations with similar loadings.

◁

See [MV] **scoreplot** for more examples of `loadingplot` and `scoreplot`.

❏ Technical Note

The loading plots and score plots we have shown were for the original unrotated factor solution. After rotating (which we will discuss next), these plots display the most-recent rotated solution. Specify option `norotated` to refer to the unrotated result. To display the plots of rotated and unrotated results at the same time, you may use either of the following two approaches. First, you may display them in different Graph windows.

```
. plotcmd, norotated name(name1)
. plotcmd, name(name2)
```

Alternatively, you may save the plots and create a combined graph:

```
. plotcmd, norotated saving(name1)
. plotcmd, saving(name2)
. graph combine name1.gph name2.gph
```

See [G] **graph combine** for details.

❏

Rotating the factor loadings

Rotation is an attempt to describe the information in several factors by re-expressing them so that loadings on a few variables are as large as possible, and loadings on the rest of the variables are as small as possible. We have this freedom to re-express because of the indeterminant nature of the factor model. For example, if you find that z_1 and z_2 are two factors, then $z_1 + z_2$ and $z_1 - z_2$ are equally valid solutions.

❏ Technical Note

Said more technically, we are trying to find a set of f factor variables such that the observed variables can be best explained by regressing them on the f factor variables. Usually, f is a small number such as 1 or 2. If $f \geq 2$, there is an inherent indeterminancy in the construction of the factors because any linear combination of the calculated factors serves equally well as a set of regressors. Rotation capitalizes on this indeterminancy to create a set of variables that looks as much like the original variables as possible.

❏

The `rotate` command modifies the results of the last `factor` or `factormat` command to create a set of loadings that are more interpretable than those produced by `factor` or `factormat`. You may perform a single factor analysis followed by several `rotate` commands, thus experimenting with different types of rotation. If you retain too few factors, the variables for several distinct concepts may be merged, as in our example below. If you retain too many factors, several factors may attempt to measure the same concept, causing the factors to get in each other's way, and suggest too many distinct concepts after rotation.

❏ Technical Note

It is possible to restrict rotation to a number of leading factors. For instance, if you extracted three factors, you may specify the option `factors(2)` to `rotate` to exclude the third factor from being rotated. The new two leading factors are combinations of the initial two leading factors and are not affected by the fixed factor.

❏

▷ Example 7

We return to our physician-cost example and perform a factor analysis using the principal-component factor method, retaining two factors. We then tell `rotate` to apply the default orthogonal varimax rotation (Kaiser 1958).

```
. use http://www.stata-press.com/data/r9/bg2
(Physician-cost data)

. quietly factor bg2cost1-bg2cost6, pcf factors(2)

. rotate
```

Factor analysis/correlation   Number of obs   =   568
    Method: principal-component factors   Retained factors =   2
    Rotation: orthogonal varimax (Horst off)   Number of params =   11

Factor	Variance	Difference	Proportion	Cumulative
Factor1	1.57170	0.03430	0.2619	0.2619
Factor2	1.53740	.	0.2562	0.5182

LR test: independent vs. saturated: chi2(15) = 269.07 Prob>chi2 = 0.0000

Rotated factor loadings (pattern matrix) and unique variances

Variable	Factor1	Factor2	Uniqueness
bg2cost1	0.6853	0.2300	0.4775
bg2cost2	-0.0126	0.7142	0.4898
bg2cost3	-0.0161	0.7818	0.3886
bg2cost4	-0.1502	0.5703	0.6521
bg2cost5	0.7292	-0.1198	0.4539
bg2cost6	0.7398	-0.1537	0.4290

Factor rotation matrix

	Factor1	Factor2
Factor1	0.7460	-0.6659
Factor2	0.6659	0.7460

In this example, the factors are rotated so that the three "negative" items are grouped together and the three "positive" items are grouped.

Look at the uniqueness column. Recall that *uniqueness* is the percentage of variance for the variable that is not explained by the common factors; we may also think of it as the variances of the specific factors for the variables. We stress that rotation involves the "common factors", so the *uniqueness* is not affected by the rotation. As we noted in [MV] **factor**, the uniqueness is relatively high in this example, placing doubt on the usefulness of the factor model here.

◁

▷ Example 8

In this example, we examine 19 variables describing various aspects of health. These variables were collected from a random selection of 9,999 visitors to doctors' offices by Tarlov et al. (1989). Factor analysis yields three clear factors. We then examine several rotations of these three factors.

```
. use http://www.stata-press.com/data/r9/sp2
. describe
Contains data from http://www.stata-press.com/data/r9/sp2.dta
  obs:         9,999
  vars:           20                          26 Jan 2005 09:26
  size:      819,918 (60.9% of memory free)   (_dta has notes)
```

variable name	storage type	display format	value label	variable label
patid	int	%9.0g		Case id
ghp31	float	%9.0g		Health excellent, very good, good, fair, poor
pf01	float	%9.0g		How long limit vigorous activity
pf02	float	%9.0g		How long limit moderate activity
pf03	float	%9.0g		How long limit walk/climb
pf04	float	%9.0g		How long limit bend/stoop
pf05	float	%9.0g		How long limit walk 1 block
pf06	float	%9.0g		How long limit eat/dress/bath
rkeep	float	%9.0g		Does health keep work-job-hse
rkind	float	%9.0g		Can't do kind/amount of work
sact0	float	%9.0g		Last month limit activities
mha01	float	%9.0g		Last month very nervous
mhp03	float	%9.0g		Last month calm/peaceful
mhd02	float	%9.0g		Last month downhearted/blue
mhp01	float	%9.0g		Last month a happy person
mhc01	float	%9.0g		Last month down in the dumps
ghp01	float	%9.0g		Somewhat ill
ghp04	float	%9.0g		Healthy as anybody I know
ghp02	float	%9.0g		Health is excellent
ghp05	float	%9.0g		Feel bad lately

```
Sorted by:  patid
```

We now perform our factorization, requesting that three factors be retained.

(*Continued on next page*)

```
. factor ghp31-ghp05, factors(3)
(obs=9999)
```

```
Factor analysis/correlation                    Number of obs    =    9999
    Method: principal factors                  Retained factors =       3
    Rotation: (unrotated)                      Number of params =      54
```

Factor	Eigenvalue	Difference	Proportion	Cumulative
Factor1	7.27086	4.90563	0.7534	0.7534
Factor2	2.36523	1.38826	0.2451	0.9985
Factor3	0.97697	1.00351	0.1012	1.0997
Factor4	-0.02654	0.00538	-0.0027	1.0970
Factor5	-0.03191	0.00378	-0.0033	1.0937
Factor6	-0.03569	0.00353	-0.0037	1.0900
Factor7	-0.03922	0.00271	-0.0041	1.0859
Factor8	-0.04193	0.00662	-0.0043	1.0815
Factor9	-0.04855	0.01015	-0.0050	1.0765
Factor10	-0.05870	0.00250	-0.0061	1.0704
Factor11	-0.06120	0.00224	-0.0063	1.0641
Factor12	-0.06344	0.00376	-0.0066	1.0575
Factor13	-0.06720	0.00345	-0.0070	1.0506
Factor14	-0.07065	0.00185	-0.0073	1.0432
Factor15	-0.07250	0.00033	-0.0075	1.0357
Factor16	-0.07283	0.00772	-0.0075	1.0282
Factor17	-0.08055	0.01190	-0.0083	1.0198
Factor18	-0.09245	0.00649	-0.0096	1.0103
Factor19	-0.09894	.	-0.0103	1.0000

```
LR test: independent vs. saturated: chi2(171) = 1.0e+05 Prob>chi2 = 0.0000
```

Factor loadings (pattern matrix) and unique variances

Variable	Factor1	Factor2	Factor3	Uniqueness
ghp31	-0.6519	-0.0562	0.3440	0.4535
pf01	0.6150	0.3226	-0.0072	0.5177
pf02	0.6867	0.3737	0.2175	0.3415
pf03	0.6712	0.3774	0.1621	0.3807
pf04	0.6540	0.3588	0.2268	0.3921
pf05	0.6209	0.3258	0.2631	0.4392
pf06	0.4370	0.1803	0.2241	0.7263
rkeep	0.6868	0.1820	0.0870	0.4876
rkind	0.7244	0.2464	0.0780	0.4085
sact0	0.6556	-0.0719	0.0461	0.5628
mha01	0.5297	-0.4773	0.1268	0.4755
mhp03	-0.4810	0.5691	-0.1238	0.4294
mhd02	0.5208	-0.5949	0.1623	0.3485
mhp01	-0.4980	0.5955	-0.1225	0.3824
mhc01	0.4927	-0.5215	0.1531	0.4618
ghp01	0.6686	0.0194	-0.3621	0.4215
ghp04	-0.6833	-0.0195	0.4089	0.3656
ghp02	-0.7398	-0.0227	0.4212	0.2748
ghp05	0.6163	-0.2760	-0.1626	0.5175

The first factor is a general health factor. (To understand that claim, compare the factor loadings with the description of the variables as shown by describe above. Also note that, just as with the physician-cost data, the sense of some of the coded responses is reversed.) The second factor loads most highly on the five "mental health" items (with names sp2mha01–sp2mhc01). The third factor loads most highly on "general health perception" items—those with names having the letters ghp in them. The other items describe "physical health". These designations are based primarily on the wording of the questions, which is summarized in the variable labels.

```
. rotate, varimax
```

```
Factor analysis/correlation                    Number of obs    =    9999
        Method: principal factors              Retained factors =       3
        Rotation: orthogonal varimax (Horst off)   Number of params =      54
```

Factor	Variance	Difference	Proportion	Cumulative
Factor1	4.20556	0.83302	0.4358	0.4358
Factor2	3.37253	0.33756	0.3495	0.7852
Factor3	3.03497	.	0.3145	1.0997

LR test: independent vs. saturated: chi2(171) = 1.0e+05 Prob>chi2 = 0.0000

Rotated factor loadings (pattern matrix) and unique variances

Variable	Factor1	Factor2	Factor3	Uniqueness
ghp31	-0.2968	-0.1647	-0.6567	0.4535
pf01	0.5872	0.0263	0.3699	0.5177
pf02	0.7740	0.0848	0.2287	0.3415
pf03	0.7386	0.0580	0.2654	0.3807
pf04	0.7484	0.0842	0.2018	0.3921
pf05	0.7256	0.1063	0.1518	0.4392
pf06	0.5023	0.1268	0.0730	0.7263
rkeep	0.6023	0.2048	0.3282	0.4876
rkind	0.6590	0.1669	0.3597	0.4085
sact0	0.4187	0.3875	0.3342	0.5628
mha01	0.1467	0.6859	0.1803	0.4755
mhp03	-0.0613	-0.7375	-0.1514	0.4294
mhd02	0.0921	0.7893	0.1416	0.3485
mhp01	-0.0570	-0.7671	-0.1612	0.3824
mhc01	0.1102	0.7124	0.1359	0.4618
ghp01	0.2783	0.1977	0.6797	0.4215
ghp04	-0.2652	-0.1908	-0.7264	0.3656
ghp02	-0.2986	-0.2116	-0.7690	0.2748
ghp05	0.1755	0.4756	0.4748	0.5175

Factor rotation matrix

	Factor1	Factor2	Factor3
Factor1	0.6658	0.4796	0.5715
Factor2	0.5620	-0.8263	0.0387
Factor3	0.4908	0.2954	-0.8197

With rotation, the structure of the data becomes much clearer. The first rotated factor is physical health, the second is mental health, and the third is general health perception. The *a priori* designation of the items is confirmed.

After rotation, physical health is the first factor. `rotate` has ordered the factors by explained variance. Still, we warn that the importance of any factor must be gauged against the number of variables that purportedly measure it. Here we included nine variables that measured physical health, five that measured mental health, and five that measured general health perception. Had we started with only one mental-health item, it would have had a high uniqueness, but we would not want to conclude that it was, therefore, largely noise.

◁

❑ Technical Note

Some people prefer specifying the option horst to apply a Horst (1965) normalization, which places equal weight on all rows of the matrix to be rotated.

❑

▷ Example 9

The literature suggests that physical health and mental health are related. In addition, general health perception may be largely a combination of the two. For these reasons, an oblique rotation of a two-factor solution is worth trying. We try the oblique oblimin rotation (Harman 1976).

```
. factor ghp31-ghp05, factors(2)
(obs=9999)
```

Factor analysis/correlation   Number of obs   =   9999
    Method: principal factors   Retained factors =   2
    Rotation: (unrotated)   Number of params =   37

Factor	Eigenvalue	Difference	Proportion	Cumulative
Factor1	7.27086	4.90563	0.7534	0.7534
Factor2	2.36523	1.38826	0.2451	0.9985
Factor3	0.97697	1.00351	0.1012	1.0997
Factor4	-0.02654	0.00538	-0.0027	1.0970
Factor5	-0.03191	0.00378	-0.0033	1.0937
Factor6	-0.03569	0.00353	-0.0037	1.0900
Factor7	-0.03922	0.00271	-0.0041	1.0859
Factor8	-0.04193	0.00662	-0.0043	1.0815
Factor9	-0.04855	0.01015	-0.0050	1.0765
Factor10	-0.05870	0.00250	-0.0061	1.0704
Factor11	-0.06120	0.00224	-0.0063	1.0641
Factor12	-0.06344	0.00376	-0.0066	1.0575
Factor13	-0.06720	0.00345	-0.0070	1.0506
Factor14	-0.07065	0.00185	-0.0073	1.0432
Factor15	-0.07250	0.00033	-0.0075	1.0357
Factor16	-0.07283	0.00772	-0.0075	1.0282
Factor17	-0.08055	0.01190	-0.0083	1.0198
Factor18	-0.09245	0.00649	-0.0096	1.0103
Factor19	-0.09894	.	-0.0103	1.0000

LR test: independent vs. saturated: chi2(171) = 1.0e+05 Prob>chi2 = 0.0000

(Continued on next page)

Factor loadings (pattern matrix) and unique variances

Variable	Factor1	Factor2	Uniqueness
ghp31	-0.6519	-0.0562	0.5718
pf01	0.6150	0.3226	0.5178
pf02	0.6867	0.3737	0.3888
pf03	0.6712	0.3774	0.4070
pf04	0.6540	0.3588	0.4435
pf05	0.6209	0.3258	0.5084
pf06	0.4370	0.1803	0.7765
rkeep	0.6868	0.1820	0.4952
rkind	0.7244	0.2464	0.4145
sact0	0.6556	-0.0719	0.5650
mha01	0.5297	-0.4773	0.4916
mhp03	-0.4810	0.5691	0.4448
mhd02	0.5208	-0.5949	0.3748
mhp01	-0.4980	0.5955	0.3974
mhc01	0.4927	-0.5215	0.4853
ghp01	0.6686	0.0194	0.5526
ghp04	-0.6833	-0.0195	0.5327
ghp02	-0.7398	-0.0227	0.4522
ghp05	0.6163	-0.2760	0.5439

. rotate, oblimin oblique

Factor analysis/correlation   Number of obs   =   9999
    Method: principal factors   Retained factors =   2
    Rotation: oblique oblimin (Horst off)   Number of params =   37

Factor	Variance	Proportion	Rotated factors are correlated
Factor1	6.58719	0.6826	
Factor2	4.65444	0.4823	

LR test: independent vs. saturated: chi2(171) = 1.0e+05 Prob>chi2 = 0.0000

Rotated factor loadings (pattern matrix) and unique variances

Variable	Factor1	Factor2	Uniqueness
ghp31	-0.5517	-0.2051	0.5718
pf01	0.7179	-0.0747	0.5178
pf02	0.8115	-0.0968	0.3888
pf03	0.8022	-0.1068	0.4070
pf04	0.7750	-0.0951	0.4435
pf05	0.7249	-0.0756	0.5084
pf06	0.4743	-0.0044	0.7765
rkeep	0.6712	0.0939	0.4952
rkind	0.7478	0.0449	0.4145
sact0	0.4608	0.3340	0.5650
mha01	0.0652	0.6869	0.4916
mhp03	0.0401	-0.7587	0.4448
mhd02	-0.0280	0.8003	0.3748
mhp01	0.0462	-0.7918	0.3974
mhc01	0.0039	0.7160	0.4853
ghp01	0.5378	0.2484	0.5526
ghp04	-0.5494	-0.2541	0.5327
ghp02	-0.5960	-0.2736	0.4522
ghp05	0.2805	0.5213	0.5439

Factor rotation matrix

	Factor1	Factor2
Factor1	0.9277	0.6831
Factor2	0.3733	-0.7303

The first factor is defined predominantly by physical health and the second by mental health. General health perception loads on both, but more on physical health than mental health. To compare the rotated and unrotated solution, it is often useful to look at both in parallel form.

```
. estat rotatecompare
```

Rotation matrix — oblique oblimin (Horst off)

Variables	Factor1	Factor2
Factor1	0.9277	0.6831
Factor2	0.3733	-0.7303

Factor loadings

Variables	Rotated Factor1	Factor2	Unrotated Factor1	Factor2
ghp31	-0.5517	-0.2051	-0.6519	-0.0562
pf01	0.7179	-0.0747	0.6150	0.3226
pf02	0.8115	-0.0968	0.6867	0.3737
pf03	0.8022	-0.1068	0.6712	0.3774
pf04	0.7750	-0.0951	0.6540	0.3588
pf05	0.7249	-0.0756	0.6209	0.3258
pf06	0.4743	-0.0044	0.4370	0.1803
rkeep	0.6712	0.0939	0.6868	0.1820
rkind	0.7478	0.0449	0.7244	0.2464
sact0	0.4608	0.3340	0.6556	-0.0719
mha01	0.0652	0.6869	0.5297	-0.4773
mhp03	0.0401	-0.7587	-0.4810	0.5691
mhd02	-0.0280	0.8003	0.5208	-0.5949
mhp01	0.0462	-0.7918	-0.4980	0.5955
mhc01	0.0039	0.7160	0.4927	-0.5215
ghp01	0.5378	0.2484	0.6686	0.0194
ghp04	-0.5494	-0.2541	-0.6833	-0.0195
ghp02	-0.5960	-0.2736	-0.7398	-0.0227
ghp05	0.2805	0.5213	0.6163	-0.2760

Look again at the `factor` output. The variances of the first and second factor of the unrotated solution are 7.27 and 2.37, respectively. After an orthogonal rotation, the explained variance of 7.27+2.37 is distributed differently over the two factors. For instance, after an orthogonal varimax rotation, the first factor has variance 5.75, and the second factor has 3.88—within rounding error 7.27+2.37 = 5.75+3.88. The situation after an oblique rotation is different. The variances of the first and second factors are 6.59 and 4.65, which add up to more than in the orthogonal case. In the oblique case, the common factors are correlated and thus "partly explain the same variance". Therefore, in this case the cumulative proportion of variance explained by the factors is not displayed.

Most researchers would not be willing to accept a solution in which the common factors are highly correlated.

```
. estat common
Correlation matrix of the Oblimin(0) rotated common factors
```

Factors	Factor1	Factor2
Factor1	1	
Factor2	.3611	1

The correlation of .36 seems acceptable, so we think that the oblique rotation was a success here.

◁

Factor scores

The `predict` command creates a set of new variables that are estimates of the first k common factors produced by `factor`, `factormat`, or `rotate`. Two types of scoring are available: regression or Thomson scoring and Bartlett scoring.

The number of variables may be less than the number of factors. If so, the first such factors will be used. If the number of variables is greater than the number of factors created or rotated, the unused factors will be filled with missing values.

▷ Example 10

Using our automobile data, we wish to develop an index of roominess based on a car's headroom, rear-seat leg room, and trunk space. We begin by extracting the factors of the three variables:

```
. use http://www.stata-press.com/data/r9/autofull
(Automobile Models)

. factor headroom rear_seat trunk
(obs=74)
```

Factor analysis/correlation			Number of obs	=	74
Method: principal factors			Retained factors	=	1
Rotation: (unrotated)			Number of params	=	3

Factor	Eigenvalue	Difference	Proportion	Cumulative
Factor1	1.71426	1.79327	1.1799	1.1799
Factor2	-0.07901	0.10329	-0.0544	1.1255
Factor3	-0.18231	.	-0.1255	1.0000

```
LR test: independent vs. saturated:  chi2(3)  =   82.93 Prob>chi2 = 0.0000
Factor loadings (pattern matrix) and unique variances
```

Variable	Factor1	Uniqueness
headroom	0.7280	0.4700
rear_seat	0.7144	0.4897
trunk	0.8209	0.3261

All the factor loadings are positive, so we have indeed obtained a "roominess" factor. The `predict` command will now create the one retained factor, which we will call `f1`:

```
. predict f1
(regression scoring assumed)
Scoring coefficients (method = regression)
```

Variable	Factor1
headroom	0.28323
rear_seat	0.26820
trunk	0.45964

The table with scoring coefficients informs us that the factor is obtained as a weighted sum of standardized versions of headroom, rear_seat, and trunk with weights 0.28, 0.26, and 0.46.

If factor had retained more than one factor, typing factor f1 would still have added only the first factor to our data. Typing predict f1 f2, however, would have added the first two factors to our data. f1 is now our "roominess" index, so we might compare the roominess of domestic and foreign cars:

```
. table foreign, c(mean f1 sd f1) row
```

Foreign	mean(f1)	sd(f1)
Domestic	.2022442	.9031404
Foreign	-.4780318	.6106609
Total	4.51e-09	.8804116

We find that domestic cars are, on average, roomier than foreign cars, at least in our data.

◁

❏ Technical Note

Are common factors not supposed to be normalized to have mean 0 and standard deviation 1? In our example above, the mean is $4.5 \cdot 10^{-9}$ and the standard deviation is .88. Why is that?

First examining the mean, the deviation from zero is due to numerical roundoff, which would diminish dramatically if we had typed predict double f1 instead. The explanation for the standard deviation of .88, on the other hand, is not numerical roundoff. At a theoretical level, the factor is supposed to have standard deviation 1, but the estimation method almost never yields that result unless an exact solution to the factor model is found. This happens for the same reason that, when you regress y on x, you do not get the same equation as if you regress x on y, unless x and y are perfectly collinear.

By the way, if you had two factors, you would expect the correlation between the two factors to be zero since that is how they are theoretically defined. The matrix algebra, however, does not usually work out that way. It is somewhat analogous to the fact that if you regress y on x and the regression assumption that the errors are uncorrelated with the dependent variable is satisfied, then it automatically cannot be satisfied if you regress x on y.

The covariance matrix of the estimated factors is

$$E(\widehat{\mathbf{f}}\widehat{\mathbf{f}}') = \mathbf{I} - (\mathbf{I} + \mathbf{\Gamma})^{-1}$$

where

$$\mathbf{\Gamma} = \mathbf{\Lambda}'\mathbf{\Psi}^{-1}\mathbf{\Lambda}$$

The columns of Λ are orthogonal to each other, but the inclusion of Ψ in the middle of the equation destroys that relationship unless all the elements of Ψ are equal.

❏

▷ Example 11

Let's pretend that we work for the K. E. Watt Company, a fictional industry group that generates statistics on automobiles. Our "roominess" index has mean 0 and standard deviation .88, but indexes we present to the public generally have mean 100 and standard deviation 10. First, we wish to rescale our index:

```
. gen roomidx = (f1/.88041161)*10 + 100
. table foreign, c(mean roomidx sd roomidx freq) row format(%9.2f)
```

Foreign	mean(roomidx)	sd(roomidx)	Freq.
Domestic	102.30	10.26	52
Foreign	94.57	6.94	22
Total	100.00	10.00	74

Now when we release our results, we can write, "The K. E. Watt index of roominess shows that domestic cars are, on average, roomier, with an index of 102 versus only 95 for foreign cars."

Now let's find the "roomiest" car in our data:

```
. sort roomidx
. list fullname roomidx in 1
```

	fullname	roomidx
74.	Merc. Marquis	116.7469

We can also write, "K. E. Watt finds that the Mercury Marquis is the roomiest automobile among those surveyed, with a roominess index of 117 versus an average of 100."

◁

❏ Technical Note

`predict` provides two methods of scoring: the default regression scoring, which we have used above, and the optional Bartlett method. An artificial example will best illustrate the use and meaning of the methods. We begin by creating a known-to-be-correct factor model in which the true loadings are 0.4, 0.6, and 0.8. The variances of the unique factors are $1 - 0.4^2 = 0.84$, $1 - 0.6^2 = 0.64$, and $1 - 0.8^2 = 0.36$, respectively. We make the sample size sufficiently large so that random fluctuations are not important.

```
. drop _all
. set seed 12345
. set obs 10000
obs was 0, now 10000
. gen ftrue = invnorm(uniform())
. gen x1 = .4*ftrue + sqrt(.84)*invnorm(uniform())
. gen x2 = .6*ftrue + sqrt(.64)*invnorm(uniform())
```

```
. gen x3 = .8*ftrue + sqrt(.36)*invnorm(uniform())

. summ x1 x2 x3
```

Variable	Obs	Mean	Std. Dev.	Min	Max
x1	10000	.0020004	1.008742	-3.717137	3.690495
x2	10000	.0063174	1.005964	-3.664064	4.297312
x3	10000	-.0022799	.9935951	-3.634207	3.489094

Having concocted our data, the iterated principal factor method reproduces the true loadings most faithfully:

```
. factor x1 x2 x3, ipf factors(1)
(obs=10000)
```

Factor analysis/correlation			Number of obs	=	10000
Method: iterated principal factors			Retained factors =		1
Rotation: (unrotated)			Number of params =		3

Factor	Eigenvalue	Difference	Proportion	Cumulative
Factor1	1.15779	1.15760	1.0000	1.0000
Factor2	0.00019	0.00042	0.0002	1.0002
Factor3	-0.00023	.	-0.0002	1.0000

LR test: independent vs. saturated: chi2(3) = 3773.10 Prob>chi2 = 0.0000

Factor loadings (pattern matrix) and unique variances

Variable	Factor1	Uniqueness
x1	0.3915	0.8467
x2	0.5976	0.6429
x3	0.8046	0.3526

Let us now compare regression and Bartlett scoring:

```
. predict freg
(regression scoring assumed)
```

Scoring coefficients (method = regression)

Variable	Factor1
x1	0.12939
x2	0.26019
x3	0.63873

```
. predict fbar, bartlett
```

Scoring coefficients (method = Bartlett)

Variable	Factor1
x1	0.17971
x2	0.36131
x3	0.88704

Comparing the two scoring vectors, we see that Bartlett scoring yields larger coefficients. The regression scoring method is biased insofar as $E(\mathrm{freg}|\mathrm{ftrue})$ is not ftrue, something we can reveal by regressing freg on ftrue:

```
. regress freg ftrue
```

Source	SS	df	MS
Model	5132.90388	1	5132.90388
Residual	2067.11409	9998	.20675276
Total	7200.01797	9999	.720073804

```
Number of obs =    10000
F(  1,  9998) =24826.29
Prob > F      =   0.0000
R-squared     =   0.7129
Adj R-squared =   0.7129
Root MSE      =    .4547
```

freg	Coef.	Std. Err.	t	P>|t|	[95% Conf. Interval]	
ftrue	.7158611	.0045433	157.56	0.000	.7069553	.7247669
_cons	.0016462	.004547	0.36	0.717	-.0072668	.0105593

Note the coefficient on `ftrue` of .736 < 1. The Bartlett scoring method, on the other hand, is unbiased:

```
. regress fbar ftrue
```

Source	SS	df	MS
Model	9899.40173	1	9899.40173
Residual	3986.67487	9998	.398747236
Total	13886.0766	9999	1.38874653

```
Number of obs =    10000
F(  1,  9998) =24826.26
Prob > F      =   0.0000
R-squared     =   0.7129
Adj R-squared =   0.7129
Root MSE      =   .63146
```

fbar	Coef.	Std. Err.	t	P>|t|	[95% Conf. Interval]	
ftrue	.9941494	.0063095	157.56	0.000	.9817815	1.006517
_cons	.0022862	.0063147	0.36	0.717	-.0100918	.0146642

The zero bias of the Bartlett method comes at the costs of less accuracy, e.g., in terms of the mean squared error.

```
. gen dbar = (fbar - ftrue)^2
. gen dreg = (freg - ftrue)^2
. summ ftrue fbar freg dbar dreg
```

Variable	Obs	Mean	Std. Dev.	Min	Max
ftrue	10000	-.0022996	1.000863	-4.238265	3.840595
fbar	10000	-1.78e-10	1.178451	-4.117305	4.208633
freg	10000	1.02e-10	.8485716	-2.964756	3.030532
dbar	10000	.3987071	.5625816	2.14e-08	6.531064
dreg	10000	.2875829	.4079801	3.83e-08	5.132763

Notice that neither estimator follows the assumption that the scaled factor has unit variance. The regression estimator has a variance less than 1, and the Bartlett estimator has a variance greater than 1.

The difference between the two scoring methods is not as important as it might seem since the bias in the regression method is only a matter of scaling and shifting.

```
. correlate freg fbar ftrue
(obs=10000)
```

	freg	fbar	ftrue
freg	1.0000		
fbar	1.0000	1.0000	
ftrue	0.8443	0.8443	1.0000

Therefore, the choice of which scoring method we apply is largely immaterial.

❏

Saved Results

Let p be the number of variables and f the number of factors.

predict, in addition to generating variables, also saves in r():

> Macros
>> r(method)   regression or Bartlett
>
> Matrices
>> r(scoef) $p \times f$ matrix of scoring coefficients

estat anti saves in r():

> Matrices
>> r(acov) $p \times p$ anti-image covariance matrix
>> r(acorr) $p \times p$ anti-image correlation matrix

estat common saves in r():

> Matrices
>> r(Phi) $f \times f$ correlation matrix of common factors

estat factors saves in r():

Matrices
> r(stats) $k \times 5$ matrix with log likelihood, degrees of freedom, AIC, and BIC
> for models with 1 to k factors estimated via maximum likelihood

estat kmo saves in r():

> Scalars
>> r(kmo)   the Kaiser–Meyer–Olkin measure of sampling adequacy
>
> Matrices
>> r(kmow)   column vector of KMO measures for each variable

estat residuals saves in r():

> Matrices
>> r(fit)   fitted matrix for the correlations, $\widehat{C}=\widehat{\Lambda}\widehat{\Phi}\widehat{\Lambda}'+\widehat{\Psi}$
>> r(res)   raw residual matrix $C-\widehat{C}$
>> r(SR)   standardized residuals (sresiduals option only)

estat smc saves in r():

Matrices
> r(smc)   vector of squared multiple correlations of variables with all other variables

estat structure saves in r():

> Matrices
>> r(st) $p \times f$ matrix of correlations between variables and common factors

See [R] **estat** for the returned results of estat summarize.

rotate after factor and factormat add to the existing e():

> Scalars
>> e(r_f)   number of factors in rotated solution
>> e(r_fmin)   rotation criterion value
>
> Macros
>> e(r_class)   orthogonal or oblique
>> e(r_criterion)   rotation criterion
>> e(r_ctitle)   title for rotation
>> e(r_normalization)   horst or none
>
> Matrices
>> e(r_L)   rotated loadings
>> e(r_T)   rotation
>> e(r_Phi)   correlations between common factors
>> e(r_Ev)   explained variance by common factors

Note that the factors in the rotated solution are in decreasing order of e(r_Ev).

Methods and Formulas

All postestimation commands listed above are implemented as ado-files.

estat

See *Methods and Formulas* of [MV] **pca postestimation** for the formulas for estat anti, estat kmo, and estat smc.

estat residuals computes the standardized residuals $\widetilde{r}_{ij}$ as

$$\widetilde{r}_{ij} = \frac{\sqrt{N}(r_{ij} - f_{ij})}{\sqrt{f_{ij}^2 + f_{ii}f_{jj}}}$$

suggested by Jöreskog and Sörbom (1986), where N is the number of observations, r_{ij} is the observed correlation of variables i and j, and f_{ij} is the fitted correlation of variables i and j. Also see Bollen (1989). Note that $\widetilde{r}_{ij} = 0$. Caution is warranted in interpretation of these residuals; see Jöreskog and Sörbom (1988).

estat structure computes the correlations of the variables and the common factors as $\mathbf{\Lambda\Phi}$.

rotate

See *Methods and Formulas* of [MV] **rotatemat** for the details of rotation.

The correlation of common factors after rotation is $\mathbf{T'T}$, where $\mathbf{T}$ is the factor rotation matrix, satisfying $\mathbf{L}_{\text{rotated}} = \mathbf{L}_{\text{unrotated}}(\mathbf{T'})^{-1}$

predict

The formula for regression scoring (Thomson 1951) in the orthogonal case is

$$\widehat{\mathbf{f}} = \mathbf{\Lambda'\Sigma}^{-1}\mathbf{x}$$

where $\mathbf{\Lambda}$ is the unrotated or orthogonally rotated loading matrix. For oblique rotation, the regression scoring is defined as

$$\widehat{\mathbf{f}} = \mathbf{\Phi\Lambda'\Sigma}^{-1}\mathbf{x}$$

where $\mathbf{\Phi}$ is the correlation matrix of the common factors.

The formula for Bartlett scoring (Bartlett 1937, 1938) is

$$\mathbf{\Gamma}^{-1}\mathbf{\Lambda'\Psi}^{-1}\mathbf{x}$$

where

$$\mathbf{\Gamma} = \mathbf{\Lambda'\Psi}^{-1}\mathbf{\Lambda}$$

See Harman (1976) and Lawley and Maxwell (1971).

References

Akaike, H. 1987. Factor analysis and AIC. *Psychometrika* 52: 317–332.

Bartlett, M. S. 1937. The statistical conception of mental factors. *British Journal of Psychology* 28: 97–104.

——. 1938. Methods of estimating mental factors. *Nature, London* 141: 609–610.

Bollen, K. A. 1989. *Structural equations with latent variables*. New York: Wiley.

Cattell, R. B. 1966. The scree test for the number of factors. *Multivariate behavioral research* 1: 245–276.

Harman, H. H. 1976. *Modern Factor Analysis*. 3rd ed. Chicago: University of Chicago Press.

Horst, P. 1965. *Factor Analysis of Data Matrices*. New York: Holt, Rinehart, and Winston.

Jöreskog, K. G. and Sörbom, D. 1986. *Analysis of linear structural relationships by the method of maximum likelihood*. Mooresville, IN: Scientific Software.

——. 1988. *PRELIS: a program for multivariate data screening and data summarization. A preprocessor for LISREL*. 2nd ed. Mooresville, IN: Scientific Software.

Kaiser, H. F. 1958. The varimax criterion for analytic rotation in factor analysis. *Psychometrika* 23: 187–200.

——. 1974. An index of factor simplicity. *Psychometrika* 39: 31–36.

Lawley, D. N. and A. E. Maxwell. 1971. *Factor Analysis as a Statistical Method*. London: Butterworths.

Schwarz, G. 1978. Estimating the dimension of a model. *Annals of Statistics* 6: 461–464.

Tarlov, A. R., J. E. Ware, Jr., S. Greenfield, E. C. Nelson, E. Perrin, and M. Zubkoff. 1989. The medical outcomes study. *Journal of the American Medical Association* 262: 925–930.

Thomson, G. H. 1951. *The Factorial Analysis of Human Ability*. London: University of London Press.

For additional references, see [MV] **factor**.

Also See

Complementary: [MV] **factor**, [MV] **rotate**, [MV] **scoreplot**, [MV] **screeplot**,
[R] **estimates**

Background: [R] **estat**, [R] **predict**

Title

hotelling — Hotelling's T-squared generalized means test

Syntax

hotelling *varlist* [*if*] [*in*] [*weight*] [, by(*varname*) <u>not</u>able]

aweights and fweights are allowed; see [U] **11.1.6 weight**.

Description

hotelling performs Hotelling's T-squared test of whether a set of means is zero or, alternatively, equal between two groups.

Options

<u>Main</u>

by(*varname*) specifies a variable identifying two groups; the test of equality of means between groups is performed. If by() is not specified, a test of means being jointly zero is performed.

notable suppresses printing a table of the means being compared.

Remarks

hotelling performs Hotelling's T-squared test of whether a set of means is zero or two sets of means are equal. It is a multivariate test that reduces to a standard t test if only one variable is specified.

▷ Example 1

You wish to test whether a new fuel additive improves gas mileage in both stop-and-go and highway situations. Taking 12 cars, you fill them with gas and run them on a highway-style track, recording their gas mileage. You then refill them and run them on a stop-and-go style track. Finally, you repeat the two runs, but this time you use fuel with the additive. Your dataset is

```
. use http://www.stata-press.com/data/r9/gasexp
. describe
Contains data from http://www.stata-press.com/data/r9/gasexp.dta
  obs:           12
  vars:           5                            15 Oct 2004 06:37
  size:         288 (99.9% of memory free)
```

variable name	storage type	display format	value label	variable label
id	float	%9.0g		car id
bmpg1	float	%9.0g		track1 before additive
ampg1	float	%9.0g		track1 after additive
bmpg2	float	%9.0g		track 2 before additive
ampg2	float	%9.0g		track 2 after additive

```
Sorted by:
```

To perform the statistical test, you jointly test whether the differences in before-and-after results are zero:

```
. gen diff1 = ampg1 - bmpg1
. gen diff2 = ampg2 - bmpg2
. hotelling diff1 diff2
```

Variable	Obs	Mean	Std. Dev.	Min	Max
diff1	12	1.75	2.70101	-3	5
diff2	12	2.083333	2.906367	-3.5	5.5

```
1-group Hotelling's T-squared = 9.6980676
F test statistic: ((12-2)/(12-1)(2)) x 9.6980676 = 4.4082126

H0: Vector of means is equal to a vector of zeros
               F(2,10) =    4.4082
        Prob > F(2,10) =    0.0424
```

The means are different at the 4.24% significance level.

◁

❏ Technical Note

We used Hotelling's T-squared test because we were testing two differences jointly. Had there been only one difference, we could have used a standard t test, which would have yielded the same results as Hotelling's test:

```
* We could have performed the test like this:
. ttest ampg1 = bmpg1
```

Paired t test

Variable	Obs	Mean	Std. Err.	Std. Dev.	[95% Conf. Interval]	
ampg1	12	22.75	.9384465	3.250874	20.68449	24.81551
bmpg1	12	21	.7881701	2.730301	19.26525	22.73475
diff	12	1.75	.7797144	2.70101	.0338602	3.46614

```
   mean(diff) = mean(ampg1 - bmpg1)                         t =   2.2444
Ho: mean(diff) = 0                        degrees of freedom =       11

Ha: mean(diff) < 0          Ha: mean(diff) != 0          Ha: mean(diff) > 0
Pr(T < t) = 0.9768        Pr(|T| > |t|) = 0.0463        Pr(T > t) = 0.0232
* Or like this:
. ttest diff1 = 0
```

One-sample t test

Variable	Obs	Mean	Std. Err.	Std. Dev.	[95% Conf. Interval]	
diff1	12	1.75	.7797144	2.70101	.0338602	3.46614

```
   mean = mean(diff1)                                       t =   2.2444
Ho: mean = 0                              degrees of freedom =       11

   Ha: mean < 0                Ha: mean != 0                Ha: mean > 0
Pr(T < t) = 0.9768        Pr(|T| > |t|) = 0.0463        Pr(T > t) = 0.0232
```

```
* Or like this:
. hotel diff1

    Variable |       Obs       Mean    Std. Dev.        Min        Max
-------------+-----------------------------------------------------------
       diff1 |        12       1.75    2.70101         -3          5

1-group Hotelling's T-squared = 5.0373832
F test statistic: ((12-1)/(12-1)(1)) x 5.0373832 = 5.0373832

H0: Vector of means is equal to a vector of zeros
              F(1,11) =     5.0374
         Prob > F(1,11) =     0.0463
```

❏

▷ Example 2

Now consider a variation on the experiment: rather than using 12 cars and running each car with and without the fuel additive, you run 24 cars, 12 with the additive and 12 without. You have the following dataset:

```
. use http://www.stata-press.com/data/r9/gasexp2, clear

. describe

Contains data from http://www.stata-press.com/data/r9/gasexp2.dta
  obs:            24
  vars:            4                          17 Oct 2004 01:43
  size:          480 (97.4% of memory free)

-------------------------------------------------------------------------------
              storage  display     value
variable name   type   format      label      variable label
-------------------------------------------------------------------------------
id            float   %9.0g                  car id
mpg1          float   %9.0g                  track 1
mpg2          float   %9.0g                  track 2
additive      float   %9.0g       yesno      additive?
-------------------------------------------------------------------------------

Sorted by:

. tabulate additive

    additive? |      Freq.      Percent        Cum.
-------------+-----------------------------------
          no |        12        50.00        50.00
         yes |        12        50.00       100.00
-------------+-----------------------------------
       Total |        24       100.00
```

This is an unpaired experiment because there is no natural pairing of the cars; you want to test that the means of mpg1 are equal for the two groups specified by additive, as are the means of mpg2:

(Continued on next page)

```
. hotelling mpg1 mpg2, by(additive)
```

```
-> additive = no
    Variable │      Obs        Mean    Std. Dev.        Min          Max
─────────────┼─────────────────────────────────────────────────────────
        mpg1 │       12          21     2.730301         17           25
        mpg2 │       12    19.91667     2.644319         16           24
```

```
-> additive = yes
    Variable │      Obs        Mean    Std. Dev.        Min          Max
─────────────┼─────────────────────────────────────────────────────────
        mpg1 │       12       22.75     3.250874         17           28
        mpg2 │       12          22     3.316625       16.5         27.5
```

```
2-group Hotelling's T-squared = 7.1347584
F test statistic: ((24-2-1)/(24-2)(2)) x 7.1347584 = 3.4052256

H0: Vectors of means are equal for the two groups
              F(2,21) =     3.4052
       Prob > F(2,21) =     0.0524
```

❑ Technical Note

As in the paired experiment, had there been only one test track, the t test would have yielded the same results as Hotelling's test:

```
. hotel mpg1, by(additive)
```

```
-> additive = no
    Variable │      Obs        Mean    Std. Dev.        Min          Max
─────────────┼─────────────────────────────────────────────────────────
        mpg1 │       12          21     2.730301         17           25
```

```
-> additive = yes
    Variable │      Obs        Mean    Std. Dev.        Min          Max
─────────────┼─────────────────────────────────────────────────────────
        mpg1 │       12       22.75     3.250874         17           28
```

```
2-group Hotelling's T-squared = 2.0390921
F test statistic: ((24-1-1)/(24-2)(1)) x 2.0390921 = 2.0390921

H0: Vectors of means are equal for the two groups
              F(1,22) =     2.0391
       Prob > F(1,22) =     0.1673
```

```
. ttest mpg1, by(additive)
```

Two-sample t test with equal variances

Group	Obs	Mean	Std. Err.	Std. Dev.	[95% Conf. Interval]	
no	12	21	.7881701	2.730301	19.26525	22.73475
yes	12	22.75	.9384465	3.250874	20.68449	24.81551
combined	24	21.875	.6264476	3.068954	20.57909	23.17091
diff		-1.75	1.225518		-4.291568	.7915684

```
      diff = mean(no) - mean(yes)                              t =   -1.4280
Ho: diff = 0                                  degrees of freedom =        22

   Ha: diff < 0                 Ha: diff != 0                 Ha: diff > 0
 Pr(T < t) = 0.0837     Pr(|T| > |t|) = 0.1673       Pr(T > t) = 0.9163
```

With more than one pair of means, however, there is no t test equivalent to Hotelling's test, although there are other logically (but not practically) equivalent solutions. One is the discriminant function: if the means of mpg1 and mpg2 are different, the discriminant function should separate the groups along that dimension.

```
. regress additive mpg1 mpg2
```

Source	SS	df	MS	
Model	1.46932917	2	.734664585	Number of obs = 24
Residual	4.53067083	21	.21574623	F(2, 21) = 3.41
				Prob > F = 0.0524
				R-squared = 0.2449
				Adj R-squared = 0.1730
Total	6	23	.260869565	Root MSE = .46448

additive	Coef.	Std. Err.	t	P>|t|	[95% Conf. Interval]	
mpg1	-.4570407	.2416657	-1.89	0.072	-.959612	.0455306
mpg2	.5014605	.2376762	2.11	0.047	.0071859	.9957352
_cons	-.0120115	.7437049	-0.02	0.987	-1.55863	1.534607

This test would declare the means different at the 5.24% level. Alternatively, you could have fitted this model using logistic regression:

```
. logit additive mpg1 mpg2

Iteration 0:   log likelihood = -16.635532
Iteration 1:   log likelihood = -13.471421
Iteration 2:   log likelihood = -13.371971
Iteration 3:   log likelihood = -13.371143
Iteration 4:   log likelihood = -13.371143
```

Logistic regression		Number of obs = 24
		LR chi2(2) = 6.53
		Prob > chi2 = 0.0382
Log likelihood = -13.371143		Pseudo R2 = 0.1962

additive	Coef.	Std. Err.	z	P>|z|	[95% Conf. Interval]	
mpg1	-2.306844	1.36139	-1.69	0.090	-4.975119	.3614307
mpg2	2.524477	1.367373	1.85	0.065	-.1555257	5.20448
_cons	-2.446527	3.689821	-0.66	0.507	-9.678443	4.78539

This test would have declared the means different at the 3.82% level.

Are the means different? Hotelling's T-squared and the discriminant function reject equality at the 5.24% level. The logistic regression rejects equality at the 3.82% level.   ❏

Saved Results

hotelling saves in r():

Scalars

r(N)	number of observations	r(T2)	Hotelling's T-squared
r(k)	number of variables	r(df)	degrees of freedom

Methods and Formulas

`hotelling` is implemented as an ado-file.

See Wilks (1962, 556–561) for a general discussion. The original formulation was by Hotelling (1931) and Mahalanobis (1930, 1936).

For the test that the means of k variables are 0, let $\overline{\mathbf{x}}$ be a $1 \times k$ matrix of the means and $\mathbf{S}$ be the estimated covariance matrix. Then $T^2 = \overline{\mathbf{x}}\mathbf{S}^{-1}\overline{\mathbf{x}}'$.

In the case of two groups, the test of equality is $T^2 = (\overline{\mathbf{x}}_1 - \overline{\mathbf{x}}_2)\mathbf{S}^{-1}(\overline{\mathbf{x}}_1 - \overline{\mathbf{x}}_2)'$.

Harold Hotelling (1895–1973) was an American economist and statistician who made many important contributions to mathematical economics, multivariate analysis, and statistical inference. After obtaining degrees in journalism and mathematics, he taught and researched at Stanford, Columbia, and the University of North Carolina. His work generalizing Student's t ratio and on principal components, canonical correlation, multivariate ANOVA, and correlation continues to be widely used.

Prasanta Chandra Mahalanobis (1893–1972) studied physics and mathematics at Calcutta and Cambridge. He became interested in statistics and on his return to India worked on applications in anthropology, meteorology, hydrology, and agriculture. Mahalanobis became the leader in Indian statistics, specializing in multivariate problems (including what is now called the Mahalanobis distance), the design of large-scale sample surveys, and the contribution of statistics to national planning.

References

Hotelling, H. 1931. The generalization of Student's ratio. *Annals of Mathematical Statistics* 2: 360–378.

Mahalanobis, P. C. 1930. On tests and measures of group divergence. *Journal Asiatic Society of Bengal* 26: 541–588.

——. 1936. On the generalized distance in statistics. *Proceedings of the National Institute of Science of India* 2: 49–55.

Olkin, I. and A. R. Sampson. 2001. Harold Hotelling. In *Statisticians of the Centuries*, ed. C. C. Heyde and E. Seneta, 454–458. New York: Springer.

Rao, C. R. 1973. Prasanta Chandra Mahalanobis, 1893–1972. *Biographical Memoirs of Fellows of The Royal Society* 19: 455–492.

Wilks, S. S. 1962. *Mathematical Statistics*. New York: Wiley.

Also See

Related: [MV] **manova**,

[R] **regress**, [R] **ttest**

Title

> **manova** — Multivariate analysis of variance and covariance

Syntax

$$\underline{\text{mano}}\text{va } \textit{depvarlist} = \textit{term} \left[\left[/ \right] \left[\textit{term} \left[/ \right] \ldots \right] \right] \left[\textit{if} \right] \left[\textit{in} \right] \left[\textit{weight} \right] \left[, \textit{options} \right]$$

where *term* has the form $\textit{varname} \left[\left\{ * \mid | \right\} \textit{varname} \left[\ldots \right] \right]$

options	description
Model	
<u>cat</u>egory(*varlist*)	names of variables in the *terms* that are categorical or class
<u>cl</u>ass(*varlist*)	synonym for **category**(*varlist*)
<u>cont</u>inuous(*varlist*)	names of variables in the *terms* that are continuous
<u>nocons</u>tant	suppress constant term
Reporting	
<u>det</u>ail	report categorical variable value mappings

bootstrap, by, jackknife, and statsby may be used with manova; see [U] **11.1.10 Prefix commands**.
aweights and fweights are allowed; see [U] **11.1.6 weight**.
See [MV] **manova postestimation** for features available after estimation.

Description

The manova command fits multivariate analysis-of-variance (MANOVA) and multivariate analysis-of-covariance (MANCOVA) models for balanced and unbalanced designs, including designs with missing cells, and for factorial, nested, or mixed designs, or designs involving repeated measures.

See [R] **anova** for univariate ANOVA and ANCOVA models.

Options

⌐ Model ⌐

category(*varlist*) indicates the names of the variables in the *terms* that are categorical or class variables. Stata ordinarily assumes that all variables are categorical variables, so, in most cases, this option need not be specified. If you specify this option, however, the variables referenced in the *terms* that are not listed in the category() option are assumed to be continuous. Also see the class() and continuous() options.

class(*varlist*) is a synonym for category(*varlist*).

continuous(*varlist*) indicates the names of the variables in the *terms* that are continuous. Stata ordinarily assumes that all variables are categorical variables. Also see the category() and class() options.

noconstant suppresses the constant term (intercept) from the model.

detail presents a table showing the actual values of the categorical variables along with their mapping into level numbers. You may specify this option at estimation or upon replay, e.g., manova, detail.

Remarks

Remarks are presented under the headings

> *One-way multivariate analysis of variance*
> *Two-way multivariate analysis of variance*
> *N-way multivariate analysis of variance*
> *Multivariate analysis of variance for Latin-square designs*
> *Multivariate analysis of variance for nested designs*
> *Multivariate analysis of variance for mixed designs*
> *Multivariate analysis of variance with repeated measures*
> *Multivariate analysis of covariance*

MANOVA is a generalization of ANOVA allowing multiple dependent variables. Several books discuss MANOVA, including Anderson (1984); Mardia, Kent, and Bibby (1979); Morrison (2005); Rencher (1998); Rencher (2002); Seber (1984); and Timm (1975). Introductory articles are provided by Pillai (1985) and Morrison (1998). Pioneering work is found in Wilks (1932), Pillai (1955), Lawley (1938), Hotelling (1951), and Roy (1939).

Four multivariate statistics are commonly computed in MANOVA: Wilks' lambda, Pillai's trace, Lawley–Hotelling trace, and Roy's largest root. See *Methods and Formulas* for details.

Why four statistics? Arnold (1981), Rencher (1998), Rencher (2002), Morrison (1998), Pillai (1985), and Seber (1984) provide guidance. All four tests are admissible, unbiased, and invariant. Asymptotically, Wilks' lambda, Pillai's trace, and the Lawley–Hotelling trace are the same, but their behavior under various violations of the null hypothesis and with small samples is different. Roy's largest root is different from the other three, even asymptotically.

None of the four multivariate criteria appears to be most powerful against all alternative hypotheses. For instance, Roy's largest root is most powerful when the null hypothesis of equal mean vectors is violated in such a way that the mean vectors tend to lie in a single line within p-dimensional space. For most other situations, Roy's largest root performs worse than the other three statistics. Pillai's trace tends to be more robust to non-normality and heteroskedasticity than the other three statistics.

The $*$ in the definition of a *term* indicates interaction. The $|$ indicates nesting (a$|$b is said: a is nested within b). A / between *terms* indicates that the *term* to the right of the slash is the error *term* for the *terms* to the left of the slash.

One-way multivariate analysis of variance

A one-way MANOVA is obtained by specifying the dependent variables followed by an equal sign, followed by the categorical variable defining the groups.

▷ Example 1: One-way MANOVA with balanced data

Rencher (2002) presents an example of a balanced one-way MANOVA using data from Andrews and Herzberg (1985, 357–360). The data from eight trees from each of six apple tree rootstocks are from table 6.2 of Rencher (2002). Four dependent variables are recorded for each tree: trunk girth at 4 years (mm $\times$ 100), extension growth at 4 years (m), trunk girth at 15 years (mm $\times$ 100), and weight of tree above ground at 15 years (lb $\times$ 1000). The grouping variable is rootstock, and the four dependent variables are y1, y2, y3, and y4, respectively.

```
. use http://www.stata-press.com/data/r9/rootstock
(Table 6.2 Rootstock Data -- Rencher (2002))

. describe

Contains data from http://www.stata-press.com/data/r9/rootstock.dta
  obs:            48                          Table 6.2 Rootstock Data --
                                                Rencher (2002)
  vars:            5                          20 Apr 2005 20:03
  size:        1,008 (99.9% of memory free)   (_dta has notes)
```

variable name	storage type	display format	value label	variable label
rootstock	byte	%9.0g		
y1	float	%4.2f		trunk girth at 4 years (mm x 100)
y2	float	%5.3f		extension growth at 4 years (m)
y3	float	%4.2f		trunk girth at 15 years (mm x 100)
y4	float	%5.3f		weight of tree above ground at 15 years (lb x 1000)

```
Sorted by:

. list in 7/10
```

	rootst~k	y1	y2	y3	y4
7.	1	1.11	3.211	3.98	1.209
8.	1	1.16	3.037	3.62	0.750
9.	2	1.05	2.074	4.09	1.036
10.	2	1.17	2.885	4.06	1.094

There are six rootstocks and four dependent variables. We test to see if the four-dimensional mean vectors of the six rootstocks are different. The null hypothesis is that the mean vectors are the same for the six rootstocks. To obtain one-way multivariate analysis-of-variance results, we type

```
. manova y1 y2 y3 y4 = rootstock
```

```
                         Number of obs =      48

                         W = Wilks' lambda      L = Lawley-Hotelling trace
                         P = Pillai's trace     R = Roy's largest root
```

Source	Statistic		df	F(df1,	df2) =	F	Prob>F	
rootstock	W	0.1540	5	20.0	130.3	4.94	0.0000	a
	P	1.3055		20.0	168.0	4.07	0.0000	a
	L	2.9214		20.0	150.0	5.48	0.0000	a
	R	1.8757		5.0	42.0	15.76	0.0000	u
Residual			42					
Total			47					

```
             e = exact, a = approximate, u = upper bound on F
```

All four multivariate tests reject the null hypothesis, indicating some kind of difference between the four-dimensional mean vectors of the six rootstocks.

Let's examine the output of manova. Above the table, it lists the number of observations used in the estimation. It also gives a key indicating that W stands for Wilks' lambda, P stands for Pillai's trace, L stands for Lawley–Hotelling trace, and R indicates Roy's largest root.

The first column of the table gives the source. In this case, we are testing the `rootstock` term (the only term in the model), and we are using residual error for the denominator of the test. Four lines of output are presented for `rootstock`, one line for each of the four multivariate tests, as indicated by the W, P, L, and R in the second column of the table.

The next column gives the multivariate statistics. In this case, Wilks' lambda is 0.1540, Pillai's trace is 1.3055, the Lawley–Hotelling trace is 2.9214, and Roy's largest root is 1.8757. Some authors report λ_1, and others (including Rencher) report $\theta = \lambda_1/(1+\lambda_1)$ for Roy's largest root. Stata reports λ_1.

The column labeled "df" gives the hypothesis degrees of freedom, the residual degrees of freedom, and the total degrees of freedom. These are just as they would be for an ANOVA. Since there are six rootstocks, we have 5 degrees of freedom for the hypothesis. There are 42 residual degrees of freedom and 47 total degrees of freedom.

The next three columns are labeled "F(df1, df2) = F ", and for each of the four multivariate tests, the degrees of freedom and F statistic are listed. The following column gives the associated p-values for the F statistics. Wilks' lambda has an F statistic of 4.94 with 20 and 130.3 degrees of freedom, which produces a p-value small enough that 0.0000 is reported. The F statistics and p-values for the other three multivariate tests follow on the three lines after Wilks' lambda.

The final column indicates if the F statistic is exact, approximate, or an upper bound. The letters e, a, and u indicate these three possibilities, as described in the footer at the bottom of the table. For this example, The F statistics (and corresponding p-values) for Wilks' lambda, Pillai's trace, and the Lawley–Hotelling trace are approximate. The F statistic for Roy's largest root is an upper bound, which means that the p-value is a lower bound.

It is easy to examine some of the underlying matrices and values used in the calculation of the four multivariate statistics. For example, you can list the sum of squares and cross products (SSCP) matrices for error and the hypothesis that are found in the `e(E)` and `e(H_m)` returned matrices, the eigenvalues of $\mathbf{E}^{-1}\mathbf{H}$ obtained from the `e(eigvals_m)` returned matrix, and the three auxiliary values (s, m, and n) that are returned in the matrix `e(aux_m)`.

```
. mat list e(E)

symmetric e(E)[4,4]
            y1          y2          y3          y4
y1   .31998754
y2   1.6965639    12.14279
y3   .55408744   4.3636123   4.2908128
y4   .21713994   2.1102135   2.4816563   1.7225248

. mat list e(H_m)

symmetric e(H_m)[4,4]
            y1          y2          y3          y4
y1   .07356042
y2   .53738525   4.1996621
y3   .33226448   2.3553887   6.1139358
y4   .20846994   1.6371084   3.7810439   2.4930912

. mat list e(eigvals_m)

e(eigvals_m)[1,4]
            c1          c2          c3          c4
r1   1.8756709   .79069412   .22904906   .02595358

. mat list e(aux_m)

e(aux_m)[3,1]
        value
s           4
m           0
n        18.5
```

The values s, m, and n are helpful when you do not want to rely on the approximate F tests, but instead want to look up critical values for the multivariate tests. Tables of critical values can be found in many multivariate texts, including Rencher (1998 and 2002).

See [MV] **manova postestimation** example 1 for an illustration of using `test` for Wald tests on expressions involving the underlying coefficients of the model and `lincom` for displaying linear combinations along with standard errors and confidence intervals from this MANOVA example.

◁

▷ Example 2: One-way MANOVA with unbalanced data

Table 4.5 of Rencher (1998) presents data reported by Allison, Zappasodi, and Lurie (1962). The dependent variables y1, recording the number of bacilli inhaled per tubercle formed, and y2, recording tubercle size (in millimeters), were measured for four groups of rabbits. Group one (unvaccinated control) and group two (infected during metabolic depression) have seven observations each, while group three (infected during heightened metabolic activity) has five observations, and group four (infected during normal activity) has only two observations.

```
. use http://www.stata-press.com/data/r9/metabolic, clear
(Table 4.5 Metabolic Comparisons of Rabbits -- Rencher (1998))
. list
```

	group	y1	y2
1.	1	24	3.5
2.	1	13.3	3.5
3.	1	12.2	4
4.	1	14	4
5.	1	22.2	3.6
6.	1	16.1	4.3
7.	1	27.9	5.2
8.	2	7.4	3.5
9.	2	13.2	3
10.	2	8.5	3
11.	2	10.1	3
12.	2	9.3	2
13.	2	8.5	2.5
14.	2	4.3	1.5
15.	3	16.4	3.2
16.	3	24	2.5
17.	3	53	1.5
18.	3	32.7	2.6
19.	3	42.8	2
20.	4	25.1	2.7
21.	4	5.9	2.3

The one-way MANOVA for testing the null hypothesis that the two-dimensional mean vectors for the four groups of rabbits are equal is

```
. manova y1 y2 = group
```

		Number of obs =	21				
		W = Wilks' lambda		L = Lawley-Hotelling trace			
		P = Pillai's trace		R = Roy's largest root			
Source		Statistic	df	F(df1,	df2) =	F	Prob>F
group	W	0.1596	3	6.0	32.0	8.02	0.0000 e
	P	1.2004		6.0	34.0	8.51	0.0000 a
	L	3.0096		6.0	30.0	7.52	0.0001 a
	R	1.5986		3.0	17.0	9.06	0.0008 u
Residual			17				
Total			20				

e = exact, a = approximate, u = upper bound on F

All four multivariate tests indicate rejection of the null hypothesis. This indicates that there are one or more differences among the two-dimensional mean vectors for the four groups. For this example, the F test for Wilks' lambda is exact because there are only two dependent variables in the model.

`manovatest` tests terms or linear combinations of the model's underlying design matrix. Example 2 of [MV] **manova postestimation** continues this example and illustrates `manovatest`.

◁

Two-way multivariate analysis of variance

You can include multiple explanatory variables with the `manova` command, and you can specify interactions by placing '*' between the variable names.

▷ Example 3: Two-way MANOVA with unbalanced data

Table 4.6 of Rencher (1998) presents unbalanced data from Woodard (1931) for a two-way MANOVA with three dependent variables (y1, y2, and y3) measured on patients with fractures of the jaw. y1 is age of patient, y2 is blood lymphocytes, and y3 is blood polymorphonuclears. The two design factors are gender (1 = male, 2 = female) and fracture (indicating the type of fracture: 1 = one compound fracture, 2 = two compound fractures, and 3 = one simple fracture). gender and fracture are numeric variables with value labels.

```
. use http://www.stata-press.com/data/r9/jaw, clear
(Table 4.6 Two-Way Unbalanced Data for Fractures of the Jaw -- Rencher (1998))
. describe
Contains data from http://www.stata-press.com/data/r9/jaw.dta
  obs:           27                          Table 4.6 Two-Way Unbalanced
                                             Data for Fractures of the Jaw
                                             -- Rencher (1998)
  vars:           5                          20 Apr 2005 14:53
  size:         243 (99.9% of memory free)   (_dta has notes)
```

variable name	storage type	display format	value label	variable label
gender	byte	%9.0g	gender	
fracture	byte	%22.0g	fractype	
y1	byte	%9.0g		age
y2	byte	%9.0g		blood lymphocytes
y3	byte	%9.0g		blood polymorphonuclears

```
Sorted by:
```

```
. list in 19/22
```

	gender	fracture	y1	y2	y3
19.	male	one simple fracture	55	32	60
20.	male	one simple fracture	30	34	62
21.	female	one compound fracture	22	56	43
22.	female	two compound fractures	22	29	68

The two-way factorial MANOVA for these data is

```
. manova y1 y2 y3 = gender fracture gender*fracture
```

```
                              Number of obs =      27

                    W = Wilks' lambda      L = Lawley-Hotelling trace
                    P = Pillai's trace     R = Roy's largest root
```

Source	Statistic		df	F(df1,	df2) =	F	Prob>F	
Model	W	0.2419	5	15.0	52.9	2.37	0.0109	a
	P	1.1018		15.0	63.0	2.44	0.0072	a
	L	1.8853		15.0	53.0	2.22	0.0170	a
	R	0.9248		5.0	21.0	3.88	0.0119	u
Residual			21					
gender	W	0.7151	1	3.0	19.0	2.52	0.0885	e
	P	0.2849		3.0	19.0	2.52	0.0885	e
	L	0.3983		3.0	19.0	2.52	0.0885	e
	R	0.3983		3.0	19.0	2.52	0.0885	e
fracture	W	0.4492	2	6.0	38.0	3.12	0.0139	e
	P	0.6406		6.0	40.0	3.14	0.0128	a
	L	1.0260		6.0	36.0	3.08	0.0155	a
	R	0.7642		3.0	20.0	5.09	0.0088	u
gender*fracture	W	0.5126	2	6.0	38.0	2.51	0.0380	e
	P	0.5245		6.0	40.0	2.37	0.0472	a
	L	0.8784		6.0	36.0	2.64	0.0319	a
	R	0.7864		3.0	20.0	5.24	0.0078	u
Residual			21					
Total			26					

```
            e = exact, a = approximate, u = upper bound on F
```

For MANOVA models with more than one term, the output of manova shows test results for the overall model, followed by results for each term in the MANOVA.

The interaction term, gender*fracture, is significant at the .05 level. Wilks' lambda for the interaction has an exact F that produces a p-value of 0.0380.

Example 3 of [MV] **manova postestimation** illustrates how the adjust postestimation command can be used to examine details of this significant interaction. It also illustrates how to obtain residuals using predict.

N-way multivariate analysis of variance

Higher-order MANOVA models are easily constructed using * to indicate the interaction terms.

▷ Example 4: MANOVA with interaction terms

Data on the wear of coated fabrics is provided by Box (1950) and is presented in table 6.20 of Rencher (2002). Variables y1, y2, and y3 are the wear after successive 1000 revolutions of an abrasive wheel. Three factors are also recorded. treatment is the surface treatment and has two levels. filler is the filler type, also with two levels. proportion is the proportion of filler and has three levels (25%, 50%, and 75%).

```
. use http://www.stata-press.com/data/r9/fabric, clear
(Table 6.20 Wear of coated fabrics -- Rencher (2002))
. describe

Contains data from http://www.stata-press.com/data/r9/fabric.dta
  obs:            24                          Table 6.20 Wear of coated
                                                fabrics -- Rencher (2002)
  vars:            6                          21 Apr 2005 02:01
  size:          312 (99.9% of memory free)  (_dta has notes)
```

variable name	storage type	display format	value label	variable label
treatment	byte	%9.0g		Surface treatment
filler	byte	%9.0g		Filler type
proportion	byte	%9.0g	prop	Proportion of filler
y1	int	%9.0g		First 1000 revolutions
y2	int	%9.0g		Second 1000 revolutions
y3	int	%9.0g		Third 1000 revolutions

```
Sorted by:

. label list prop
prop:
           1 25%
           2 50%
           3 75%
```

(Continued on next page)

. list

	treatm~t	filler	propor~n	y1	y2	y3
1.	0	1	25%	194	192	141
2.	0	1	50%	233	217	171
3.	0	1	75%	265	252	207
4.	0	1	25%	208	188	165
5.	0	1	50%	241	222	201
6.	0	1	75%	269	283	191
7.	0	2	25%	239	127	90
8.	0	2	50%	224	123	79
9.	0	2	75%	243	117	100
10.	0	2	25%	187	105	85
11.	0	2	50%	243	123	110
12.	0	2	75%	226	125	75
13.	1	1	25%	155	169	151
14.	1	1	50%	198	187	176
15.	1	1	75%	235	225	166
16.	1	1	25%	173	152	141
17.	1	1	50%	177	196	167
18.	1	1	75%	229	270	183
19.	1	2	25%	137	82	77
20.	1	2	50%	129	94	78
21.	1	2	75%	155	76	92
22.	1	2	25%	160	82	83
23.	1	2	50%	98	89	48
24.	1	2	75%	132	105	67

proportion is a numeric variable taking on values 1, 2, and 3 and is value-labeled with labels 25%, 50%, and 75%. treatment takes on values of 0 and 1, while filler is either 1 or 2.

First, we examine these data, ignoring the repeated measures aspects of y1, y2, and y3. In example 10, we will take it into account.

(Continued on next page)

```
. manova y1 y2 y3 = p t p*t f p*f t*f p*t*f, detail
Factor        Value            Value           Value          Value

proportion    1 1              2 2             3 3
treatment     1 0              2 1
filler        1 1              2 2
```

```
                         Number of obs =      24

                         W = Wilks' lambda      L = Lawley-Hotelling trace
                         P = Pillai's trace     R = Roy's largest root
           Source  |  Statistic    df    F(df1,    df2) =    F    Prob>F
```

Source		Statistic	df	F(df1,	df2) =	F	Prob>F	
Model	W	0.0007	11	33.0	30.2	10.10	0.0000	a
	P	2.3030		33.0	36.0	3.60	0.0001	a
	L	74.4794		33.0	26.0	19.56	0.0000	a
	R	59.1959		11.0	12.0	64.58	0.0000	u
Residual			12					
proportion	W	0.1375	2	6.0	20.0	5.65	0.0014	e
	P	0.9766		6.0	22.0	3.50	0.0139	a
	L	5.4405		6.0	18.0	8.16	0.0002	a
	R	5.2834		3.0	11.0	19.37	0.0001	u
treatment	W	0.0800	1	3.0	10.0	38.34	0.0000	e
	P	0.9200		3.0	10.0	38.34	0.0000	e
	L	11.5032		3.0	10.0	38.34	0.0000	e
	R	11.5032		3.0	10.0	38.34	0.0000	e
proportion*treatment	W	0.7115	2	6.0	20.0	0.62	0.7134	e
	P	0.2951		6.0	22.0	0.63	0.7013	a
	L	0.3962		6.0	18.0	0.59	0.7310	a
	R	0.3712		3.0	11.0	1.36	0.3055	u
filler	W	0.0192	1	3.0	10.0	170.60	0.0000	e
	P	0.9808		3.0	10.0	170.60	0.0000	e
	L	51.1803		3.0	10.0	170.60	0.0000	e
	R	51.1803		3.0	10.0	170.60	0.0000	e
proportion*filler	W	0.1785	2	6.0	20.0	4.56	0.0046	e
	P	0.9583		6.0	22.0	3.37	0.0164	a
	L	3.8350		6.0	18.0	5.75	0.0017	a
	R	3.6235		3.0	11.0	13.29	0.0006	u
treatment*filler	W	0.3552	1	3.0	10.0	6.05	0.0128	e
	P	0.6448		3.0	10.0	6.05	0.0128	e
	L	1.8150		3.0	10.0	6.05	0.0128	e
	R	1.8150		3.0	10.0	6.05	0.0128	e
proportion*treatment*	W	0.7518	2	6.0	20.0	0.51	0.7928	e
filler	P	0.2640		6.0	22.0	0.56	0.7589	a
	L	0.3092		6.0	18.0	0.46	0.8260	a
	R	0.2080		3.0	11.0	0.76	0.5381	u
Residual			12					
Total			23					

```
        e = exact, a = approximate, u = upper bound on F
```

The detail option of manova presents a table showing the mapping between the internal values used by Stata and the actual values of each factor. Here we see that a treatment value of 0 is represented internally as level 1 of treatment and that value 1 is level 2 of treatment. This is probably only important to you if you are specifying particular coefficients in commands such as lincom (see [R] **lincom**) or test after manova (see [MV] **manova postestimation**). When in doubt, specify detail with the manova command or upon replay. You can type manova to replay the current MANOVA. Typing manova, detail adds the detail table to the top of the replay.

The MANOVA table indicates that all the terms are significant, except for proportion*treatment and proportion*treatment*filler.

◁

Multivariate analysis of variance for Latin-square designs

▷ Example 5: MANOVA with Latin-square data

Exercise 5.11 from Timm (1975) presents data from a multivariate Latin-square design. Two dependent variables are measured in a 4×4 Latin square. W is the student's score on determining distances within the solar system. B is the student's score on determining distances beyond the solar system. The three variables comprising the square are machine, ability, and treatment, each at four levels.

```
. use http://www.stata-press.com/data/r9/solardistance, clear
(Multivariate Latin Square, Timm (1975), Exercise 5.11 #1)

. describe

Contains data from http://www.stata-press.com/data/r9/solardistance.dta
  obs:            16                       Multivariate Latin Square, Timm
                                             (1975), Exercise 5.11 #1
  vars:            5                       23 Apr 2005 03:27
  size:          144 (99.9% of memory free)  (_dta has notes)

              storage  display   value
variable name   type   format    label      variable label

machine         byte   %9.0g                teaching machine
ability         byte   %9.0g                ability tracks
treatment       byte   %9.0g                method of measuring
                                              astronomical distances
W               byte   %9.0g                Solar system distances (within)
B               byte   %9.0g                Solar system distances (beyond)

Sorted by:
```

(Continued on next page)

```
. list
```

	machine	ability	treatm~t	W	B
1.	1	1	2	33	15
2.	1	2	1	40	4
3.	1	3	3	31	16
4.	1	4	4	37	10
5.	2	1	4	25	20
6.	2	2	3	30	18
7.	2	3	1	22	6
8.	2	4	2	25	18
9.	3	1	1	10	5
10.	3	2	4	20	16
11.	3	3	2	17	16
12.	3	4	3	12	4
13.	4	1	3	24	15
14.	4	2	2	20	13
15.	4	3	4	19	14
16.	4	4	1	29	20

```
. manova W B = machine ability treatment
```

Number of obs = 16

W = Wilks' lambda   L = Lawley-Hotelling trace
P = Pillai's trace   R = Roy's largest root

Source	Statistic		df	F(df1,	df2) =	F	Prob>F	
Model	W	0.0378	9	18.0	10.0	2.30	0.0898	e
	P	1.3658		18.0	12.0	1.44	0.2645	a
	L	14.7756		18.0	8.0	3.28	0.0455	a
	R	14.0137		9.0	6.0	9.34	0.0066	u
Residual			6					
machine	W	0.0561	3	6.0	10.0	5.37	0.0101	e
	P	1.1853		6.0	12.0	2.91	0.0545	a
	L	12.5352		6.0	8.0	8.36	0.0043	a
	R	12.1818		3.0	6.0	24.36	0.0009	u
ability	W	0.4657	3	6.0	10.0	0.78	0.6070	e
	P	0.5368		6.0	12.0	0.73	0.6322	a
	L	1.1416		6.0	8.0	0.76	0.6199	a
	R	1.1367		3.0	6.0	2.27	0.1802	u
treatment	W	0.4697	3	6.0	10.0	0.77	0.6137	e
	P	0.5444		6.0	12.0	0.75	0.6226	a
	L	1.0988		6.0	8.0	0.73	0.6378	a
	R	1.0706		3.0	6.0	2.14	0.1963	u
Residual			6					
Total			15					

e = exact, a = approximate, u = upper bound on F

We find that machine is a significant factor in the model, while ability and treatment are not.

◁

Multivariate analysis of variance for nested designs

Nested terms are specified using a vertical bar. A|B is read as A nested within B. A|B|C is read as A nested within B, which is nested within C. A|B*C is read as A is nested within the interaction of B and C. A*B|C is read as the interaction of A and B, which is nested within C.

Different error terms can be specified for different parts of the model. The forward slash is used to indicate that the next term in the model is the error term for what precedes it. For instance, `manova y1 y2 = A / B|A` indicates that the multivariate tests for A are to be tested using the SSCP matrix from B|A in the denominator. Error terms (terms following the slash) are generally not tested unless they are themselves followed by a slash. The residual error SSCP matrix is the default error-term matrix.

For example, consider T_1 / T_2 / T_3, where T_1, T_2, and T_3 may be arbitrarily complex terms. `manova` will report T_1 tested by T_2 and T_2 tested by T_3. If we add one more slash on the end to form T_1 / T_2 / T_3 /, `manova` will also report T_3 tested by the residual error.

▷ Example 6: MANOVA with nested data

A medical researcher comes to you for help in analyzing some data he has collected. Two skin-rash treatment protocols were tested at eight clinics (four clinics for each protocol). Three doctors were selected at random from each of the clinics to administer the particular protocol to four of their patients. Each patient was treated for four separate rash patches, and two response variables, `response` and `response2`, were measured. This Stata dataset is called `rash2.dta`. The same data minus the `response2` variable is known as `rash.dta` and is used as example 10 in [R] **anova**. The rash2 data are described below.

```
. use http://www.stata-press.com/data/r9/rash2, clear
(skin rash data)

. describe
Contains data from http://www.stata-press.com/data/r9/rash2.dta
  obs:           384                          skin rash data
 vars:             6                          22 Apr 2005 12:57
 size:         3,840 (99.6% of memory free)   (_dta has notes)

              storage  display    value
variable name   type   format     label      variable label

response        byte   %9.0g
response2       byte   %9.0g
treatment       byte   %9.0g                  2 treatment protocols
clinic          byte   %9.0g                  4 clinics per treatment
doctor          byte   %9.0g                  3 doctors per clinic
patient         byte   %9.0g                  4 patients per doctor

Sorted by:  treatment  clinic  doctor  patient
```

In this fully nested design, `treatment` is a fixed factor, while the remaining terms are random factors.

Abbreviation of variable names makes the `manova` model statement easier to read.

```
. set matsize 262
. manova response response2 = t / c|t / d|c|t / p|d|c|t /
```

		Number of obs =		384				

| | | W = Wilks' lambda | | L = Lawley-Hotelling trace | | | | |
| | | P = Pillai's trace | | R = Roy's largest root | | | | |

Source		Statistic	df	F(df1,	df2) =	F	Prob>F	
Model	W	0.3438	95	190.0	574.0	2.13	0.0000	e
	P	0.8053		190.0	576.0	2.04	0.0000	a
	L	1.4751		190.0	572.0	2.22	0.0000	a
	R	1.0696		95.0	288.0	3.24	0.0000	u
Residual			288					
treatment	W	0.1036	1	2.0	5.0	21.64	0.0035	e
	P	0.8964		2.0	5.0	21.64	0.0035	e
	L	8.6553		2.0	5.0	21.64	0.0035	e
	R	8.6553		2.0	5.0	21.64	0.0035	e
clinic|treatment			6					
clinic|treatment	W	0.4546	6	12.0	30.0	1.21	0.3226	e
	P	0.5763		12.0	32.0	1.08	0.4085	a
	L	1.1320		12.0	28.0	1.32	0.2619	a
	R	1.0685		6.0	16.0	2.85	0.0440	u
doctor|clinic|treatment			16					
doctor|clinic|treatment	W	0.6708	16	32.0	142.0	0.98	0.5049	e
	P	0.3589		32.0	144.0	0.98	0.4996	a
	L	0.4465		32.0	140.0	0.98	0.5106	a
	R	0.2979		16.0	72.0	1.34	0.1974	u
patient|doctor|clinic| treatment			72					
patient|doctor|clinic| treatment	W	0.4899	72	144.0	574.0	1.71	0.0000	e
	P	0.5856		144.0	576.0	1.66	0.0000	a
	L	0.8869		144.0	572.0	1.76	0.0000	a
	R	0.6498		72.0	288.0	2.60	0.0000	u
Residual			288					
Total			383					

e = exact, a = approximate, u = upper bound on F

As with the univariate ANOVA (see [R] **anova**), you conclude that the two treatment protocols are significantly different, that the clinic and doctor effects are not significant, and that the patient effect is significant.

See example 4 of [MV] **manova postestimation** for a continuation of this example. It illustrates how to test pooled terms against nonresidual error terms using the `manovatest` postestimation command. In that example, `clinic` and `doctor` are pooled with `patient` from the original fully specified MANOVA. Another way of pooling is to refit the model, discarding the higher level terms. Be careful in doing this to ensure that the remaining lower-level terms have a numbering scheme that will not mistakenly consider different subjects as being the same. The `rash2` dataset has `patient` numbered uniquely, so that we can simply type

```
. manova response response2 = t / p|t /
```

	Number of obs =	384

W = Wilks' lambda   L = Lawley-Hotelling trace
P = Pillai's trace   R = Roy's largest root

Source		Statistic	df	F(df1,	df2) =	F	Prob>F	
Model	W	0.3438	95	190.0	574.0	2.13	0.0000	e
	P	0.8053		190.0	576.0	2.04	0.0000	a
	L	1.4751		190.0	572.0	2.22	0.0000	a
	R	1.0696		95.0	288.0	3.24	0.0000	u
Residual			288					
treatment	W	0.6219	1	2.0	93.0	28.27	0.0000	e
	P	0.3781		2.0	93.0	28.27	0.0000	e
	L	0.6080		2.0	93.0	28.27	0.0000	e
	R	0.6080		2.0	93.0	28.27	0.0000	e
patient|treatment			94					
patient|treatment	W	0.4037	94	188.0	574.0	1.75	0.0000	e
	P	0.7052		188.0	576.0	1.67	0.0000	a
	L	1.2071		188.0	572.0	1.84	0.0000	a
	R	0.9109		94.0	288.0	2.79	0.0000	u
Residual			288					
Total			383					

e = exact, a = approximate, u = upper bound on F

and get the same results that we obtained using `manovatest` to get a pooled test after the full MANOVA; see [MV] **manova postestimation** example 4.

◁

❏ Technical Note

You may have noticed that we typed `set matsize 262` before performing the `manova` on the skin-rash data. Where did this number come from?

For an ANOVA, the design matrix would have one column for the overall mean, two columns for the levels of `treatment`, eight columns for the levels of `clinic` nested in `treatment`, 24 columns for the levels of `doctor` nested in `clinic`, and 96 columns for the levels of `patient` nested in `doctor`. This gives a total of 131 columns.

MANOVA uses the same design matrix as ANOVA. The need for a `matsize` of $262 = 131 \times 2$ comes when `manova` saves the full variance–covariance matrix and coefficient vector. These need a dimension equal to the dimension of the design matrix times the number of variables in the *depvarlist*.

With the fabric-wear data of example 4, we could have typed `set matsize 108` ($108 = 36 \times 3$) since there are 3 dependent variables and the design matrix has 36 columns. The 36 columns comprise a column for the overall mean, 3 columns for p, 2 columns for t, 6 columns for p*t, 2 columns for f, 6 columns for p*f, 4 columns for t*f, and 12 columns for p*t*f. Since the default maximum matrix size is larger than 108, we did not bother to set `matsize` to 108.

❏

Multivariate analysis of variance for mixed designs

▷ Example 7: Split-plot MANOVA

reading2.dta has data from an experiment involving two reading programs and three skill-enhancement techniques. Ten classes of first-grade students were randomly assigned so that five classes were taught with one reading program and another five classes were taught with the other. The 30 students in each class were divided into six groups with five students each. Within each class, the six groups were divided randomly so that each of the three skill-enhancement techniques was taught to two of the groups within each class. At the end of the school year, a reading assessment test was administered to all the students. Two scores were recorded. The first was a reading score (score), and the second was a comprehension score (comprehension).

Example 11 of [R] **anova** uses reading.dta to illustrate mixed designs for ANOVA. reading2.dta is the same as reading.dta, with the exception that the comprehension variable is added.

```
. use http://www.stata-press.com/data/r9/reading2, clear
(Reading experiment data)

. describe
Contains data from http://www.stata-press.com/data/r9/reading2.dta
  obs:           300                          Reading experiment data
  vars:            6                          22 Apr 2005 15:46
  size:        3,000 (99.7% of memory free)   (_dta has notes)

              storage  display    value
variable name  type   format     label      variable label

score          byte   %9.0g                  reading score
comprehension  byte   %9.0g                  comprehension score
program        byte   %9.0g                  reading program
class          byte   %9.0g                  class nested in program
skill          byte   %9.0g                  skill enhancement technique
group          byte   %9.0g                  group nested in class and skill

Sorted by:
```

In this split-plot MANOVA, the whole-plot treatment is the two reading programs, and the split-plot treatment is the three skill-enhancement techniques.

For this split-plot MANOVA, the error term for program is class nested within program. The error term for skill and the program by skill interaction is the class by skill interaction nested within program. Other terms are also involved in the model and can be seen below.

```
. set matsize 224
. manova score comp = pr / cl|pr sk pr*sk / cl*sk|pr / gr|cl*sk|pr /
                      Number of obs =      300
                      W = Wilks' lambda     L = Lawley-Hotelling trace
                      P = Pillai's trace    R = Roy's largest root
             Source | Statistic    df   F(df1,    df2) =    F   Prob>F
            --------+----------------------------------------------------
              Model | W  0.5234    59   118.0    478.0   1.55  0.0008  e
                    | P  0.5249          118.0    480.0   1.45  0.0039  a
                    | L  0.8181          118.0    476.0   1.65  0.0001  a
                    | R  0.6830           59.0    240.0   2.78  0.0000  u
            --------+----------------------------------------------------
           Residual |              240
            --------+----------------------------------------------------
            program | W  0.4543     1     2.0      7.0   4.20  0.0632  e
                    | P  0.5457            2.0      7.0   4.20  0.0632  e
                    | L  1.2010            2.0      7.0   4.20  0.0632  e
                    | R  1.2010            2.0      7.0   4.20  0.0632  e
            --------+----------------------------------------------------
      class|program |                8
            --------+----------------------------------------------------
              skill | W  0.6754     2     4.0     30.0   1.63  0.1935  e
                    | P  0.3317            4.0     32.0   1.59  0.2008  a
                    | L  0.4701            4.0     28.0   1.65  0.1908  a
                    | R  0.4466            2.0     16.0   3.57  0.0522  u
            --------+----------------------------------------------------
      program*skill | W  0.3955     2     4.0     30.0   4.43  0.0063  e
                    | P  0.6117            4.0     32.0   3.53  0.0171  a
                    | L  1.5100            4.0     28.0   5.29  0.0027  a
                    | R  1.4978            2.0     16.0  11.98  0.0007  u
            --------+----------------------------------------------------
class*skill|program |               16
            --------+----------------------------------------------------
class*skill|program | W  0.4010    16    32.0     58.0   1.05  0.4265  e
                    | P  0.7324           32.0     60.0   1.08  0.3860  a
                    | L  1.1609           32.0     56.0   1.02  0.4688  a
                    | R  0.6453           16.0     30.0   1.21  0.3160  u
            --------+----------------------------------------------------
 group|class*skill| |               30
            program |
            --------+----------------------------------------------------
 group|class*skill| | W  0.7713    30    60.0    478.0   1.10  0.2844  e
            program | P  0.2363           60.0    480.0   1.07  0.3405  a
                    | L  0.2867           60.0    476.0   1.14  0.2344  a
                    | R  0.2469           30.0    240.0   1.98  0.0028  u
            --------+----------------------------------------------------
           Residual |              240
            --------+----------------------------------------------------
              Total |              299
```

                e = exact, a = approximate, u = upper bound on F

The program*skill interaction is significant. Here is a look at the average reading score and the comprehension score for each combination of program and skill.

```
. table prog skill, c(mean score mean comp) row col f(%8.2f)
```

reading program	skill enhancement technique			
	1	2	3	Total
1	68.16	52.86	61.54	60.85
	64.80	54.54	58.74	59.36
2	50.70	56.54	52.10	53.11
	53.30	58.52	54.40	55.41
Total	59.43	54.70	56.82	56.98
	59.05	56.53	56.57	57.38

The first reading program combined with the first skill-enhancement technique provide the best average scores.

◁

Multivariate analysis of variance with repeated measures

One approach to analyzing repeated measures in an analysis-of-variance setting is to use correction factors for terms in an ANOVA that involve the repeated measures. These correction factors attempt to correct for the violated assumption of independence of observations; see [R] **anova**. In this approach, the data are in long form; see [D] **reshape**.

Another approach to repeated measures is to use MANOVA with the repeated measures appearing as dependent variables, followed by tests involving linear combinations of these repeated measures. This approach involves fewer assumptions than the repeated measures ANOVA approach.

The simplest possible repeated-measures design has no between-subjects factors and only one within-subjects factor (the repeated measures).

▷ Example 8: MANOVA with repeated measures data

Here are data on five subjects, each of whom took three tests.

```
. use http://www.stata-press.com/data/r9/nobetween, clear
. list
```

	subject	test1	test2	test3
1.	1	68	69	95
2.	2	50	74	69
3.	3	72	89	71
4.	4	61	64	61
5.	5	60	71	90

manova must be tricked into fitting a constant-only model. To do this, you generate a variable equal to one, use that variable as the single *term* in your manova, and then specify the noconstant option. From the resulting MANOVA, you then test the repeated measures using the ytransform() option of manovatest; see [MV] **manova postestimation** for syntax details.

```
. generate mycons = 1
. manova test1 test2 test3 = mycons, noconstant
                        Number of obs =      5
                        W = Wilks' lambda       L = Lawley-Hotelling trace
                        P = Pillai's trace      R = Roy's largest root
              Source |  Statistic    df   F(df1,    df2) =    F    Prob>F
              -------+-----------------------------------------------------
              mycons | W   0.0076      1     3.0     2.0    86.91  0.0114 e
                     | P   0.9924            3.0     2.0    86.91  0.0114 e
                     | L 130.3722            3.0     2.0    86.91  0.0114 e
                     | R 130.3722            3.0     2.0    86.91  0.0114 e
              -------+-----------------------------------------------------
            Residual |             4
              -------+-----------------------------------------------------
               Total |             5
              --------------------------------------------------------------
                        e = exact, a = approximate, u = upper bound on F
. mat c = (1,0,-1\0,1,-1)
. manovatest mycons, ytransform(c)
Transformations of the dependent variables
 (1)    test1 - test3
 (2)    test2 - test3
                        W = Wilks' lambda       L = Lawley-Hotelling trace
                        P = Pillai's trace      R = Roy's largest root
              Source |  Statistic    df   F(df1,    df2) =    F    Prob>F
              -------+-----------------------------------------------------
              mycons | W   0.2352      1     2.0     3.0     4.88  0.1141 e
                     | P   0.7648            2.0     3.0     4.88  0.1141 e
                     | L   3.2509            2.0     3.0     4.88  0.1141 e
                     | R   3.2509            2.0     3.0     4.88  0.1141 e
              -------+-----------------------------------------------------
            Residual |             4
              --------------------------------------------------------------
                        e = exact, a = approximate, u = upper bound on F
```

The test produced directly with manova is not interesting. It is testing the hypothesis that the three test score means are zero. The test produced by manovatest is of interest. Based on the contrasts in the matrix c, you produce a test that there is a difference between the test1, test2, and test3 scores. In this case, the test produces a p-value of 0.1141, and you fail to reject the null hypothesis of equality between the test scores.

You can compare this with the results obtained from a repeated-measures ANOVA:

```
. reshape long test, i(subject) j(testnum)
. anova test subject testnum, repeated(testnum)
```

which produced an uncorrected p-value of 0.1160 and corrected p-values of 0.1181, 0.1435, and 0.1665 using the Huynh–Feldt, Greenhouse–Geisser, and Box's conservative correction, respectively.

◁

▷ Example 9: Randomized block design with repeated measures

Milliken and Johnson (1984) demonstrate using manova to analyze repeated measures from a randomized block design used in studying the differences among varieties of sorghum. Table 31.1 of Milliken and Johnson (1984) provides the data. Four sorghum varieties were each planted in five blocks. A leaf-area index measurement was recorded for each of five weeks, starting two weeks after emergence.

The tests of interest include a test for equal variety marginal means, equal time marginal means, and a test for the interaction of variety and time. The MANOVA below does not directly provide these tests. manovatest after the manova gives the three tests of interest.

```
. use http://www.stata-press.com/data/r9/sorghum, clear
(Leaf area index on 4 sorghum varieties, Milliken & Johnson (1984))
. manova time1 time2 time3 time4 time5 = variety block
```

```
                         Number of obs =      20

                         W = Wilks' lambda     L = Lawley-Hotelling trace
                         P = Pillai's trace    R = Roy's largest root
```

Source	Statistic	df	F(df1,	df2) =	F	Prob>F	
Model	W 0.0001	7	35.0	36.1	9.50	0.0000	a
	P 3.3890		35.0	60.0	3.61	0.0000	a
	L 126.2712		35.0	32.0	23.09	0.0000	a
	R 109.7360		7.0	12.0	188.12	0.0000	u
Residual		12					
variety	W 0.0011	3	15.0	22.5	16.11	0.0000	a
	P 2.5031		15.0	30.0	10.08	0.0000	a
	L 48.3550		15.0	20.0	21.49	0.0000	a
	R 40.0068		5.0	10.0	80.01	0.0000	u
block	W 0.0047	4	20.0	27.5	5.55	0.0000	a
	P 1.7518		20.0	44.0	1.71	0.0681	a
	L 77.9162		20.0	26.0	25.32	0.0000	a
	R 76.4899		5.0	11.0	168.28	0.0000	u
Residual		12					
Total		19					

```
          e = exact, a = approximate, u = upper bound on F
```

Two matrices are needed for transformations of the time# variables. m1 is a row vector containing five ones. m2 provides contrasts for time#. The manovatest, showorder command lists the underlying ordering of columns for constructing two additional matrices used to obtain linear combinations from the design matrix. Matrix c1 provides contrasts on variety. Matrix c2 is used to collapse to the overall margin of the design matrix in order to obtain time marginal means.

```
. matrix m1 = J(1,5,1)
. matrix m2 = (1,-1,0,0,0 \ 1,0,-1,0,0 \ 1,0,0,-1,0 \ 1,0,0,0,-1)
. manovatest, showorder
Order of columns in the design matrix
      1: _cons
      2: (variety==1)
      3: (variety==2)
      4: (variety==3)
      5: (variety==4)
      6: (block==1)
      7: (block==2)
      8: (block==3)
      9: (block==4)
     10: (block==5)
. matrix c1 = (0,1,-1,0,0,0,0,0,0,0\0,1,0,-1,0,0,0,0,0,0\0,1,0,0,-1,0,0,0,0,0)
. matrix c2 = (1,.25,.25,.25,.25,.2,.2,.2,.2,.2)
```

The test for equal variety marginal means uses matrix m1 to obtain the sum of the time# variables, and matrix c1 to provide the contrasts on variety. The second test uses m2 to provide contrasts on time#, and matrix c2 to collapse to the appropriate margin for the test of time marginal means. The final test uses m2 for contrasts on time#, and c1 for contrasts on variety to test the variety by time interaction.

```
. manovatest, test(c1) ytransform(m1)

Transformation of the dependent variables
(1)     time1 + time2 + time3 + time4 + time5

Test constraints
(1)     variety[1] - variety[2] = 0
(2)     variety[1] - variety[3] = 0
(3)     variety[1] - variety[4] = 0
```

	W = Wilks' lambda		L = Lawley-Hotelling trace			
	P = Pillai's trace		R = Roy's largest root			
Source	Statistic	df	F(df1,	df2) =	F	Prob>F
manovatest	W 0.0435	3	3.0	12.0	88.05	0.0000 e
	P 0.9565		3.0	12.0	88.05	0.0000 e
	L 22.0133		3.0	12.0	88.05	0.0000 e
	R 22.0133		3.0	12.0	88.05	0.0000 e
Residual		12				

e = exact, a = approximate, u = upper bound on F

```
. manovatest, test(c2) ytransform(m2)

Transformations of the dependent variables
(1)     time1 - time2
(2)     time1 - time3
(3)     time1 - time4
(4)     time1 - time5

Test constraint
(1)     _cons + .25 variety[1] + .25 variety[2] + .25 variety[3] + .25
        variety[4] + .2 block[1] + .2 block[2] + .2 block[3] + .2 block[4] + .2
        block[5] = 0
```

	W = Wilks' lambda		L = Lawley-Hotelling trace			
	P = Pillai's trace		R = Roy's largest root			
Source	Statistic	df	F(df1,	df2) =	F	Prob>F
manovatest	W 0.0050	1	4.0	9.0	445.62	0.0000 e
	P 0.9950		4.0	9.0	445.62	0.0000 e
	L 198.0544		4.0	9.0	445.62	0.0000 e
	R 198.0544		4.0	9.0	445.62	0.0000 e
Residual		12				

e = exact, a = approximate, u = upper bound on F

```
. manovatest, test(c1) ytransform(m2)

Transformations of the dependent variables
(1)     time1 - time2
(2)     time1 - time3
(3)     time1 - time4
(4)     time1 - time5

Test constraints
(1)     variety[1] - variety[2] = 0
(2)     variety[1] - variety[3] = 0
(3)     variety[1] - variety[4] = 0
```

		W = Wilks' lambda		L = Lawley-Hotelling trace				
		P = Pillai's trace		R = Roy's largest root				
Source		Statistic	df	F(df1,	df2) =	F	Prob>F	
manovatest	W	0.0143	3	12.0	24.1	8.00	0.0000	a
	P	2.1463		12.0	33.0	6.91	0.0000	a
	L	12.1760		12.0	23.0	7.78	0.0000	a
	R	8.7953		4.0	11.0	24.19	0.0000	u
Residual			12					

```
e = exact, a = approximate, u = upper bound on F
```

All three tests are significant, indicating differences in variety, in time, and in the variety by time interaction.

◁

▷ Example 10: MANOVA and dependent variable effects

Recall the fabric-data example from Rencher (2002) that we used in example 4 to illustrate a 3-way MANOVA. Rencher has as an additional exercise to test the period effect (the y1, y2, and y3 repeated measures variables) and the interaction of period with the other factors in the model. The ytransform() option of manovatest provides a method to do this; see [MV] **manova postestimation**. Here are the tests of the period effect interacted with each term in the model. We create the matrix c with rows corresponding to the linear and quadratic contrasts for the three dependent variables.

```
. quietly manova y1 y2 y3 = p t p*t f p*f t*f p*t*f

. matrix c = (-1,0,1 \ -1,2,-1)

. manovatest p, ytransform(c)

Transformations of the dependent variables
(1)      - y1 + y3
(2)      - y1 + 2 y2 - y3
```

		W = Wilks' lambda		L = Lawley-Hotelling trace				
		P = Pillai's trace		R = Roy's largest root				
Source		Statistic	df	F(df1,	df2) =	F	Prob>F	
proportion	W	0.4749	2	4.0	22.0	2.48	0.0736	e
	P	0.5454		4.0	24.0	2.25	0.0936	a
	L	1.0631		4.0	20.0	2.66	0.0630	a
	R	1.0213		2.0	12.0	6.13	0.0147	u
Residual			12					

```
e = exact, a = approximate, u = upper bound on F
```

```
. manovatest t, ytransform(c)
```

Transformations of the dependent variables
```
(1)    - y1 + y3
(2)    - y1 + 2 y2 - y3
```

		W = Wilks' lambda		L = Lawley-Hotelling trace				
		P = Pillai's trace		R = Roy's largest root				
Source		Statistic	df	F(df1,	df2) =	F	Prob>F	
treatment	W	0.1419	1	2.0	11.0	33.27	0.0000	e
	P	0.8581		2.0	11.0	33.27	0.0000	e
	L	6.0487		2.0	11.0	33.27	0.0000	e
	R	6.0487		2.0	11.0	33.27	0.0000	e
Residual			12					

e = exact, a = approximate, u = upper bound on F

```
. manovatest p*t, ytransform(c)
```

Transformations of the dependent variables
```
(1)    - y1 + y3
(2)    - y1 + 2 y2 - y3
```

		W = Wilks' lambda		L = Lawley-Hotelling trace				
		P = Pillai's trace		R = Roy's largest root				
Source		Statistic	df	F(df1,	df2) =	F	Prob>F	
proportion*treatment	W	0.7766	2	4.0	22.0	0.74	0.5740	e
	P	0.2276		4.0	24.0	0.77	0.5550	a
	L	0.2824		4.0	20.0	0.71	0.5972	a
	R	0.2620		2.0	12.0	1.57	0.2476	u
Residual			12					

e = exact, a = approximate, u = upper bound on F

```
. manovatest f, ytransform(c)
```

Transformations of the dependent variables
```
(1)    - y1 + y3
(2)    - y1 + 2 y2 - y3
```

		W = Wilks' lambda		L = Lawley-Hotelling trace				
		P = Pillai's trace		R = Roy's largest root				
Source		Statistic	df	F(df1,	df2) =	F	Prob>F	
filler	W	0.0954	1	2.0	11.0	52.17	0.0000	e
	P	0.9046		2.0	11.0	52.17	0.0000	e
	L	9.4863		2.0	11.0	52.17	0.0000	e
	R	9.4863		2.0	11.0	52.17	0.0000	e
Residual			12					

e = exact, a = approximate, u = upper bound on F

(Continued on next page)

```
. manovatest p*f, ytransform(c)
```

Transformations of the dependent variables
(1)   - y1 + y3
(2)   - y1 + 2 y2 - y3

		W = Wilks' lambda P = Pillai's trace		L = Lawley-Hotelling trace R = Roy's largest root			
Source		Statistic	df	F(df1,	df2) =	F	Prob>F
proportion*filler	W	0.6217	2	4.0	22.0	1.48 0.2436	e
	P	0.3870		4.0	24.0	1.44 0.2515	a
	L	0.5944		4.0	20.0	1.49 0.2439	a
	R	0.5698		2.0	12.0	3.42 0.0668	u
Residual			12				

e = exact, a = approximate, u = upper bound on F

```
. manovatest t*f, ytransform(c)
```

Transformations of the dependent variables
(1)   - y1 + y3
(2)   - y1 + 2 y2 - y3

		W = Wilks' lambda P = Pillai's trace		L = Lawley-Hotelling trace R = Roy's largest root			
Source		Statistic	df	F(df1,	df2) =	F	Prob>F
treatment*filler	W	0.3867	1	2.0	11.0	8.72 0.0054	e
	P	0.6133		2.0	11.0	8.72 0.0054	e
	L	1.5857		2.0	11.0	8.72 0.0054	e
	R	1.5857		2.0	11.0	8.72 0.0054	e
Residual			12				

e = exact, a = approximate, u = upper bound on F

```
. manovatest p*t*f, ytransform(c)
```

Transformations of the dependent variables
(1)   - y1 + y3
(2)   - y1 + 2 y2 - y3

		W = Wilks' lambda P = Pillai's trace		L = Lawley-Hotelling trace R = Roy's largest root			
Source		Statistic	df	F(df1,	df2) =	F	Prob>F
proportion*treatment*	W	0.7812	2	4.0	22.0	0.72 0.5857	e
filler	P	0.2290		4.0	24.0	0.78 0.5518	a
	L	0.2671		4.0	20.0	0.67 0.6219	a
	R	0.2028		2.0	12.0	1.22 0.3303	u
Residual			12				

e = exact, a = approximate, u = upper bound on F

The first test, manovatest p, ytransform(c), provides the test of proportion interacted with the period effect. The F tests for Wilks' lambda, Pillai's trace, and the Lawley–Hotelling trace do not reject the null hypothesis using a significance level of .05 (p-values of 0.0736, 0.0936, and 0.0630). The F test for Roy's largest root is an upper bound, so the p-value of 0.0147 is a lower bound.

The tests of treatment interacted with the period effect, filler interacted with the period effect, and treatment*filler interacted with the period effect are significant. The remaining tests are not.

To test the period effect, we call manovatest with both the ytransform() and test() options. The showorder option guides us in constructing the matrix for the test() option.

```
. manovatest, showorder
Order of columns in the design matrix
      1: _cons
      2: (proportion==1)
      3: (proportion==2)
      4: (proportion==3)
      5: (treatment==0)
      6: (treatment==1)
      7: (filler==1)
      8: (filler==2)
      9: (proportion==1)*(treatment==0)
     10: (proportion==1)*(treatment==1)
     11: (proportion==2)*(treatment==0)
     12: (proportion==2)*(treatment==1)
     13: (proportion==3)*(treatment==0)
     14: (proportion==3)*(treatment==1)
     15: (proportion==1)*(filler==1)
     16: (proportion==1)*(filler==2)
     17: (proportion==2)*(filler==1)
     18: (proportion==2)*(filler==2)
     19: (proportion==3)*(filler==1)
     20: (proportion==3)*(filler==2)
     21: (treatment==0)*(filler==1)
     22: (treatment==0)*(filler==2)
     23: (treatment==1)*(filler==1)
     24: (treatment==1)*(filler==2)
     25: (proportion==1)*(treatment==0)*(filler==1)
     26: (proportion==1)*(treatment==0)*(filler==2)
     27: (proportion==1)*(treatment==1)*(filler==1)
     28: (proportion==1)*(treatment==1)*(filler==2)
     29: (proportion==2)*(treatment==0)*(filler==1)
     30: (proportion==2)*(treatment==0)*(filler==2)
     31: (proportion==2)*(treatment==1)*(filler==1)
     32: (proportion==2)*(treatment==1)*(filler==2)
     33: (proportion==3)*(treatment==0)*(filler==1)
     34: (proportion==3)*(treatment==0)*(filler==2)
     35: (proportion==3)*(treatment==1)*(filler==1)
     36: (proportion==3)*(treatment==1)*(filler==2)
```

We create a row vector, m, that has a 1 for the first column (corresponding to the constant in the model), followed by 1/3 for three columns (corresponding to proportion), followed by 1/2 for four columns (corresponding to the treatment and filler terms), followed by twelve columns of 1/6 (for the proportion*treatment and proportion*filler terms), followed by four columns of 1/4 (for treatment*filler), and, finally, followed by 1/12 for the last 12 columns (corresponding to the proportion*treatment*filler term). The test of period effect then uses this m matrix and the c matrix previously defined as the basis of the test for the period effect.

(Continued on next page)

```
. matrix m = (1), J(1,3,1/3), J(1,4,1/2), J(1,12,1/6), J(1,4,1/4), J(1,12,1/12)
. manovatest, test(m) ytrans(c)
```
Transformations of the dependent variables
```
(1)     - y1 + y3
(2)     - y1 + 2 y2 - y3
```
Test constraint
```
(1)     _cons + .3333333 proportion[1] + .3333333 proportion[2] + .3333333
        proportion[3] + .5 treatment[1] + .5 treatment[2] + .5 filler[1] + .5
        filler[2] + .1666667 proportion[1]*treatment[1] + .1666667
        proportion[1]*treatment[2] + .1666667 proportion[2]*treatment[1] +
        .1666667 proportion[2]*treatment[2] + .1666667
        proportion[3]*treatment[1] + .1666667 proportion[3]*treatment[2] +
        .1666667 proportion[1]*filler[1] + .1666667 proportion[1]*filler[2] +
        .1666667 proportion[2]*filler[1] + .1666667 proportion[2]*filler[2] +
        .1666667 proportion[3]*filler[1] + .1666667 proportion[3]*filler[2] +
        .25 treatment[1]*filler[1] + .25 treatment[1]*filler[2] + .25
        treatment[2]*filler[1] + .25 treatment[2]*filler[2] + .0833333
        proportion[1]*treatment[1]*filler[1] + .0833333
        proportion[1]*treatment[1]*filler[2] + .0833333
        proportion[1]*treatment[2]*filler[1] + .0833333
        proportion[1]*treatment[2]*filler[2] + .0833333
        proportion[2]*treatment[1]*filler[1] + .0833333
        proportion[2]*treatment[1]*filler[2] + .0833333
        proportion[2]*treatment[2]*filler[1] + .0833333
        proportion[2]*treatment[2]*filler[2] + .0833333
        proportion[3]*treatment[1]*filler[1] + .0833333
        proportion[3]*treatment[1]*filler[2] + .0833333
        proportion[3]*treatment[2]*filler[1] + .0833333
        proportion[3]*treatment[2]*filler[2] = 0
```

```
                          W = Wilks' lambda        L = Lawley-Hotelling trace
                          P = Pillai's trace       R = Roy's largest root

          Source  |  Statistic    df   F(df1,    df2) =    F    Prob>F
        ----------+------------------------------------------------------
        manovatest|  W    0.0208    1    2.0     11.0    259.04 0.0000 e
                  |  P    0.9792         2.0     11.0    259.04 0.0000 e
                  |  L   47.0988         2.0     11.0    259.04 0.0000 e
                  |  R   47.0988         2.0     11.0    259.04 0.0000 e
        ----------+------------------------------------------------------
          Residual|                12
```

```
            e = exact, a = approximate, u = upper bound on F
```

This agrees with the answers provided by Rencher (2002).

◁

In the previous three examples, one factor has been encoded within the dependent variables. We have seen that the `ytransform()` option of `manovatest` provides the method for testing this factor and its interactions with the factors that appear on the right-hand side of the MANOVA.

More than one factor could be encoded within the dependent variables. Again the `ytransform()` option of `manovatest` allows us to perform multivariate tests of interest:

▷ Example 11: MANOVA and multiple dependent variable effects

Table 6.14 of Rencher (2002) provides an example with two within-subjects factors represented in the dependent variables and one between-subjects factor.

```
. use http://www.stata-press.com/data/r9/table614, clear
(Table 6.14, Rencher (2002))
. list in 9/12, noobs compress
```

c	sub~t	ab11	ab12	ab13	ab21	ab22	ab23	ab31	ab32	ab33
1	9	41	32	23	37	51	39	27	28	30
1	10	39	32	24	30	35	31	26	29	32
2	1	47	36	25	31	36	29	21	24	27
2	2	53	43	32	40	48	47	46	50	54

There are a total of 20 observations. Factors a and b are encoded in the names of the nine dependent variables. variable name ab23, for instance, indicates factor a at level 2 and factor b at level 3. Factor c is the between-subjects factor.

We first compute a MANOVA using the dependent variables and our one between-subjects term.

```
. manova ab11 ab12 ab13 ab21 ab22 ab23 ab31 ab32 ab33 = c
```

```
                      Number of obs =      20
                      W = Wilks' lambda     L = Lawley-Hotelling trace
                      P = Pillai's trace    R = Roy's largest root
            Source |  Statistic    df   F(df1,   df2) =    F   Prob>F
          ---------+---------------------------------------------------
                c  | W  0.5330      1     9.0    10.0    0.97 0.5114 e
                   | P  0.4670            9.0    10.0    0.97 0.5114 e
                   | L  0.8762            9.0    10.0    0.97 0.5114 e
                   | R  0.8762            9.0    10.0    0.97 0.5114 e
          ---------+---------------------------------------------------
          Residual |                18
          ---------+---------------------------------------------------
             Total |                19

          e = exact, a = approximate, u = upper bound on F
```

This provides the basis for computing tests on all terms of interest. We use the ytransform() and test() options of manovatest using the following matrices to obtain the tests of interest.

```
. mat a = (2,2,2,-1,-1,-1,-1,-1,-1 \ 0,0,0,1,1,1,-1,-1,-1)
. mat b = (2,-1,-1,2,-1,-1,2,-1,-1 \ 0,1,-1,0,1,-1,0,1,-1)
. forvalues i = 1/2 {
  2.          forvalues j = 1/2 {
  3.                  mat g = nullmat(g) \ vecdiag(a['i',1...]'*b['j',1...])
  4.          }
  5. }
. mat list g
g[4,9]
     c1  c2  c3  c4  c5  c6  c7  c8  c9
r1    4  -2  -2  -2   1   1  -2   1   1
r1    0   2  -2   0  -1   1   0  -1   1
r1    0   0   0   2  -1  -1  -2   1   1
r1    0   0   0   0   1  -1   0  -1   1
. mat j = J(1,9,1/9)
. mat xall = (1,.5,.5)
```

Matrices a and b correspond to factors a and b. Matrix g is the elementwise multiplication of each row of a with each row of b and corresponds to the a*b interaction. Matrix j is used to average the dependent variables, while matrix xall collapses over factor c.

Here are the tests for a, b, and a*b.

```
. manovatest, test(xall) ytrans(a)
```

Transformations of the dependent variables
```
(1)    2 ab11 + 2 ab12 + 2 ab13 - ab21 - ab22 - ab23 - ab31 - ab32 - ab33
(2)    ab21 + ab22 + ab23 - ab31 - ab32 - ab33
```

Test constraint
```
(1)    _cons + .5 c[1] + .5 c[2] = 0
```

		W = Wilks' lambda		L = Lawley-Hotelling trace			
		P = Pillai's trace		R = Roy's largest root			
Source		Statistic	df	F(df1,	df2) =	F	Prob>F
manovatest	W	0.6755	1	2.0	17.0	4.08	0.0356 e
	P	0.3245		2.0	17.0	4.08	0.0356 e
	L	0.4803		2.0	17.0	4.08	0.0356 e
	R	0.4803		2.0	17.0	4.08	0.0356 e
Residual			18				

e = exact, a = approximate, u = upper bound on F

```
. manovatest, test(xall) ytrans(b)
```

Transformations of the dependent variables
```
(1)    2 ab11 - ab12 - ab13 + 2 ab21 - ab22 - ab23 + 2 ab31 - ab32 - ab33
(2)    ab12 - ab13 + ab22 - ab23 + ab32 - ab33
```

Test constraint
```
(1)    _cons + .5 c[1] + .5 c[2] = 0
```

		W = Wilks' lambda		L = Lawley-Hotelling trace			
		P = Pillai's trace		R = Roy's largest root			
Source		Statistic	df	F(df1,	df2) =	F	Prob>F
manovatest	W	0.3247	1	2.0	17.0	17.68	0.0001 e
	P	0.6753		2.0	17.0	17.68	0.0001 e
	L	2.0799		2.0	17.0	17.68	0.0001 e
	R	2.0799		2.0	17.0	17.68	0.0001 e
Residual			18				

e = exact, a = approximate, u = upper bound on F

```
. manovatest, test(xall) ytrans(g)
```

Transformations of the dependent variables
```
(1)    4 ab11 - 2 ab12 - 2 ab13 - 2 ab21 + ab22 + ab23 - 2 ab31 + ab32 + ab33
(2)    2 ab12 - 2 ab13 - ab22 + ab23 - ab32 + ab33
(3)    2 ab21 - ab22 - ab23 - 2 ab31 + ab32 + ab33
(4)    ab22 - ab23 - ab32 + ab33
```

Test constraint
```
(1)    _cons + .5 c[1] + .5 c[2] = 0
```

		W = Wilks' lambda		L = Lawley-Hotelling trace			
		P = Pillai's trace		R = Roy's largest root			
Source		Statistic	df	F(df1,	df2) =	F	Prob>F
manovatest	W	0.2255	1	4.0	15.0	12.88	0.0001 e
	P	0.7745		4.0	15.0	12.88	0.0001 e
	L	3.4347		4.0	15.0	12.88	0.0001 e
	R	3.4347		4.0	15.0	12.88	0.0001 e
Residual			18				

e = exact, a = approximate, u = upper bound on F

Factors a, b, and a*b are significant with p-values of 0.0356, 0.0001, and 0.0001, respectively. The multivariate statistics are equivalent to the T^2 values Rencher reports using the relationship $T^2 = (n_1 + n_2 - 2) \times (1 - \Lambda)/\Lambda$ that applies in this situation. For instance, Wilks' lambda for factor a is reported as 0.6755 (and the actual value recorded in r(stat) is 0.67554286) so that $T^2 = (10 + 10 - 2) \times (1 - 0.67554286)/0.67554286 = 8.645$, as reported by Rencher.

We now compute the tests for c and the interactions of c with the other terms in the model.

```
. manovatest c, ytrans(j)
```

Transformation of the dependent variables
(1)   .1111111 ab11 + .1111111 ab12 + .1111111 ab13 + .1111111 ab21 +
         .1111111 ab22 + .1111111 ab23 + .1111111 ab31 + .1111111 ab32 +
         .1111111 ab33

		W = Wilks' lambda P = Pillai's trace		L = Lawley-Hotelling trace R = Roy's largest root			
Source		Statistic	df	F(df1,	df2) =	F	Prob>F
c	W	0.6781	1	1.0	18.0	8.54	0.0091 e
	P	0.3219		1.0	18.0	8.54	0.0091 e
	L	0.4747		1.0	18.0	8.54	0.0091 e
	R	0.4747		1.0	18.0	8.54	0.0091 e
Residual			18				

e = exact, a = approximate, u = upper bound on F

```
. manovatest c, ytrans(a)
```

Transformations of the dependent variables
(1)   2 ab11 + 2 ab12 + 2 ab13 - ab21 - ab22 - ab23 - ab31 - ab32 - ab33
(2)   ab21 + ab22 + ab23 - ab31 - ab32 - ab33

		W = Wilks' lambda P = Pillai's trace		L = Lawley-Hotelling trace R = Roy's largest root			
Source		Statistic	df	F(df1,	df2) =	F	Prob>F
c	W	0.9889	1	2.0	17.0	0.10	0.9097 e
	P	0.0111		2.0	17.0	0.10	0.9097 e
	L	0.0112		2.0	17.0	0.10	0.9097 e
	R	0.0112		2.0	17.0	0.10	0.9097 e
Residual			18				

e = exact, a = approximate, u = upper bound on F

```
. manovatest c, ytrans(b)
```

Transformations of the dependent variables
(1)   2 ab11 - ab12 - ab13 + 2 ab21 - ab22 - ab23 + 2 ab31 - ab32 - ab33
(2)   ab12 - ab13 + ab22 - ab23 + ab32 - ab33

		W = Wilks' lambda P = Pillai's trace		L = Lawley-Hotelling trace R = Roy's largest root			
Source		Statistic	df	F(df1,	df2) =	F	Prob>F
c	W	0.9718	1	2.0	17.0	0.25	0.7845 e
	P	0.0282		2.0	17.0	0.25	0.7845 e
	L	0.0290		2.0	17.0	0.25	0.7845 e
	R	0.0290		2.0	17.0	0.25	0.7845 e
Residual			18				

e = exact, a = approximate, u = upper bound on F

```
. manovatest c, ytrans(g)
Transformations of the dependent variables
(1)    4 ab11 - 2 ab12 - 2 ab13 - 2 ab21 + ab22 + ab23 - 2 ab31 + ab32 + ab33
(2)    2 ab12 - 2 ab13 - ab22 + ab23 - ab32 + ab33
(3)    2 ab21 - ab22 - ab23 - 2 ab31 + ab32 + ab33
(4)    ab22 - ab23 - ab32 + ab33
```

		Statistic	df	F(df1,	df2) =	F	Prob>F	
		W = Wilks' lambda			L = Lawley-Hotelling trace			
		P = Pillai's trace			R = Roy's largest root			
Source								
c	W	0.9029	1	4.0	15.0	0.40	0.8035	e
	P	0.0971		4.0	15.0	0.40	0.8035	e
	L	0.1075		4.0	15.0	0.40	0.8035	e
	R	0.1075		4.0	15.0	0.40	0.8035	e
Residual			18					

```
                e = exact, a = approximate, u = upper bound on F
```

The test of c is equivalent to an ANOVA using the sum or average of the dependent variables as the dependent variable. The test of c produces an F of 8.54 with a p-value of 0.0091, which agrees with the results of Rencher (2002).

The tests of a*c, b*c, and a*b*c produce p-values of 0.9097, 0.7845, and 0.8035, respectively.

In summary, the factors that are significant are a, b, a*b, and c.

◁

Multivariate analysis of covariance

Multivariate analysis of covariance (MANCOVA) models are specified by including the covariates as *term*s in the manova and then also listing them in the continuous() option to indicate that they are to be treated as continuous instead of categorical variables.

▷ Example 12: MANCOVA

Table 4.9 of Rencher (1998) provides biochemical measurements on four weight groups. Rencher extracted the data from Brown and Beerstecher (1951) and Smith, Gnanadesikan, and Hughes (1962). Three dependent variables and two covariates are recorded for eight subjects within each of the four groups. The first two groups are underweight, and the last two groups are overweight. The dependent variables are modified creatinine coefficient (y1), pigment creatinine (y2), and phosphate in mg/ml (y3). The two covariates are volume in ml (x1) and specific gravity (x2).

```
. use http://www.stata-press.com/data/r9/biochemical
(Table 4.9, Rencher (1998))

. describe

Contains data from http://www.stata-press.com/data/r9/biochemical.dta
  obs:            32                          Table 4.9, Rencher (1998)
  vars:            6                          22 Apr 2005 21:48
  size:          640 (99.9% of memory free)  (_dta has notes)
```

variable name	storage type	display format	value label	variable label
group	byte	%9.0g		
y1	float	%9.0g		modified creatinine coefficient
y2	float	%9.0g		pigment creatinine
y3	float	%9.0g		phosphate (mg/ml)
x1	int	%9.0g		volume (ml)
x2	byte	%9.0g		specific gravity

```
Sorted by:
```

Rencher performs three tests on these data. The first is a test of equality of group effects adjusted for the covariates. The second is a test that the coefficients for the covariates are jointly equal to zero. The third test is that the coefficients for the covariates are equal across groups.

```
. manova y1 y2 y3 = group x1 x2, continuous(x1 x2)

                    Number of obs =        32

                    W = Wilks' lambda     L = Lawley-Hotelling trace
                    P = Pillai's trace    R = Roy's largest root
```

Source	Statistic		df	F(df1,	df2) =	F	Prob>F	
Model	W	0.0619	5	15.0	66.7	7.73	0.0000	a
	P	1.4836		15.0	78.0	5.09	0.0000	a
	L	6.7860		15.0	68.0	10.25	0.0000	a
	R	5.3042		5.0	26.0	27.58	0.0000	u
Residual			26					
group	W	0.1491	3	9.0	58.6	7.72	0.0000	a
	P	0.9041		9.0	78.0	3.74	0.0006	a
	L	5.3532		9.0	68.0	13.48	0.0000	a
	R	5.2872		3.0	26.0	45.82	0.0000	u
x1	W	0.6841	1	3.0	24.0	3.69	0.0257	e
	P	0.3159		3.0	24.0	3.69	0.0257	e
	L	0.4617		3.0	24.0	3.69	0.0257	e
	R	0.4617		3.0	24.0	3.69	0.0257	e
x2	W	0.9692	1	3.0	24.0	0.25	0.8576	e
	P	0.0308		3.0	24.0	0.25	0.8576	e
	L	0.0318		3.0	24.0	0.25	0.8576	e
	R	0.0318		3.0	24.0	0.25	0.8576	e
Residual			26					
Total			31					

```
                e = exact, a = approximate, u = upper bound on F
```

The test of equality of group effects adjusted for the covariates is shown in the MANCOVA table above. Rencher reports a Wilks' lambda value of 0.1491, which agrees with the value shown for the group term above. group is found to be significant.

The test that the coefficients for the covariates are jointly equal to zero is obtained by using manovatest.

```
. manovatest x1 x2
                    W = Wilks' lambda      L = Lawley-Hotelling trace
                    P = Pillai's trace     R = Roy's largest root
            Source |  Statistic    df   F(df1,    df2) =   F    Prob>F
      -------------+----------------------------------------------------
             x1 x2 | W    0.4470     2     6.0     48.0      3.97 0.0027 e
                   | P    0.5621           6.0     50.0      3.26 0.0088 a
                   | L    1.2166           6.0     46.0      4.66 0.0009 a
                   | R    1.1995           3.0     25.0     10.00 0.0002 u
      -------------+----------------------------------------------------
          Residual |                26
      --------------------------------------------------------------------
                    e = exact, a = approximate, u = upper bound on F
```

Wilks' lambda of 0.4470 agrees with the value reported by Rencher. With a p-value of .0027, we reject the null hypothesis that the coefficients for the covariates are jointly zero.

To test that the coefficients for the covariates are equal across groups, we perform a MANCOVA that includes our covariates (x1 and x2) interacted with group. We then use manovatest to obtain the combined test of equal coefficients for x1 and x2 across groups.

Samuel Stanley Wilks (1906–1964) was born in Texas. He gained degrees in architecture, mathematics, and statistics from North Texas Teachers' College and the Universities of Texas and Iowa. After periods in Columbia and England, he moved to Princeton in 1933. Wilks published various widely-used texts, was founding Editor of the *Annals of Mathematical Statistics*, and made many key contributions to multivariate statistics. Wilks' lambda is named for him.

(Continued on next page)

```
. manova y1 y2 y3 = group x1 x2 group*x1 group*x2, cont(x1 x2)

                    Number of obs =      32

                    W = Wilks' lambda      L = Lawley-Hotelling trace
                    P = Pillai's trace     R = Roy's largest root
```

Source		Statistic	df	F(df1,	df2) =	F	Prob>F	
Model	W	0.0205	11	33.0	53.7	4.47	0.0000	a
	P	1.9571		33.0	60.0	3.41	0.0000	a
	L	10.6273		33.0	50.0	5.37	0.0000	a
	R	7.0602		11.0	20.0	12.84	0.0000	u
Residual			20					
group	W	0.4930	3	9.0	44.0	1.65	0.1317	a
	P	0.5942		9.0	60.0	1.65	0.1226	a
	L	0.8554		9.0	50.0	1.58	0.1458	a
	R	0.5746		3.0	20.0	3.83	0.0256	u
x1	W	0.7752	1	3.0	18.0	1.74	0.1947	e
	P	0.2248		3.0	18.0	1.74	0.1947	e
	L	0.2900		3.0	18.0	1.74	0.1947	e
	R	0.2900		3.0	18.0	1.74	0.1947	e
x2	W	0.8841	1	3.0	18.0	0.79	0.5169	e
	P	0.1159		3.0	18.0	0.79	0.5169	e
	L	0.1311		3.0	18.0	0.79	0.5169	e
	R	0.1311		3.0	18.0	0.79	0.5169	e
group*x1	W	0.4590	3	9.0	44.0	1.84	0.0873	a
	P	0.6378		9.0	60.0	1.80	0.0869	a
	L	0.9702		9.0	50.0	1.80	0.0923	a
	R	0.6647		3.0	20.0	4.43	0.0152	u
group*x2	W	0.5275	3	9.0	44.0	1.47	0.1899	a
	P	0.5462		9.0	60.0	1.48	0.1747	a
	L	0.7567		9.0	50.0	1.40	0.2130	a
	R	0.4564		3.0	20.0	3.04	0.0527	u
Residual			20					
Total			31					

```
             e = exact, a = approximate, u = upper bound on F
. manovatest group*x1 group*x2

                    W = Wilks' lambda      L = Lawley-Hotelling trace
                    P = Pillai's trace     R = Roy's largest root
```

Source		Statistic	df	F(df1,	df2) =	F	Prob>F	
group*x1 group*x2	W	0.3310	6	18.0	51.4	1.37	0.1896	a
	P	0.8600		18.0	60.0	1.34	0.1973	a
	L	1.4629		18.0	50.0	1.35	0.1968	a
	R	0.8665		6.0	20.0	2.89	0.0341	u
Residual			20					

```
             e = exact, a = approximate, u = upper bound on F
```

Rencher reports 0.3310 for Wilks' lambda, which agrees with the results of manovatest above. In this case, we fail to reject the null hypothesis.

◁

Saved Results

manova saves in e():

Scalars

e(N)	number of observations
e(df_m)	model degrees of freedom
e(df_r)	residual degrees of freedom
e(k_eq)	number of equations
e(df_#)	degrees of freedom for term #

Macros

e(cmd)	manova
e(depvars)	names of dependent variables
e(indepvars)	names of the right-hand-side variables
e(term_#)	term #
e(errorterm_#)	error term for term # (defined for terms using nonresidual error)
e(wtype)	weight type
e(wexp)	weight expression
e(predict)	program used to implement predict

Matrices

e(b)	coefficient vector (a stacked version of e(B))
e(V)	variance–covariance matrix of the estimators
e(B)	coefficient matrix
e(E)	residual-error SSCP matrix
e(xpxinv)	generalized inverse of $\mathbf{X'X}$
e(H_m)	hypothesis SSCP matrix for the overall model
e(stat_m)	multivariate statistics for the overall model
e(eigvals_m)	eigenvalues of $\mathbf{E}^{-1}\mathbf{H}$ for the overall model
e(aux_m)	s, m, and n values for the overall model
e(H_#)	hypothesis SSCP matrix for term #
e(stat_#)	multivariate statistics for term # (if computed)
e(eigvals_#)	eigenvalues of $\mathbf{E}^{-1}\mathbf{H}\_\#$ for term # (if computed)
e(aux_#)	s, m, and n values for term # (if computed)

Functions

e(sample)	marks estimation sample

Methods and Formulas

Let $\mathbf{Y}$ denote the matrix of observations on the left-hand-side (lhs) variables. Let $\mathbf{X}$ denote the design matrix based on the right-hand-side (rhs) variables. The first column of $\mathbf{X}$ is equal to all ones (unless the noconstant option was specified). Categorical rhs variables are placed in $\mathbf{X}$ as a set of indicator (sometimes called dummy) variables, while continuous variables enter as is. Columns of $\mathbf{X}$ corresponding to interactions are formed by multiplying the various combinations of columns for the variables involved in the interaction.

The multivariate model

$$\mathbf{Y} = \mathbf{X}\boldsymbol{\beta} + \boldsymbol{\epsilon}$$

leads to multivariate hypotheses of the form

$$\mathbf{C}\beta\mathbf{A}' = 0$$

where β is a matrix of parameters, $\mathbf{C}$ specifies constraints on the design matrix $\mathbf{X}$ for a particular hypothesis, and $\mathbf{A}$ provides a transformation of $\mathbf{Y}$. $\mathbf{A}$ is often the identity matrix.

An estimate of β is provided by

$$\mathbf{B} = (\mathbf{X}'\mathbf{X})^-\mathbf{X}'\mathbf{Y}$$

The error sum of squares and cross products (SSCP) matrix is

$$\mathbf{E} = \mathbf{A}(\mathbf{Y}'\mathbf{Y} - \mathbf{B}'\mathbf{X}'\mathbf{X}\mathbf{B})\mathbf{A}'$$

and the SSCP matrix for the hypothesis is

$$\mathbf{H} = \mathbf{A}(\mathbf{CB})'\{\mathbf{C}(\mathbf{X}'\mathbf{X})^-\mathbf{C}'\}^{-1}(\mathbf{CB})\mathbf{A}'$$

The inclusion of weights, if specified, enters the formulas in a manner similar to that shown in [R] **regress**.

Let $\lambda_1 > \lambda_2 > \cdots > \lambda_s$ represent the nonzero eigenvalues of $\mathbf{E}^{-1}\mathbf{H}$. $s = \min(p, \nu_h)$, where p is the number of columns of $\mathbf{YA}'$ (i.e., the number of y-variables or number of resultant transformed lhs variables), and ν_h is the hypothesis degrees of freedom.

Wilks' (1932) lambda statistic is

$$\Lambda = \prod_{i=1}^{s} \frac{1}{1 + \lambda_i} = \frac{|\mathbf{E}|}{|\mathbf{H} + \mathbf{E}|}$$

and is a likelihood-ratio test. This statistic is distributed as the Wilks' Λ-distribution if $\mathbf{E}$ has the Wishart distribution, $\mathbf{H}$ has the Wishart distribution under the null hypothesis, and $\mathbf{E}$ and $\mathbf{H}$ are independent. The null hypothesis is rejected for small values of Λ.

Pillai's (1955) trace is

$$V = \sum_{i=1}^{s} \frac{\lambda_i}{1 + \lambda_i} = \mathrm{tr}\left\{(\mathbf{E} + \mathbf{H})^{-1}\mathbf{H}\right\}$$

and the Lawley–Hotelling trace (Lawley 1938 and Hotelling 1951) is

$$U = \sum_{i=1}^{s} \lambda_i = \mathrm{tr}(\mathbf{E}^{-1}\mathbf{H})$$

and is also known as Hotelling's generalized T^2-statistic.

Roy's largest root is taken as λ_1, though some report $\theta = \lambda_1/(1 + \lambda_1)$, which is bounded between zero and one. Roy's largest root provides a test based on the union-intersection approach to test construction introduced by Roy (1939).

Tables providing critical values for these four multivariate statistics are found in many of the books that discuss MANOVA, including Rencher (1998 and 2002).

Let p be the number of columns of $\mathbf{YA'}$ (i.e., the number of y variables or the number of resultant transformed y-variables), ν_h be the hypothesis degrees of freedom, ν_e be the error degrees of freedom, $s = \min(\nu_h, p)$, $m = (|\nu_h - p| - 1)/2$, and $n = (\nu_e - p - 1)/2$. Transformations of these four multivariate statistics to F statistics are as follows.

For Wilks' lambda, an approximate F statistic (Rao 1951) with df_1 and df_2 degrees of freedom is

$$F = \frac{(1 - \Lambda^{1/t})df_1}{(\Lambda^{1/t})df_2}$$

where

$$df_1 = p\nu_h \qquad df_2 = wt - 1 - p\nu_h/2$$

$$w = \nu_e + \nu_h - (p + \nu_h + 1)/2$$

$$t = \left(\frac{p^2\nu_h^2 - 4}{p^2 + \nu_h^2 - 5} \right)^{1/2}$$

t is set to one if either the numerator or the denominator equals zero. This F statistic is exact when p equals 1 or 2 or when ν_h equals 1 or 2.

An approximate F statistic for Pillai's trace (Pillai 1954 and 1956a) with $s(2m + s + 1)$ and $s(2n + s + 1)$ degrees of freedom is

$$F = \frac{(2n + s + 1)V}{(2m + s + 1)(s - V)}$$

An approximate F statistic for the Lawley–Hotelling trace (Pillai 1954 and 1956b) with $s(2m+s+1)$ and $2sn + 2$ degrees of freedom is

$$F = \frac{2(sn + 1)U}{s^2(2m + s + 1)}$$

When p or ν_h are 1, an exact F statistic for Roy's largest root is

$$F = \lambda_1 \frac{\nu_e - p + 1}{p}$$

with $|\nu_h - p| + 1$ and $\nu_e - p + 1$ degrees of freedom. In other cases, an upper bound F statistic (providing a lower bound on the p-value) for Roy's largest root is

$$F = \lambda_1 \frac{\nu_e - d + \nu_h}{d}$$

with d and $\nu_e - d + \nu_h$ degrees of freedom, where $d = \max(p, \nu_h)$.

References

Allison, M. J., P. Zappasodi, and M. B. Lurie. 1962. The correlation of a biphasic metabolic response with a biphasic response in resistance to tuberculosis in rabbits. *Journal of Experimental Medicine* 115: 881–890.

Anderson, T. W. 1965. Samuel Stanley Wilks, 1906–1964. *Annals of Mathematical Statistics* 36: 1–23.

——. 1984. *Introduction to Multivariate Statistical Analysis.* New York: Wiley.

Andrews, D. F. and A. M. Herzberg (eds.). 1985. *Data.* New York: Springer.

Box, G. E. P. 1950. Problems in the analysis of growth and linear curves. *Biometrics.* 6: 362–389.

Brown, J. D. and E. Beerstecher. 1951. Metabolic patterns of underweight and overweight individuals. *Biochemical Institute Studies IV, No. 5109.* Austin: University of Texas Press.

Hilbe, J. 1992. smv4: One-way multivariate analysis of variance (MANOVA). *Stata Technical Bulletin* 6: 5–7. Reprinted in *Stata Technical Bulletin Reprints*, vol. 1, pp. 138–139.

Hotelling, H. 1951. A generalized t^2 test and measure of multivariate dispersion. *Proceedings of the Second Berkeley Symposium on Mathematical Statistics and Probability* 1: 23–41.

Lawley, D. N. 1938. A generalization of Fisher's z-test. *Biometrika* 30: 180–187.

Mardia, K. V., J. T. Kent, and J. M. Bibby. 1979. *Multivariate Analysis.* New York: Academic Press.

Milliken, G. A. and D. E. Johnson. 1984. *Analysis of Messy Data, Volume 1: Designed Experiments.* New York: Nostrand Reinhold.

Morrison, D. F. 1998. Multivariate analysis of variance. In *Encyclopedia of Biostatistics*, Vol. 4, ed. P. Armitage and T. Colton, 2820–2825. New York: Wiley.

——. 2005. *Multivariate Statistical Methods.* 4th ed. Belmont, CA: Duxbury.

Pillai, K. C. S. 1954. On some distribution problems in multivariate analysis. *Mimeograph Series No. 88.* Institute of Statistics, University of North Carolina, Chapel Hill.

——. 1955. Some new test criteria in multivariate analysis. *Annals of Mathematical Statistics* 26: 117–121.

——. 1956a. On the distribution of the largest or the smallest root of a matrix in multivariate analysis. *Biometrika* 43: 122–127.

——. 1956b. Some results useful in multivariate analysis. *Annals of Mathematical Statistics* 27: 1106–1113.

——. 1985. Multivariate analysis of variance (MANOVA). In *Encyclopedia of Statistical Sciences*, Vol. 6, ed. S. Kotz, N. L. Johnson, and C. B. Read, 20–29. New York: Wiley.

Rao, C. R. 1951. An asymptotic expansion of the distribution of Wilks' criterion. *Bulletin of the International Statistical Institute* 33: 177–180.

Rencher, A. C. 1998. *Multivariate Statistical Inference and Applications.* New York: Wiley.

——. 2002. *Methods of Multivariate Analysis.* 2nd ed. New York: Wiley.

Roy, S. N. 1939. p-Statistics or some generalizations in analysis of variance appropriate to multivariate problems. *Sankhyā* 4: 381–396.

Seber, G. A. F. 1984. *Multivariate Observations.* New York: Wiley.

Smith, H., R. Gnanadesikan, and J. B. Hughes. 1962. Multivariate analysis of variance (MANOVA). *Biometrics* 18: 22–41.

Timm, N. H. 1975. *Multivariate Analysis with Applications in Education and Psychology.* Pacific Grove, CA: Brooks/Cole.

Wilks, S. S. 1932. Certain generalizations in the analysis of variance. *Biometrika* 24: 471–494.

Woodard, D. E. 1931. Healing time of fractures of the jaw in relation to delay before reduction, infection, syphilis and blood calcium and phosphorus content. *Journal of the American Dental Association* 18: 419–442.

(Continued on next page)

Also See

Complementary:	[MV] **manova postestimation**,
	[D] **encode**, [D] **reshape**
Related:	[MV] **hotelling**,
	[R] **anova**, [R] **mvreg**
Background:	[U] **11.1.10 Prefix commands**,
	[U] **13.5 Accessing coefficients and standard errors**,
	[U] **20 Estimation and postestimation commands**

Title

manova postestimation — Postestimation tools for manova

Description

The following postestimation command is of special interest after `manova`:

command	description
manovatest	multivariate tests after `manova`

For information about `manovatest`, see below.

In addition, the following standard postestimation commands are available:

command	description
adjust	adjusted predictions of $\mathbf{x}\beta$; see [R] **adjust**
* estat	VCE and estimation sample summary; see [R] **estat**
estimates	cataloging estimation results; see [R] **estimates**
lincom	point estimates, standard errors, testing, and inference for linear combinations of coefficients; see [R] **lincom**
predict	predictions, residuals, and standard errors; see below
test	Wald tests for simple and composite linear hypotheses; see below

* `estat ic` is not available after `manova`.

Special-interest postestimation commands

`manovatest` provides multivariate tests involving *term*s or linear combinations of the underlying design matrix from the most-recently fitted `manova`. The four multivariate test statistics are Wilks' lambda, Pillai's trace, Lawley–Hotelling trace, and Roy's largest root. The format of the output is similar to that shown by `manova`.

`test` has specialized syntax for use after `manova`; see below. `test` performs Wald tests of expressions involving the coefficients of the underlying regression model. Simple and composite linear hypotheses are possible.

(*Continued on next page*)

Syntax for predict

predict [*type*] *newvar* [*if*] [*in*] [, equation(*eqno*[, *eqno*]) *statistic*]

statistic	description
xb	$x_j b$, fitted values (the default)
stdp	standard error of the prediction
residuals	residuals
difference	difference between the linear predictions of two equations
stddp	standard error of the difference in linear predictions

These statistics are available both in and out of sample; type predict ... if e(sample) ... if wanted only for the estimation sample.

Options for predict

equation(*eqno*[, *eqno*]) specifies the equation to which you are referring.

equation() is filled in with one *eqno* for options xb, stdp, and residuals. equation(#1) would mean that the calculation is to be made for the first equation, equation(#2) would mean the second, and so on. Alternatively, you could refer to the equations by their names. equation(income) would refer to the equation named income and equation(hours) to the equation named hours.

If you do not specify equation(), results are the same as if you specified equation(#1).

difference and stddp refer to between-equation concepts. To use these options, you must specify two equations, e.g., equation(#1,#2) or equation(income,hours). When two equations must be specified, equation() is required. With equation(#1,#2), difference computes the prediction of equation(#1) minus the prediction of equation(#2).

xb, the default, calculates the fitted values—the prediction of $x_j b$ for the specified equation.

stdp calculates the standard error of the prediction for the specified equation (the standard error of the estimated expected value or mean for the observation's covariate pattern). This is also referred to as the standard error of the fitted value.

residuals calculates the residuals.

difference calculates the difference between the linear predictions of two equations in the system.

stddp calculates the standard error of the difference in linear predictions $(x_{1j} b - x_{2j} b)$ between equations 1 and 2.

For more information on using predict after multiple-equation estimation commands, see [R] **predict**.

Syntax for manovatest

manovatest *term* [*term* ...] [/ *term* [*term* ...]] [, ytransform(*matname*)]

manovatest , test(*matname*) [ytransform(*matname*)]

manovatest , showorder

where *term* has the form *varname*[{ * | | }*varname*[...]]

Options for manovatest

ytransform(*matname*) specifies a matrix for transforming the y variables (the *depvarlist* from manova) as part of the test. The multivariate tests are based on $(\mathbf{AEA}')^{-1}(\mathbf{AHA}')$. By default, $\mathbf{A}$ is the identity matrix. ytransform() is how you specify an $\mathbf{A}$ matrix to be used in the multivariate tests. Specifying ytransform() provides the same results as first transforming the y variables with $\mathbf{YA}'$, where $\mathbf{Y}$ is the matrix formed by binding the y variables by column, and $\mathbf{A}$ is the matrix stored in *matname*; performing manova on the transformed ys; and then running manovatest without ytransform().

The number of columns of *matname* must equal the number of variables in the *depvarlist* from manova. The number of rows must be less than or equal to the number of variables in the *depvarlist* from manova. *matname* should have columns in the same order as the *depvarlist* from manova. The column and row names of *matname* are ignored.

When ytransform() is specified, a listing of the transformations is presented before the table containing the multivariate tests. You should examine this table to verify that you have applied the transformation you desired.

test(*matname*) is required with the second syntax of manovatest. The rows of *matname* specify linear combinations of the underlying design matrix of the MANOVA that are to be jointly tested. The columns correspond to the underlying design matrix (including the constant if it has not been suppressed). The column and row names of *matname* are ignored.

A listing of the constraints imposed by the test() option is presented before the table containing the multivariate tests. You should examine this table to verify that you have applied the linear combinations you desired. Typing manovatest, showorder allows you to examine the ordering of the columns for the design matrix from the MANOVA.

showorder causes manovatest to list the definition of each column in the design matrix. showorder is not allowed with any other option or when *term*s are specified.

Syntax for test following manova

test , test(*matname*) [mtest[(*opt*)] matvlc(*matname2*)]

test [*exp* = *exp*] [, accumulate notest matvlc(*matname2*)]

test , showorder

where *exp* must have references to coefficients enclosed in [*eqno*]_coef[] (or synonym [*eqno*]_b[]). *eqno* is either ## or *name*. Omitting [*eqno*] is equivalent to specifying [#1].

Options for test after manova

test(*matname*) is required with the first syntax of test. The rows of *matname* specify linear combinations of the underlying design matrix of the MANOVA that are to be jointly tested. The columns correspond to the underlying design matrix (including the constant if it has not been suppressed). The column and row names of *matname* are ignored.

A listing of the constraints imposed by the test() option is presented before the table containing the tests. You should examine this table to verify that you have applied the linear combinations you desired. Typing test, showorder allows you to examine the ordering of the columns for the design matrix from the MANOVA.

matname should have as many columns as the number of dependent variables times the number of columns in the basic design matrix. The design matrix is repeated for each dependent variable.

mtest [(*opt*)] specifies that tests be performed for each condition separately. *opt* specifies the method for adjusting *p*-values for multiple testing. Valid values for *opt* are

<u>b</u>onferroni   Bonferroni's method
<u>h</u>olm   Holm's method
<u>s</u>idak   Sidak's method
<u>noadj</u>ust   no adjustment is to be made

Specifying mtest without an argument is equivalent to specifying mtest(noadjust).

accumulate allows a hypothesis to be tested jointly with the previously tested hypotheses.

notest suppresses the output. This option is useful when you are interested only in the joint test with additional hypotheses specified in a subsequent call to test, accumulate.

matvlc(*matname*), a programmer's option, saves the variance–covariance matrix of the linear combinations involved in the suite of tests. For the test of $Ho : \mathbf{Lb} = \mathbf{c}$, what is returned in *matname* is $\mathbf{LVL}'$, where $\mathbf{V}$ is the estimated variance–covariance matrix of $\mathbf{b}$.

showorder causes test to list the definition of each column in the design matrix. showorder is not allowed with any other option.

Remarks

Several postestimation tools are available after manova. We demonstrate these tools by extending examples 1, 2, 3, and 6 of [MV] **manova**.

▷ Example 1

Example 1 of [MV] **manova** presented a balanced one-way MANOVA on the rootstock data.

```
. use http://www.stata-press.com/data/r9/rootstock
(Table 6.2 Rootstock Data -- Rencher (2002))
. manova y1 y2 y3 y4 = rootstock
(output omitted)
```

test provides Wald tests on expressions involving the underlying coefficients of the model, and lincom provides linear combinations along with standard errors and confidence intervals.

```
. test [y3]_b[rootstock[3]] = ([y3]_b[rootstock[1]] + [y3]_b[rootstock[2]])/2
( 1) - .5 [y3]rootstock[1] - .5 [y3]rootstock[2] + [y3]rootstock[3] = 0

       F( 1,    42) =    5.62
            Prob > F =    0.0224

. lincom [y3]_b[rootstock[4]] - [y1]_b[rootstock[4]]
( 1) - [y1]rootstock[4] + [y3]rootstock[4] = 0
```

	Coef.	Std. Err.	t	P>|t|	[95% Conf. Interval]	
(1)	.24875	.1443917	1.72	0.092	-.0426442	.5401442

If the equation portion of the expression is omitted, the first equation (first dependent variable) is assumed.

◁

The `manovatest` postestimation command provides multivariate tests of terms or linear combinations of the underlying design matrix from the most recent MANOVA model.

▷ Example 2

In example 2 of [MV] **manova**, a one-way MANOVA on the metabolic dataset was shown.

```
. use http://www.stata-press.com/data/r9/metabolic
(Table 4.5 Metabolic Comparisons of Rabbits -- Rencher (1998))
. manova y1 y2 = group
  (output omitted )
```

`manovatest` can test terms from the preceding `manova`. In this case, we test the `group` term from our one-way MANOVA.

```
. manovatest group
```

	W = Wilks' lambda		L = Lawley-Hotelling trace			
	P = Pillai's trace		R = Roy's largest root			
Source	Statistic	df	F(df1,	df2) =	F	Prob>F
group	W 0.1596	3	6.0	32.0	8.02	0.0000 e
	P 1.2004		6.0	34.0	8.51	0.0000 a
	L 3.0096		6.0	30.0	7.52	0.0001 a
	R 1.5986		3.0	17.0	9.06	0.0008 u
Residual		17				

e = exact, a = approximate, u = upper bound on F

Using `manovatest` to test model terms is not interesting in this example. It merely repeats information already presented by `manova`. Later we will see useful applications of term testing via `manovatest`.

`manovatest` can also be used to test linear combinations of the underlying design matrix of the MANOVA model. While the MANOVA indicates that there are differences in the groups, it does not indicate the nature of those differences. Rencher discusses three linear contrasts of interest for this example: group one (the control) versus the rest, group four versus group two and three, and group two versus group three. The `test()` option of `manovatest` allows us to test these hypotheses.

Since we did not use the `noconstant` option with our `manova`, the underlying parameterization of the design matrix has the first column corresponding to the constant in the model, while the next four columns correspond to the four groups of rabbits. The `showorder` option of `manovatest` illustrates this point. The tests on the three contrasts of interest follow.

```
. manovatest, showorder
Order of columns in the design matrix
      1: _cons
      2: (group==1)
      3: (group==2)
      4: (group==3)
      5: (group==4)
. matrix c1 = (0,3,-1,-1,-1)
```

```
. manovatest, test(c1)

Test constraint
 (1)     3 group[1] - group[2] - group[3] - group[4] = 0
```

				W = Wilks' lambda P = Pillai's trace		L = Lawley-Hotelling trace R = Roy's largest root		
Source		Statistic	df	F(df1,	df2) =	F	Prob>F	
manovatest	W	0.4063	1	2.0	16.0	11.69	0.0007	e
	P	0.5937		2.0	16.0	11.69	0.0007	e
	L	1.4615		2.0	16.0	11.69	0.0007	e
	R	1.4615		2.0	16.0	11.69	0.0007	e
Residual			17					

```
                       e = exact, a = approximate, u = upper bound on F
. matrix c2 = (0,0,-1,-1,2)

. manovatest, test(c2)

Test constraint
 (1)      - group[2] - group[3] + 2 group[4] = 0
```

				W = Wilks' lambda P = Pillai's trace		L = Lawley-Hotelling trace R = Roy's largest root		
Source		Statistic	df	F(df1,	df2) =	F	Prob>F	
manovatest	W	0.9567	1	2.0	16.0	0.36	0.7018	e
	P	0.0433		2.0	16.0	0.36	0.7018	e
	L	0.0453		2.0	16.0	0.36	0.7018	e
	R	0.0453		2.0	16.0	0.36	0.7018	e
Residual			17					

```
                       e = exact, a = approximate, u = upper bound on F
. matrix c3 = (0,0,1,-1,0)

. manovatest, test(c3)

Test constraint
 (1)     group[2] - group[3] = 0
```

				W = Wilks' lambda P = Pillai's trace		L = Lawley-Hotelling trace R = Roy's largest root		
Source		Statistic	df	F(df1,	df2) =	F	Prob>F	
manovatest	W	0.4161	1	2.0	16.0	11.23	0.0009	e
	P	0.5839		2.0	16.0	11.23	0.0009	e
	L	1.4033		2.0	16.0	11.23	0.0009	e
	R	1.4033		2.0	16.0	11.23	0.0009	e
Residual			17					

```
                       e = exact, a = approximate, u = upper bound on F
```

Since there is only one degree of freedom for each of the hypotheses, the F tests are exact (and identical for the four multivariate methods). The first test indicates that the mean vector for the control group is significantly different from the mean vectors for the other three groups. The second test, with a p-value of 0.7018, fails to reject the null hypothesis that group four equals groups two and three. The third test, with a p-value of 0.0009, indicates differences between the mean vectors of groups two and three.

Rencher also tests using weighted orthogonal contrasts. manovatest can provide these tests as well.

```
. matrix c1w = (0,14,-7,-5,-2)

. manovatest, test(c1w)

Test constraint
(1)    14 group[1] - 7 group[2] - 5 group[3] - 2 group[4] = 0
                           W = Wilks' lambda       L = Lawley-Hotelling trace
                           P = Pillai's trace      R = Roy's largest root
            Source  | Statistic    df    F(df1,    df2) =    F    Prob>F
        ------------+--------------------------------------------------------
         manovatest | W    0.3866    1     2.0     16.0    12.70 0.0005 e
                    | P    0.6134          2.0     16.0    12.70 0.0005 e
                    | L    1.5869          2.0     16.0    12.70 0.0005 e
                    | R    1.5869          2.0     16.0    12.70 0.0005 e
        ------------+--------------------------------------------------------
           Residual |              17
        ------------+--------------------------------------------------------

                    e = exact, a = approximate, u = upper bound on F

. matrix c2w = (0,0,-7,-5,12)

. manovatest, test(c2w)

Test constraint
(1)     -7 group[2] - 5 group[3] + 12 group[4] = 0
                           W = Wilks' lambda       L = Lawley-Hotelling trace
                           P = Pillai's trace      R = Roy's largest root
            Source  | Statistic    df    F(df1,    df2) =    F    Prob>F
        ------------+--------------------------------------------------------
         manovatest | W    0.9810    1     2.0     16.0    0.15 0.8580 e
                    | P    0.0190          2.0     16.0    0.15 0.8580 e
                    | L    0.0193          2.0     16.0    0.15 0.8580 e
                    | R    0.0193          2.0     16.0    0.15 0.8580 e
        ------------+--------------------------------------------------------
           Residual |              17
        ------------+--------------------------------------------------------

                    e = exact, a = approximate, u = upper bound on F
```

These two weighted contrasts do not lead to different conclusions compared with their unweighted counterparts.

◁

❑ Technical Note

`manovatest, test(`*matname*`)` displays the linear combination (labeled "`Test constraint`") indicated by *matname*. You should examine this listing to make sure that the matrix you specify in `test()` provides the test you desire.

❑

The `adjust` postestimation command provides tables of predicted means and confidence intervals based on the most recently estimated model.

▷ Example 3

Example 3 of [MV] **manova** presented a two-way MANOVA model on the jaw data.

```
. use http://www.stata-press.com/data/r9/jaw
(Table 4.6 Two-Way Unbalanced Data for Fractures of the Jaw -- Rencher (1998))
. manova y1 y2 y3 = gender fracture gender*fracture
  (output omitted)
```

The interaction term, gender*fracture, was significant. adjust may be used to examine the interaction; see [R] **adjust**.

. adjust, by(fracture gender) xb ci equation(y1)

```
Equation: y1     Command: manova
```

| | gender | |
fracture	male	female
one compound fracture	39.5	22
	[30.8251,48.1749]	[.750989,43.249]
two compound fractures	26.875	30.75
	[19.3623,34.3877]	[20.1255,41.3745]
one simple fracture	45.1667	36.5
	[36.4918,53.8415]	[21.4747,51.5253]

```
Key:   Linear Prediction
       [95% Confidence Interval]
```

. adjust, by(fracture gender) xb ci equation(y2)

```
Equation: y2     Command: manova
```

| | gender | |
fracture	male	female
one compound fracture	35.5	56
	[31.0268,39.9732]	[45.043,66.957]
two compound fractures	32.375	33.25
	[28.5011,36.2489]	[27.7715,38.7285]
one simple fracture	36.1667	33
	[31.6935,40.6398]	[25.2522,40.7478]

```
Key:   Linear Prediction
       [95% Confidence Interval]
```

. adjust, by(fracture gender) xb ci equation(y3)

```
Equation: y3     Command: manova
```

| | gender | |
fracture	male	female
one compound fracture	61.1667	43
	[56.9271,65.4063]	[32.6151,53.3849]
two compound fractures	62.25	64
	[58.5784,65.9216]	[58.8076,69.1924]
one simple fracture	58.1667	63.5
	[53.9271,62.4063]	[56.1568,70.8432]

```
Key:   Linear Prediction
       [95% Confidence Interval]
```

The first adjust table shows the predicted mean and confidence interval of y1 (age of patient) for each combination of fracture and gender. The second and third adjust tables provide this information for y2 (blood lymphocytes) and y3 (blood polymorphonuclears). These three tables of predictions are the same as those you would obtain from adjust after running anova for each of the three dependent variables separately.

Note that the predicted y2 value is larger than the predicted y3 value for females with one compound fracture. For the other five combinations of gender and fracture, the relationship is reversed. Note that there is only one observation for the combination of female and one compound fracture.

Let's examine the residuals using the predict command.

```
. predict y1res, residual equation(y1)
. predict y2res, residual equation(y2)
. predict y3res, residual equation(y3)
. list gender fracture y1res y2res y3res
```

	gender	fracture	y1res	y2res	y3res
1.	male	one compound fracture	2.5	-.5	-.1666667
2.	male	one compound fracture	2.5	7.5	-6.166667
3.	male	one compound fracture	8.5	-.5	2.833333
4.	male	one compound fracture	-4.5	-2.5	3.833333
5.	male	one compound fracture	-14.5	-4.5	2.833333
6.	male	one compound fracture	5.5	.5	-3.166667
7.	male	two compound fractures	-3.875	-5.375	1.75
8.	male	two compound fractures	-4.875	-.375	1.75
9.	male	two compound fractures	-1.875	-2.375	1.75
10.	male	two compound fractures	1.125	6.625	-6.25
11.	male	two compound fractures	-2.875	-1.375	6.75
12.	male	two compound fractures	25.125	-4.375	-2.25
13.	male	two compound fractures	-9.875	-2.375	1.75
14.	male	two compound fractures	-2.875	9.625	-5.25
15.	male	one simple fracture	-13.16667	.8333333	-4.166667
16.	male	one simple fracture	6.833333	-2.166667	3.833333
17.	male	one simple fracture	7.833333	8.833333	-7.166667
18.	male	one simple fracture	3.833333	-1.166667	1.833333
19.	male	one simple fracture	9.833333	-4.166667	1.833333
20.	male	one simple fracture	-15.16667	-2.166667	3.833333
21.	female	one compound fracture	7.11e-15	-1.42e-14	1.42e-14
22.	female	two compound fractures	-8.75	-4.25	4
23.	female	two compound fractures	7.25	-8.25	9
24.	female	two compound fractures	-9.75	3.75	-5
25.	female	two compound fractures	11.25	8.75	-8
26.	female	one simple fracture	6.5	-3	3.5
27.	female	one simple fracture	-6.5	3	-3.5

The single observation for a female with one compound fracture has residuals that are within roundoff of zero. With only one observation for that cell of the design, this MANOVA model is forced to fit to that point. The largest residual (in absolute value) appears for observation 12, which has an age 25.125 higher than the model prediction for a male with two compound fractures. ◁

▷ Example 4

Example 6 of [MV] **manova** presents a nested MANOVA on the rash2 data.

```
. use http://www.stata-press.com/data/r9/rash2
(skin rash data)

. set matsize 262

. manova response response2 = t / c|t / d|c|t / p|d|c|t /
 (output omitted )
```

The MANOVA indicated that the treatment and patient terms were significant while clinic and doctor terms were not.

You decide to follow the rule of thumb that says to pool terms whose p-values are larger than 0.25. Wilks' lambda reports a p-value of 0.3226 for the test of clinic|treatment (see example 6 of [MV] **manova**). You decide to pool the doctor and clinic terms in the MANOVA in order to gain power for the test of treatment. The forward-slash notation of manova is also allowed with manovatest to indicate nonresidual error terms. Here is the multivariate test of treatment using the pooled clinic and doctor terms and then the multivariate test of the pooled term using patient as the error term.

```
. manovatest t / c|t d|c|t
```

| | | W = Wilks' lambda | | L = Lawley-Hotelling trace | | | |
| | | P = Pillai's trace | | R = Roy's largest root | | | |
Source		Statistic	df	F(df1,	df2) =	F	Prob>F
treatment	W	0.2762	1	2.0	21.0	27.52	0.0000 e
	P	0.7238		2.0	21.0	27.52	0.0000 e
	L	2.6210		2.0	21.0	27.52	0.0000 e
	R	2.6210		2.0	21.0	27.52	0.0000 e
clinic|treatment doctor| clinic|treatment		22					

e = exact, a = approximate, u = upper bound on F

```
. manovatest c|t d|c|t / p|d|c|t
```

| | | W = Wilks' lambda | | L = Lawley-Hotelling trace | | | |
| | | P = Pillai's trace | | R = Roy's largest root | | | |
Source		Statistic	df	F(df1,	df2) =	F	Prob>F
clinic|treatment doctor| clinic|treatment	W	0.5713	22	44.0	142.0	1.04	0.4156 e
	P	0.4843		44.0	144.0	1.05	0.4103 a
	L	0.6529		44.0	140.0	1.04	0.4214 a
	R	0.4225		22.0	72.0	1.38	0.1533 u
patient|doctor|clinic| treatment		72					

e = exact, a = approximate, u = upper bound on F

Pooling clinic with doctor helps increase the power for the test of treatment. The pooled clinic with doctor term is not significant, having a p-value of 0.4156 from Wilks' lambda. You decide, based on this, to pool both clinic and doctor with patient.

Note that you can show the univariate analysis for one of your dependent variables by using the ytransform() option of manovatest.

```
. mat resp1only = (1,0)

. manovatest t / c|t d|c|t, ytransform(resp1only)

Transformation of the dependent variables
(1)     response
```

		W = Wilks' lambda		L = Lawley-Hotelling trace			
		P = Pillai's trace		R = Roy's largest root			
Source		Statistic	df	F(df1,	df2) =	F	Prob>F
treatment	W	0.7029	1	1.0	22.0	9.30	0.0059 e
	P	0.2971		1.0	22.0	9.30	0.0059 e
	L	0.4228		1.0	22.0	9.30	0.0059 e
	R	0.4228		1.0	22.0	9.30	0.0059 e
clinic|treatment doctor| clinic|treatment			22				

```
                   e = exact, a = approximate, u = upper bound on F

. mat list r(H)

symmetric r(H)[1,1]
           response
response   4240.0417

. mat list r(E)

symmetric r(E)[1,1]
           response
response   10029.073

. manovatest c|t d|c|t / p|d|c|t, ytransform(resp1only)

Transformation of the dependent variables
(1)     response
```

		W = Wilks' lambda		L = Lawley-Hotelling trace			
		P = Pillai's trace		R = Roy's largest root			
Source		Statistic	df	F(df1,	df2) =	F	Prob>F
clinic|treatment doctor| clinic|treatment	W	0.7156	22	22.0	72.0	1.30	0.2014 e
	P	0.2844		22.0	72.0	1.30	0.2014 e
	L	0.3974		22.0	72.0	1.30	0.2014 e
	R	0.3974		22.0	72.0	1.30	0.2014 e
patient|doctor|clinic| treatment			72				

```
                   e = exact, a = approximate, u = upper bound on F

. mat list r(H)

symmetric r(H)[1,1]
           response
response   10029.073

. mat list r(E)

symmetric r(E)[1,1]
           response
response   25236.875
```

Compare the results above with those shown in example 3 of [R] **anova postestimation**, which discusses the use of test in the pooling of terms after ANOVA. The F statistics, p-values, degrees of freedom, and sums of squares from the manovatests above are in agreement with the univariate analysis.

Continuing with the multivariate analysis, you can now pool `clinic` and `doctor` with `patient`.

```
. manovatest t / c|t d|c|t p|d|c|t
```

| | W = Wilks' lambda | | L = Lawley-Hotelling trace | | | |
| | P = Pillai's trace | | R = Roy's largest root | | | |
Source	Statistic	df	F(df1,	df2) =	F	Prob>F
treatment	W 0.6219	1	2.0	93.0	28.27	0.0000 e
	P 0.3781		2.0	93.0	28.27	0.0000 e
	L 0.6080		2.0	93.0	28.27	0.0000 e
	R 0.6080		2.0	93.0	28.27	0.0000 e
clinic|treatment doctor| clinic|treatment patient|doctor|clinic| treatment		94				

```
                     e = exact, a = approximate, u = upper bound on F
. manovatest c|t d|c|t p|d|c|t
```

| | W = Wilks' lambda | | L = Lawley-Hotelling trace | | | |
| | P = Pillai's trace | | R = Roy's largest root | | | |
Source	Statistic	df	F(df1,	df2) =	F	Prob>F
clinic|treatment doctor| clinic|treatment patient|doctor|clinic| treatment	W 0.4037	94	188.0	574.0	1.75	0.0000 e
	P 0.7052		188.0	576.0	1.67	0.0000 a
	L 1.2071		188.0	572.0	1.84	0.0000 a
	R 0.9109		94.0	288.0	2.79	0.0000 u
Residual		288				

```
                     e = exact, a = approximate, u = upper bound on F
```

See the second `manova` run from example 6 of [MV] **manova** for an alternate way of pooling the terms by refitting the MANOVA model.

◁

See examples 8 through 12 of [MV] **manova** for further examples of `manovatest`, including examples involving both the `test()` and `ytransform()` options.

Saved Results

`manovatest` saves in `r()`:

Scalars
  r(df)   hypothesis degrees of freedom
  r(df_r)   residual degrees of freedom

Matrices
  r(H)   hypothesis SSCP matrix
  r(E)   residual error SSCP matrix
  r(stat)   multivariate statistics
  r(eigvals)   eigenvalues of $E^{-1}H$
  r(aux) s, m, and n values

`test` after `manova` saves in `r()`:

Scalars

`r(p)`	two sided p-value
`r(F)`	F statistic
`r(df)`	hypothesis degrees of freedom
`r(df_r)`	residual degrees of freedom
`r(drop)`	0 if no constraints dropped, 1 otherwise
`r(dropped_#)`	index of #th constraint dropped

Macros

`r(mtmethod)`	method of adjustment for multiple testing

Matrices

`r(mtest)`	multiple test results

Methods and Formulas

All postestimation commands listed above are implemented as ado-files.

See [MV] **manova** for methods and formulas for the multivariate tests performed by `manovatest`.

Also See

Complementary:	[MV] **manova**;
	[R] **adjust**, [R] **estimates**, [R] **lincom**, [R] **test**
Background:	[R] **estat**, [R] **predict**

Title

> **matrix dissimilarity** — Compute similarity or dissimilarity measures

Syntax

matrix dissimilarity *matname* = [*varlist*] [*if*] [*in*] [, *options*]

options	description
measure	similarity or dissimilarity measure; default is L2 (Euclidean)
observations	compute similarities or dissimilarities between observations; default
variables	compute similarities or dissimilarities between variables
names(*varname*)	row/column names for *matname* (allowed with observations)
allbinary	check that all values are 0, 1, or missing
proportions	interpret values as proportions of binary values
dissim(*method*)	change similarity measure to dissimilarity measure

where *method* transforms similarities to dissimilarities using

$$\text{oneminus} \quad d_{ij} = 1 - s_{ij}$$
$$\text{standard} \quad d_{ij} = \sqrt{s_{ii} + s_{jj} - 2s_{ij}}$$

Description

matrix dissimilarity computes a similarity, dissimilarity, or distance matrix.

Options

measure specifies one of the similarity or dissimilarity measures allowed by Stata. The default is L2, Euclidean distance. Numerous similarity and dissimilarity measures are provided for continuous data and for binary data; see [MV] ***measure_option***.

observations and variables specify whether similarities or dissimilarities are computed between observations or variables. The default is observations.

names(*varname*) provides row and column names for *matname*. *varname* must be a string variable with a length of 32 or less. You will want to pick a *varname* that yields unique values for the row and column names. Uniqueness of values is not checked by matrix dissimilarity. names() is not allowed with the variables option. The default row and column names when the similarities or dissimilarities are computed between observations is obs#, where # is the observation number corresponding to that row or column.

allbinary checks that all values are 0, 1, or missing. Stata treats nonzero values as one (excluding missing values) when dealing with what are supposed to be binary data (including binary similarity *measures*). allbinary causes matrix dissimilarity to exit with an error message if the values are not truly binary. allbinary is not allowed with proportions.

proportions is for use with binary similarity *measure*s. It specifies that values be interpreted as proportions of binary values. The default action treats all nonzero values as one (excluding missing values). With proportions, the values are confirmed to be between zero and one, inclusive. See [MV] ***measure_option*** for a discussion of the use of proportions with binary *measure*s. proportions is not allowed with allbinary.

dissim(*method*) specifies that similarity measures be transformed into dissimilarity measures. *method* may be oneminus or standard. oneminus transforms similarities to dissimilarities using $d_{ij} = 1 - s_{ij}$. standard uses $d_{ij} = \sqrt{s_{ii} + s_{jj} - 2s_{ij}}$. dissim() does nothing when the *measure* is already a dissimilarity or distance. See [MV] ***measure_option*** to see which *measure*s are similarities.

Remarks

Commands such as cluster singlelinkage, cluster completelinkage, and mds (see [MV] **cluster** and [MV] **mds**) have options allowing the user to select the similarity or dissimilarity measure to use for its computation. If you are developing a command that requires a similarity or dissimilarity matrix, the matrix dissimilarity command provides a convenient way to obtain it.

The similarity or dissimilarity between each observation (or variable if the variables option is specified) and the others is placed in *matname*. The element in the *i*th row and *j*th column gives either the similarity or dissimilarity between the *i*th and *j*th observation (or variable). Whether you get a similarity or a dissimilarity depends upon the requested *measure*; see [MV] ***measure_option***.

If there are a large number of observations (variables when the variables option is specified), you may need to increase the maximum matrix size; see [R] **matsize**. If the number of observations (or variables) is so large that storing the results in a matrix is not practical, you may wish to consider using the cluster measures command, which stores similarities or dissimilarities in variables; see [MV] **cluster programming utilities**.

When computing similarities or dissimilarities between observations, the default row and column names of *matname* are set to obs#, where # is the observation number. The names() option allows you to override this default. For similarities or dissimilarities between variables, the row and column names of *matname* are set to the appropriate variable names.

The order of the rows and columns corresponds with the order of your observations when you are computing similarities or dissimilarities between observations. Warning: if you reorder your data (e.g., using sort or gsort) after running matrix dissimilarity, the row and column ordering will no longer match your data.

An additional use of matrix dissimilarity is in performing a cluster analysis on variables instead of observations. The cluster command performs a cluster analysis of the observations; see [MV] **cluster**. If you instead wish to cluster variables, you can use the variables option of matrix dissimilarity to obtain a dissimilarity matrix that can then be used with clustermat; see [MV] **clustermat** and example 2 below.

▷ Example 1

Example 1 of [MV] **cluster singlelinkage** introduces data with 4 chemical laboratory measurements on 50 different samples of a particular plant. Let's find the Canberra distance between the measurements performed by lab technician Bill found among the first 25 observations of the labtech dataset.

```
. use http://www.stata-press.com/data/r9/labtech

. matrix dissim D = x1 x2 x3 x4 if labtech=="Bill" in 1/25, canberra

. matrix list D

symmetric D[6,6]
              obs7       obs18       obs20       obs22       obs23       obs25
 obs7            0
obs18    1.3100445           0
obs20    1.1134916     .87626565           0
obs22     1.452748    1.0363077    1.0621064           0
obs23    1.0380665    1.4952796     .81602718    1.6888123           0
obs25    1.4668898    1.5139834    1.4492336    1.0668425    1.1252514           0
```

Notice that, by default, the row and column names of the matrix indicate the observations involved. The Canberra distance between the 23rd observation and the 18th observation is 1.4952796. See [MV] *measure_option* for a description of the Canberra distance.

◁

▷ Example 2

Example 2 from [MV] **cluster singlelinkage** presents a dataset with 30 observations of 60 binary variables, a1, a2, ..., a30. In [MV] **cluster singlelinkage** the observations were clustered. Here we instead cluster the variables by computing the dissimilarity matrix using `matrix dissimilarity` with the `variables` option followed by the `clustermat` command.

We use the `matching` option to obtain the simple matching similarity coefficient but then specify `dissim(oneminus)` to transform the similarities to dissimilarities using the transformation $d_{ij} = 1 - s_{ij}$. The `allbinary` option checks that the variables really are binary (0/1) data.

```
. use http://www.stata-press.com/data/r9/homework

. matrix dissim Avars = a*, variables matching dissim(oneminus) allbinary

. matrix subA = Avars[1..5,1..5]

. matrix list subA

symmetric subA[5,5]
          a1          a2          a3          a4          a5
a1         0
a2        .4           0
a3        .4   .46666667           0
a4        .3          .3   .36666667           0
a5        .4          .4   .13333333          .3           0
```

We listed the first 5 rows and columns of the 60×60 matrix. Notice that the matrix row and column names correspond to the variable names.

To perform an average-linkage cluster analysis on the 60 variables, we supply the `Avars` matrix created by `matrix dissimilarity` to the `clustermat averagelinkage` command; see [MV] **cluster averagelinkage**.

```
. clustermat averagelinkage Avars, clear
obs was 0, now 60
cluster name: _cl_1

. cluster generate g5 = groups(5)
```

```
. table g5
```

g5	Freq.
1	21
2	9
3	25
4	4
5	1

We generated a variable, g5, indicating the five-group cluster solution and then tabulated to show how many variables were clustered into each of the five groups. Group five has only 1 member.

```
. list g5 if g5==5
```

	g5
13.	5

The member corresponds to the 13th observation in the current dataset, which in turn corresponds to variable a13 from the original dataset. It appears that a13 is not like the other variables.

◁

Also See

Complementary:	[MV] **cluster**, [MV] **clustermat**, [MV] **mdslong**
Related:	[MV] **cluster programming utilities**
Background:	[MV] *measure_option*; [P] **matrix**

Title

mds — Multidimensional scaling for two-way data

Syntax

mds *varlist* [*if*] [*in*], id(*varname*) [*options*]

options	description
Model	
* id(*varname*)	identify observations
<u>unit</u>[(*varlist₂*)]	scale variables to min = 0 and max = 1
<u>std</u>[(*varlist₃*)]	scale variables to mean = 0 and sd = 1
<u>mea</u>sure(*measure*)	similarity or dissimilarity measure (see [MV] *measure_option*); default is L2 (Euclidean)
s2d(<u>st</u>andard)	convert similarity to dissimilarity: $d_{ij} = \sqrt{s_{ii} + s_{jj} - 2s_{ij}}$, the default
s2d(<u>one</u>minus)	convert similarity to dissimilarity: $d_{ij} = 1 - s_{ij}$
<u>dim</u>ension(#)	configuration dimensions; default is dimension(2)
<u>add</u>constant	make distance matrix positive definite
Reporting	
<u>neig</u>en(#)	maximum number of eigenvalues to display; default is neigen(10)
<u>config</u>	display table with configuration coordinates
<u>noplot</u>	suppress configuration plot

* id(*varname*) is required.

bootstrap, by, jackknife, rolling, statsby, and xi may be used with mds; see [U] **11.1.10 Prefix commands**. The maximum number of observations allowed in mds is the maximum matrix size; see [R] **matsize**.

See [MV] **mds postestimation** for features available after estimation.

Description

mds performs classical metric multidimensional scaling (MDS) for dissimilarity between observations with respect to the variables in *varlist*. A wide selection of similarity and dissimilarity measures is available, see the measure() option.

While mds computes dissimilarities from the observations, mdslong and mdsmat are for use when you already have similarity or dissimilarity information. mdslong and mdsmat offer the same statistical features but require different data organizations. mdslong expects the proximity information in a "long format" (pairwise or dyadic form), whereas mdsmat performs MDS on a symmetric dissimilarity matrix; see [MV] **mdslong** and [MV] **mdsmat**.

Options

> Model

id(*varname*) is required; it specifies a variable that identifies observations. A warning message is displayed if *varname* has duplicate values.

unit$\left[\,(varlist_2)\,\right]$ specifies variables that are transformed to min = 0 and max = 1 before entering in the computation of similarities or dissimilarities. unit by itself, without an argument, is a shorthand for unit(_all). Variables in unit() should not be included in std().

std$\left[\,(varlist_3)\,\right]$ specifies variables that are transformed to mean = 0 and sd = 1 before entering in the computation of similarities or dissimilarities. std by itself, without an argument, is a shorthand for std(_all). Variables in std() should not be included in unit().

measure(*measure*) specifies the similarity or dissimilarity measure. The default is measure(L2), Euclidean distance. See [MV] *measure_option* for detailed descriptions of the supported measures.

If a similarity measure is selected, the computed similarities will first be transformed into dissimilarities, before proceeding with the scaling; see the s2d() option below.

Classical metric MDS with Euclidean distance is equivalent to principal component analysis (see [MV] **pca**); the MDS configuration coordinates are the principal components.

s2d(*conversion*) specifies how measures in similarity form are converted to dissimilarities. The following conversions are available:

$$
\begin{array}{lll}
\text{standard} & d_{ij} = \sqrt{s_{ii} + s_{jj} - 2s_{ij}} & \text{the default} \\
\text{oneminus} & d_{ij} = 1 - s_{ij} &
\end{array}
$$

Obviously, s2d() should only be specified with measures in similarity form.

dimension(#) specifies the dimension of the approximating configuration. # defaults to 2 and should not exceed the number of positive eigenvalues of the centered distance matrix.

addconstant specifies that if the double-centered distance matrix is not positive semidefinite (psd), a constant should be added to the squared distances to make it psd and, hence, Euclidean.

> Reporting

neigen(#) specifies the number of eigenvalues to be included in the table. The default is neigen(10). Specifying neigen(0) suppresses the table.

config displays the table with the coordinates of the approximating configuration. This table may also be displayed using the postestimation command estat config; see [MV] **mds postestimation**.

noplot suppresses the graph of the approximating configuration. Note that the graph can still be produced later via mdsconfig, which also allows the standard graphics options for fine-tuning the plot; see [MV] **mds postestimation**.

Remarks

Remarks are presented under the headings

> *Introduction*
> *Euclidean distances*
> *Non-Euclidean dissimilarity measures*

Introduction

Multidimensional scaling (MDS) is a dimension-reduction and visualization technique. Dissimilarities (for instance, Euclidean distances) between observations in a high-dimensional space are represented in a lower-dimensional space (typically two dimensions) so that the Euclidean distance in the lower-dimensional space approximates the dissimilarities in the higher-dimensional space. See Kruskal and Wish (1978) for a brief nontechnical introduction to MDS. Young and Hamer (1994) and Borg and Groenen (1996) offer more advanced textbook-sized treatments.

If you already have the similarities or dissimilarities of the n objects, you should continue by reading [MV] **mdsmat**.

In many applications of MDS, however, the similarity or dissimilarity of objects is not measured, but *defined* by the researcher in terms of variables ("attributes") $x_1, \ldots, x_k$ that are measured on the objects. The pairwise dissimilarity of objects can be expressed using a variety of similarity or dissimilarity measures in the attributes (e.g., Mardia, Kent, and Bibby 1979, section 13.4; Cox and Cox 2001, section 1.3). A common measure is the Euclidean distance L2 between the attributes of the objects i and j:

$$L2_{ij} = \sqrt{(x_{i1} - x_{j1})^2 + (x_{i2} - x_{j2})^2 + \cdots + (x_{ik} - x_{jk})^2}$$

A popular alternative is the L1 distance, also known as the `cityblock` or `Manhattan` distance. In comparison to L2, L1 gives less influence to larger differences in attributes:

$$L1_{ij} = |x_{i1} - x_{j1}| + |x_{i2} - x_{j2}| + \cdots + |x_{ik} - x_{jk}|$$

In contrast, we may also define the extent of dissimilarity between two observations as the maximum absolute difference in the attributes and give a larger influence to larger differences:

$$\texttt{Linfinity}_{ij} = \max(|x_{i1} - x_{j1}|, |x_{i2} - x_{j2}|, \cdots, |x_{ik} - x_{jk}|)$$

These three measures are special cases of the Minkowski distance $L(q)$, for $q = 2$ (L2), $q = 1$ (L1), and $q = \infty$ (`Linfinity`), respectively. Minkowski distances with other values of q may be used, as well. Stata supports a wide variety of other similarity and dissimilarity measures, both for continuous variables and for binary variables. See [MV] *measure_option* for details.

Multidimensional scaling constructs approximations for dissimilarities, not for similarities. Thus if a similarity measure is specified, `mds` first transforms the similarities into dissimilarities. Two methods to do this are available. The default `standard` method,

$$\text{dissim}_{ij} = \sqrt{\text{sim}_{ii} - 2\text{sim}_{ij} + \text{sim}_{jj}}$$

has a useful property: if the similarity matrix is positive semidefinite, a property satisfied by most similarity measures, the standard dissimilarities are Euclidean.

In many applications, the number of observations exceeds the number of variables on which the observations are compared, but this is not a requirement for MDS. MDS creates an $n \times n$ dissimilarity matrix $\mathbf{D}$ from the n observations on k variables. It then constructs an approximation of $\mathbf{D}$ by the Euclidean distances in a matching configuration $\mathbf{Y}$ of n points in p-dimensional space:

$$\text{dissimilarity}(x_i, x_j) \approx L2(y_i, y_j) \quad \text{for all } i, j$$

Typically, of course, $p \ll k$, and most often $p = 1$, 2, or 3.

A wide variety of MDS methods have been proposed. mds performs classical scaling that has its roots in Young and Householder (1938) and Torgerson (1952). Classical scaling requires complete and symmetric dissimilarity interval-level data.

Euclidean distances

▷ Example 1

The most popular dissimilarity measure is Euclidean distance. We illustrate with data from table 7.1 of Yang and Trewn (2004, 182). This dataset consists of 8 variables with nutrition data on 25 breakfast cereals.

```
. use http://www.stata-press.com/data/r9/cerealnut
(Cereal Nutrition)

. describe
Contains data from http://www.stata-press.com/data/r9/cerealnut.dta
  obs:           25                          Cereal Nutrition
  vars:           9                          24 Feb 2005 17:19
  size:        1,150 (99.9% of memory free)  (_dta has notes)
```

variable name	storage type	display format	value label	variable label
brand	str25	%25s		Cereal Brand
calories	int	%9.0g		Calories (Cal/oz)
protein	byte	%9.0g		Protein (g)
fat	byte	%9.0g		Fat (g)
Na	int	%9.0g		Na (mg)
fiber	float	%9.0g		Fiber (g)
carbs	float	%9.0g		Carbs (g)
sugar	byte	%9.0g		Sugar (g)
K	int	%9.0g		K (mg)

```
Sorted by:

. summarize calories-K, sep(4)
```

Variable	Obs	Mean	Std. Dev.	Min	Max
calories	25	109.6	21.30728	50	160
protein	25	2.68	1.314027	1	6
fat	25	.92	.7593857	0	2
Na	25	195.8	71.32204	0	320
fiber	25	1.7	2.056494	0	9
carbs	25	15.3	4.028544	7	22
sugar	25	7.4	4.609772	0	14
K	25	90.6	77.5043	15	320

```
. replace brand = subinstr(brand," ","_",.)
(20 real changes made)
```

We replaced spaces in the cereal brand names with underscores to avoid confusing which words in the brand names are associated with which points in the graphs we are about to produce. Removing spaces is not required.

The default dissimilarity measure used by mds is the Euclidean distance L2 computed on the raw data (unstandardized). The summary of the 8 nutrition variables shows that K, Na, and calories—having much larger standard deviations—will largely determine the Euclidean distances.

```
. mds calories-K, id(brand)
```

```
Classical metric multidimensional scaling
    dissimilarity: L2, computed on 8 variables
```

			Number of obs	=	25
Eigenvalues > 0	=	8	Mardia fit measure 1 =		0.9603
Retained dimensions	=	2	Mardia fit measure 2 =		0.9970

Dimension	Eigenvalue	abs(eigenvalue)		(eigenvalue)^2	
		Percent	Cumul.	Percent	Cumul.
1	158437.92	56.95	56.95	67.78	67.78
2	108728.77	39.08	96.03	31.92	99.70
3	10562.645	3.80	99.83	0.30	100.00
4	382.67849	0.14	99.97	0.00	100.00
5	69.761715	0.03	99.99	0.00	100.00
6	12.520822	0.00	100.00	0.00	100.00
7	5.7559984	0.00	100.00	0.00	100.00
8	2.2243244	0.00	100.00	0.00	100.00

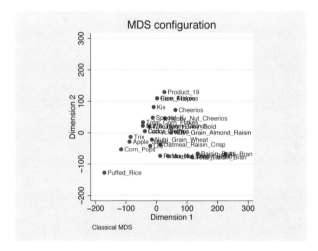

The default MDS configuration graph can be improved upon using the `mdsconfig` postestimation command. We will demonstrate this in a moment. But first, we explain the output of `mds`.

`mds` has extracted two dimensions, which is the default action. To assess goodness of fit, the two statistics proposed by Mardia are reported (see Mardia, Kent, and Bibby 1979, section 14.4). The statistics are defined in terms of the eigenvalues of the double-centered distance matrix. If the dissimilarities are truly Euclidean, all eigenvalues are non-negative. Look at the eigenvalues. We may interpret these as the extent to which the dimensions account for dissimilarity between the cereals. Depending on whether you look at the eigenvalues or squared eigenvalues, it takes two or three dimensions to account for over 99% of the dissimilarity.

We can produce a prettier configuration plot using the `mdsconfig` command; see [MV] **mds postestimation** for details.

```
. generate place = 3
. replace place = 9 if inlist(brand,"Rice_Krispies","Nut_&_Honey_Crunch",
>                              "Special_K","Raisin_Nut_Bran","Lucky_Charms")
(5 real changes made)
. replace place = 12 if inlist(brand,"Mueslix_Crispy_Blend")
(1 real change made)
. mdsconfig, autoaspect mlabvpos(place)
```

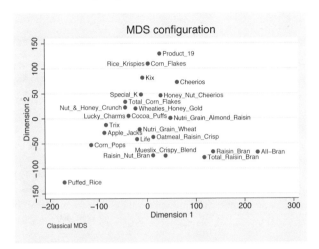

The *marker_label_option* `mlabvpos()` allowed fine control over the placement of the cereal brand names. We created a variable called `place` giving clock positions where the cereal names were to appear in relation to the plotted point. We set these to minimize overlap of the names. We also requested the `autoaspect` option to obtain better use of the graphing region while preserving the scale of the x- and y-axes.

MDS has placed the cereals so that all the brands fall within a triangle defined by Product 19, All-Bran, and Puffed Rice. You can examine the graph to see how close your favorite cereal is to the other cereals.

But, as we saw from the variable summary, three of the eight variables are controlling the distances. If we want to provide for a more equal footing for the eight variables, we can request that `mds` compute the Euclidean distances on standardized variables. Euclidean distance based on standardized variables is also known as the *Karl Pearson distance*. We obtain standardized measures with the option `std`.

(Continued on next page)

```
. mds calories-K, id(brand) std noplot
Classical metric multidimensional scaling
      dissimilarity: L2, computed on 8 variables
```

			Number of obs =	25
Eigenvalues > 0 =	8		Mardia fit measure 1 =	0.5987
Retained dimensions =	2		Mardia fit measure 2 =	0.7697

Dimension	Eigenvalue	abs(eigenvalue)		(eigenvalue)^2	
		Percent	Cumul.	Percent	Cumul.
1	65.645395	34.19	34.19	49.21	49.21
2	49.311416	25.68	59.87	27.77	76.97
3	38.826608	20.22	80.10	17.21	94.19
4	17.727805	9.23	89.33	3.59	97.78
5	11.230087	5.85	95.18	1.44	99.22
6	8.2386231	4.29	99.47	0.78	99.99
7	.77953426	0.41	99.87	0.01	100.00
8	.24053137	0.13	100.00	0.00	100.00

Notice that it now takes more MDS retained dimensions to account for over 99% of the underlying distances. For this example, we have still retained only two dimensions. We specified the `noplot` option because we wanted to exercise control over the configuration plot using the `mdsconfig` command. We generated a variable named `pos` that will help minimize cereal brand name overlap. The `xnegate` option negates the data in the first dimension to give a more natural left to right orientation.

```
. generate pos = 3
. replace pos = 5 if inlist(brand,"Honey_Nut_Cheerios","Raisin_Nut_Bran",
>                       "Nutri_Grain_Almond_Raisin")
(3 real changes made)
. replace pos = 8 if inlist(brand,"Oatmeal_Raisin_Crisp")
(1 real change made)
. replace pos = 9 if inlist(brand,"Corn_Pops","Trix","Nut_&_Honey_Crunch",
>                       "Rice_Krispies","Wheaties_Honey_Gold")
(5 real changes made)
. replace pos = 12 if inlist(brand,"Life")
(1 real change made)
```

. mdsconfig, autoaspect xnegate mlabvpos(pos)

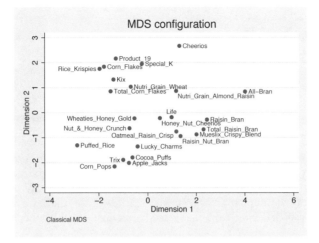

This configuration plot, based on the standardized variables, better incorporates all the nutrition data. If you are familiar with these cereal brands, it is easy to spot groups of similar cereals appearing near each other. The bottom left corner has several of the most sweetened cereals. The brands containing the word "Bran" all appear to the right of center. Rice Krispies and Puffed Rice are the farthest to the left.

Classical multidimensional scaling based on standardized Euclidean distances is actually equivalent to a principal component analysis of the correlation matrix of the variables. See Mardia, Kent, and Bibby (1979, section 14.3) for details.

We now demonstrate this property by doing a principal component analysis extracting the leading two principal components. See [MV] **pca** for details.

. pca calories-K, comp(2)

Principal components/correlation			Number of obs	=	25
			Number of comp.	=	2
			Trace	=	8
Rotation: (unrotated = principal)			Rho	=	0.5987

Component	Eigenvalue	Difference	Proportion	Cumulative
Comp1	2.73522	.680583	0.3419	0.3419
Comp2	2.05464	.436867	0.2568	0.5987
Comp3	1.61778	.879117	0.2022	0.8010
Comp4	.738659	.270738	0.0923	0.8933
Comp5	.46792	.124644	0.0585	0.9518
Comp6	.343276	.310795	0.0429	0.9947
Comp7	.0324806	.0224585	0.0041	0.9987
Comp8	.0100221	.	0.0013	1.0000

Principal components (eigenvectors)

Variable	Comp1	Comp2	Unexplained
calories	0.1992	-0.0632	.8832
protein	0.3376	0.4203	.3253
fat	0.3811	-0.0667	.5936
Na	0.0962	0.5554	.3408
fiber	0.5146	0.0913	.2586
carbs	-0.2574	0.4492	.4043
sugar	0.2081	-0.5426	.2765
K	0.5635	0.0430	.1278

Notice that the proportion and cumulative proportion of the eigenvalues in the PCA match the percentages from MDS. We will ignore the interpretation of the principal components, but move directly to the principal coordinates, also known as the scores of the PCA. We make a plot of the first and second scores, using the `scoreplot` command; see [MV] **scoreplot**. We specify the `mlabel()` option to label the cereals and the `mlabvpos()` option for fine control over placement of the brand names.

```
. replace pos = 11 if inlist(brand,"All-Bran")
(1 real change made)
. scoreplot, mlabel(brand) mlabvpos(pos)
```

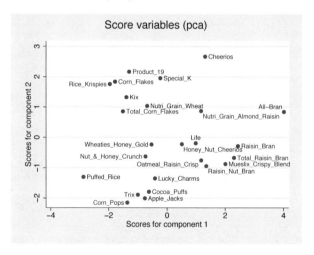

Compare this PCA score plot with the MDS configuration plot. Apart from some differences in how the graphs were rendered, they are the same.

◁

Non-Euclidean dissimilarity measures

With non-Euclidean dissimilarity measures, the parallel between PCA and MDS no longer holds.

▷ Example 2

To illustrate MDS with non-Euclidean distance measures, we will analyze books on multivariate statistics. Gifi (1990) reports on the number of pages devoted to six topics in 20 textbooks on multivariate statistics. We added similar data on five more recent books.

```
. use http://www.stata-press.com/data/r9/mvstatsbooks
. describe
Contains data from http://www.stata-press.com/data/r9/mvstatsbooks.dta
  obs:            25
  vars:            8                         15 Mar 2005 16:27
  size:          825 (99.9% of memory free)  (_dta has notes)
```

variable name	storage type	display format	value label	variable label
author	str17	%17s		
math	int	%9.0g		math other than statistics (e.g., linear algebra)
corr	int	%9.0g		correlation and regression, including linear structural and functional equations
fact	byte	%9.0g		factor analysis and principal component analysis
cano	byte	%9.0g		canonical correlation analysis
disc	int	%9.0g		discriminant analysis, classification, and cluster analysis
stat	int	%9.0g		statistics, incl. dist. theory, hypothesis testing & est.; categorical data
mano	int	%9.0g		manova and the general linear model

```
Sorted by:
```

A brief description of the topics is given in the variable labels. For more details, we refer to Gifi (1990, 15). Here are the data:

(*Continued on next page*)

. list, noobs

author	math	corr	fact	cano	disc	stat	mano
Roy57	31	0	0	0	0	164	11
Kendall57	0	16	54	18	27	13	14
Kendall75	0	40	32	10	42	60	0
Anderson58	19	0	35	19	28	163	52
CooleyLohnes62	14	7	35	22	17	0	56
CooleyLohnes71	20	69	72	33	55	0	32
Morrison67	74	0	86	14	0	84	48
Morrison76	78	0	80	5	17	105	60
VandeGeer67	74	19	33	12	26	0	0
VandeGeer71	80	68	67	15	29	0	0
Dempster69	108	48	4	10	46	108	0
Tasuoka71	109	13	5	17	39	32	46
Harris75	16	35	69	24	0	26	41
Dagnelie75	26	86	60	6	48	48	28
GreenCaroll76	290	10	6	0	8	0	2
CailliezPages76	184	48	82	42	134	0	0
Giri77	29	0	0	0	41	211	32
Gnanadesikan77	0	19	56	0	39	75	0
Kshirsagar78	0	22	45	42	60	230	59
Thorndike78	30	128	90	28	48	0	0
MardiaKentBibby79	34	28	68	19	67	131	55
Seber84	16	0	59	13	116	129	101
Stevens96	23	87	67	21	30	43	249
EverittDunn01	0	54	65	0	56	20	30
Rencher02	38	0	71	19	105	135	131

For instance, the 1979 book by Mardia, Kent, and Bibby has 34 pages on mathematics (mostly linear algebra), 28 pages on correlation, regression, and related topics (in this particular case, simultaneous equations), etc. In most of these books, some pages are not classified. Anyway, the number of pages and the amount of information per page vary widely among the books. A Euclidean distance measure is not appropriate here. Standardization does not help us here—the problem is not differences in the scales of the variables, but those in the observations. One possibility is to transform the data into *compositional data* by dividing the variables by the total number of classified pages. See Mardia, Kent, and Bibby (1979, 377–380) for a discussion of specialized dissimilarity measures for compositional data. However, we can also use the correlation between observations (not between variables) as the similarity measure. The higher the correlation between the attention given to the various topics, the more similar two textbooks are. We do a classical MDS, suppressing the plot to first assess the quality of a two-dimensional representation.

```
. mds math-mano, id(author) measure(corr) noplot

Classical metric multidimensional scaling
      similarity: correlation, computed on 7 variables
    dissimilarity: sqrt(2(1-similarity))
                                           Number of obs      =        25
    Eigenvalues > 0       =        6       Mardia fit measure 1 =    0.6680
    Retained dimensions   =        2       Mardia fit measure 2 =    0.8496
```

Dimension	Eigenvalue	abs(eigenvalue) Percent	Cumul.	(eigenvalue)^2 Percent	Cumul.
1	8.469821	38.92	38.92	56.15	56.15
2	6.0665813	27.88	66.80	28.81	84.96
3	3.8157101	17.53	84.33	11.40	96.35
4	1.6926956	7.78	92.11	2.24	98.60
5	1.2576053	5.78	97.89	1.24	99.83
6	.45929376	2.11	100.00	0.17	100.00

Again the quality of a two-dimensional approximation is somewhat unsatisfactory with 67% and 85% of the variation accounted for according to the two Mardia criteria. Still, let's look at the plot, using a title that refers to the self-referential aspect of the analysis (Smullyan 1986). We reposition some of the author labels to enhance readability using the mlabvpos() option.

```
. gen spot = 3
. replace spot = 2 if inlist(author,"Seber84","Kshirsagar78","Kendall75")
(3 real changes made)
. replace spot = 5 if author=="MardiaKentBibby79"
(1 real change made)
. replace spot = 9 if inlist(author, "Dagnelie75","Rencher02",
>   "GreenCaroll76","EverittDunn01","CooleyLohnes62","Morrison67")
(6 real changes made)
. mdsconfig, mlabvpos(spot) title(This plot needs no title)
```

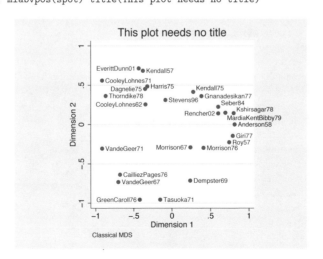

A striking characteristic of the plot is that the textbooks seem to be located on a circle. This is a phenomenon that is regularly encountered in multidimensional scaling and was labeled the "horse shoe effect" by Kendall (1971, 215–251). This phenomenon seems to occur especially in situations in which a one-dimensional representation of objects needs to be constructed, e.g., in *seriation*

applications, from data in which small dissimilarities were measured accurately but moderate and larger dissimilarities are "lumped together".

◁

❏ Technical Note

These data could also be analyzed in differently. A particularly interesting method is correspondence analysis (CA), which seeks a simultaneous geometric representation of the rows (textbooks) and columns (topics). We used `camat` to analyze these data. The results for the textbooks were not very different. Textbooks that were mapped as similar using MDS were also mapped this way by CA. The Green and Carroll book that appeared much different from the rest was also displayed away from the rest by CA. In the CA biplot, it was immediately clear that this book was so different because its pages were predominantly classified by Gifi (1990) as mathematical. But CA also located the topics in this space. The pattern was easy to interpret and not unexpected. The seven topics were mapped in three groups. `math` and `stat` appear as two groups by themselves, and the five applied topics were mapped close together. See [MV] **ca** for information on the `ca` command.

❏

Saved Results

mds saves in e():

Scalars
e(N)	number of observations
e(p)	number of dimensions in the approximating configuration
e(np)	number of strictly positive eigenvalues
e(addcons)	constant added to squared dissimilarities to force positive semidefiniteness
e(mardia1)	Mardia measure 1
e(mardia2)	Mardia measure 2

Macros
e(cmd)	mds
e(method)	classical
e(varlist)	variables used in computing similarities or dissimilarities
e(id)	observation identifier variable
e(numid)	1 if e(id) is numeric, 0 otherwise
e(dname)	similarity or dissimilarity measure name
e(dtype)	similarity or dissimilarity
e(s2d)	standard or oneminus
e(unique)	1 if eigenvalues are distinct, 0 otherwise
e(properties)	nob noV eigen
e(predict)	mds_p
e(estat_cmd)	mds_estat

Matrices

e(D)	dissimilarity matrix
e(Y)	approximating configuration coordinates
e(Ev)	eigenvalues
e(coding)	variable standardization values; first column has value to subtract and second column has divisor
e(linearf)	two element vector defining the linear transformation; distance equals first element plus second element times dissimilarity

Functions

e(sample)	marks estimation sample

Methods and Formulas

mds is implemented as an ado-file.

mds creates a dissimilarity matrix D according to the *measure* specified in option measure(). See [MV] *measure_option* for descriptions of these measures. Subsequently, mds uses mdsmat to compute the classical MDS solution for D. See [MV] **mdsmat** for information on the methods and formulas.

References

Borg, I. and P. Groenen. 1996. *Modern Multidimensional Scaling: Theory and Applications*. New York: Springer.

Cox, T. F. and M. A. A. Cox. 2001. *Multidimensional Scaling*. 2nd ed. Boca Raton, FL: Chapman & Hall.

Gifi, A. 1990. *Nonlinear Multivariate Analysis*. New York: Wiley.

Kendall, D. G. 1971. Seriation from abundance matrices. *Mathematics in the Archaeological and Historical Sciences*, Edinburgh: Edinburgh University Press.

Kruskal, J. B. and M. Wish. 1978. *Multidimensional Scaling*. Newbury Park, CA: Sage.

Lingoes, J. C. 1971. Some boundary conditions for a monotone analysis of symmetric matrices. *Psychometrika* 36: 195–203.

Mardia, K. V., J. T. Kent, and J. M. Bibby. 1979. *Multivariate Analysis*. New York: Academic Press.

Smullyan, R. 1986. *This Book Needs No Title: A Budget of Living Paradoxes*. New York: Touchstone.

Torgerson, W. S. 1952. Multidimensional scaling: I. Theory and method. *Psychometrika* 17: 401–419.

Yang, K. and J. Trewn. 2004. *Multivariate Statistical Methods in Quality Management*. New York: McGraw–Hill.

Young, F. W. and R. M. Hamer. 1994. *Theory and Applications of Multidimensional Scaling*. Hillsdale, NJ: Erlbaum Associates.

Young, G. and A. S. Householder. 1938. Discussion of a set of points in terms of their mutual distances. *Psychometrika* 3: 19–22.

Also See

Complementary:	[MV] **mds postestimation**
Related:	[MV] **biplot**, [MV] **ca**, [MV] **factor**, [MV] **mdslong**, [MV] **mdsmat**, [MV] **pca**
Background:	[U] **11.1.10 Prefix commands**, [U] **20 Estimation and postestimation commands**

Title

> **mds postestimation** — Postestimation tools for mds, mdsmat, and mdslong

Description

The following postestimation commands are of special interest after `mds`, `mdslong`, and `mdsmat`:

command	description
estat config	coordinates of the approximating configuration
estat correlations	correlations between dissimilarities and approximating distances
estat pairwise	pairwise dissimilarities, approximating distances, and raw residuals
estat quantiles	quantiles of the residuals per object
estat stress	Kruskal stress (loss) measure
* estat summarize	estimation sample summary
mdsconfig	plot of approximating configuration
mdsshepard	Shepard diagram

* `estat summarize` is not available after `mdsmat`.

For more information on these commands, see below.

In addition, the following standard postestimation commands are available:

command	description
* estimates	cataloging estimation results
predict	approximating configuration, dissimilarities, distances, and residuals

* All `estimates` subcommands except `table` and `stats` are available.

See the corresponding entries in the *Stata Base Reference Manual* for more information.

Special-interest postestimation commands

`estat config` lists the coordinates of the approximating configuration.

`estat correlations` lists the Pearson and Spearman correlations between the dissimilarities and the approximating distances for each object.

`estat pairwise` lists the pairwise statistics: the dissimilarities, the approximating distances, and the raw residuals.

`estat quantiles` lists the quantiles of the residuals per object.

`estat stress` displays the Kruskal stress (loss) measure between the (transformed) dissimilarities and fitted distances per object.

`estat summarize` summarizes the variables in the MDS over the estimation sample. After `mds`, `estat summarize` also reports whether and how variables were transformed before computing similarities or dissimilarities.

mdsconfig produces a plot of the approximating Euclidean configuration. The first two approximating dimensions are plotted.

mdsshepard produces a Shepard diagram of the dissimilarities against the approximating Euclidean distances. Ideally, the points in the plot should be close to the "$x=y$ line". Optionally, separate plots are generated for each "row" (value of id()).

Syntax for predict

predict [*type*] *newvarlist* [*if*] [*in*] [, *statistic options*]

statistic	description
Main	
config	approximating configuration; specify dimension() or fewer variables
pairwise	dissimilarity, distance, and raw residuals; specify 3 variables

options	description
Main	
* saving(*filename*)	save results to *filename*
replace	overwrite *filename*
notransform	suppress linear transformation of the dissimilarities; pairwise only
full	generate all pairs; default is only the ordered pairs, excluding the diagonal; pairwise only

  * saving() is required after mdsmat, after mds if pairwise is selected, and after mdslong if config is selected.

Options for predict

⌐ Main ⌐

config generates variables containing the approximating configuration in Euclidean space. Specify as many new variables as approximating dimensions (as determined by the dimension() option of mds, mdsmat, or mdslong), though you may specify fewer. Note that estat config displays the same information but does not store the information in variables. After mdsmat and mdslong, you must also specify the saving() option.

pairwise generates three new variables containing pairwise comparison statistics: dissimilarities, approximating Euclidean distances, and raw residuals. Specify three new variable names. Note that estat pairwise displays the same information but does not store the information in variables.

After mds and mdsmat, you must also specify the saving() option. With n objects, the pairwise dataset has $n(n-1)/2$ observations. In addition to the three requested variables, predict produces variables *id*1 and *id*2, which identify pairs of objects. With mds, *id* is the name of the identification variable (id() option), and with mdsmat it is "Category".

saving(*filename*) is required after mdsmat, after mds if pairwise is selected, and after mdslong if config is selected. saving() indicates that the generated variables are to be created in a new Stata dataset and saved in the file named *filename*. Unless saving() is specified, the variables are generated in the current dataset. In many cases saving() is required; see config and pairwise for details.

replace indicates that *filename* specified in saving() may be overwritten.

notransform suppresses the linear transformation of the dissimilarities to match the fitted Euclidean distances. notransform may only be specified with pairwise.

full creates predictions for all pairs of objects (j_1, j_2). The default is to generate predictions only for pairs (j_1, j_2) with $j_1 > j_2$. full may only be specified with pairwise.

Syntax for estat

List the coordinates of the approximating configuration

    estat config $\big[$, maxlength(#) format(%fmt) $\big]$

List the Pearson and Spearman correlations

    estat correlations $\big[$, maxlength(#) format(%fmt) nototal $\big]$

List the pairwise statistics

    estat pairwise $\big[$, maxlength(#) notransform full separator $\big]$

List the quantiles of the residuals

    estat quantiles $\big[$, maxlength(#) format(%fmt) nototal notransform $\big]$

Display the Kruskal stress (loss) measure

    estat stress $\big[$, maxlength(#) format(%fmt) nototal notransform $\big]$

Summarize the variables in MDS

    estat summarize $\big[$, labels $\big]$

options	description
Main	
maxlength(#)	maximum number of characters for displaying object names; default is 12
format(%fmt)	display format
nototal	suppress display of overall summary statistics
notransform	suppress linear transform of the dissimilarities
full	display all pairs (j_1, j_2); default is $(j_1 > j_2)$ only
separator	draw separating lines
labels	display variable labels

Options for estat

maxlength(#), an option used with all but estat summarize, specifies the maximum number of characters of the object names to be displayed; the default is maxlength(12).

format(%fmt), an option used with estat config, estat correlations, estat quantiles, and estat stress, specifies the display format; the default differs between the subcommands.

nototal, an option used with estat correlations, estat quantiles, and estat stress, suppresses the overall summary statistics.

notransform, an option used with estat pairwise, estat quantiles, and estat stress, suppresses the linear transformation of the dissimilarities to match the fitted Euclidean distances.

full, an option used with estat pairwise, displays a row for all pairs (j_1, j_2). The default is to display rows only for pairs with $j_1 > j_2$.

separator, an option used with estat pairwise, draws separating lines between blocks of rows corresponding to changes in the first of the pair of objects.

labels, an option used with estat summarize, displays variable labels.

Syntax for mdsconfig

mdsconfig [, *options*]

options	description
Main	
<u>dim</u>ensions(# #)	two dimensions to be displayed; default is dimensions(2 1)
<u>x</u>negate	negate data relative to the x-axis
<u>y</u>negate	negate data relative to the y-axis
<u>auto</u>aspect	adjust aspect ratio based on the data; default aspect ratio is 1
<u>max</u>length(#)	maximum number of characters used in marker labels
cline_options	affect rendition of the lines connecting points
marker_options	change look of markers (color, size, etc.)
marker_label_options	change look or position of marker labels
Y-Axis, X-Axis, Title, Caption, Overall	
twoway_options	any options other than by() documented in [G] ***twoway_options***

Options for mdsconfig

dimensions(# #) identifies the dimensions to be displayed. For instance, dimensions(3 2) plots the third dimension (vertically) versus the second dimension (horizontally). The dimension number cannot exceed the number of extracted dimensions. The default is dimensions(2 1).

xnegate specifies that the data be negated relative to the x-axis.

ynegate specifies that the data be negated relative to the y-axis.

autoaspect specifies that the aspect ratio be automatically adjusted based on the range of the data to be plotted. This option can make some plots more readable. By default, mdsconfig uses an aspect ratio of one, producing a square plot. Some plots will have little variation in the y-axis direction, and use of the autoaspect option will better fill the available graph space and preserve the equivalence of distance in the x- and y-axes.

As an alternative to autoaspect, the *twoway_option* aspectratio() can be used to override the default aspect ratio. mdsconfig accepts the aspectratio() option as a suggestion only and will override it when necessary to produce plots with balanced axes; i.e., distance on the x-axis equals distance on the y-axis.

twoway_options, such as xlabel(), xscale(), ylabel(), and yscale(), should be used with caution. These options are accepted but may have unintended side effects on the aspect ratio.

maxlength(#) specifies the maximum number of characters for object names used to mark the points; the default is maxlength(12).

cline_options affect the rendition of the lines connecting the plotted points; see [G] ***cline_options***. Note that if you are drawing connected lines, the appearance of the plot depends on the sort order of the data.

marker_options affect the rendition of the markers drawn at the plotted points, including their shape, size, color, and outline; see [G] ***marker_options***.

marker_label_options specify if and how the markers are to be labeled; see [G] ***marker_label_options***.

Y-Axis, X-Axis, Title, Caption, Overall
---

twoway_options are any of the options documented in [G] ***twoway_options***, excluding by(). These include options for titling the graph (see [G] ***title_options***) and options for saving the graph to disk (see [G] ***saving_option***). See autoaspect above for a warning against using options such as xlabel(), xscale(), ylabel(), and yscale().

Syntax for mdsshepard

mdsshepard [, *mdsshepard_options*]

mdsshepard_options	description
Main	
separate	draw separate Shepard diagrams for each object
notransform	suppress the linear transformation of the dissimilarities
autoaspect	adjust aspect ratio based on the data; default aspect ratio is 1
cline_options	affect rendition of the lines connecting points
marker_options	change look of markers (color, size, etc.)
marker_label_options	change look or position of marker labels
Y-Axis, X-Axis, Title, Caption, Overall	
twoway_options	any options other than by() documented in [G] ***twoway_options***
By	
byopts(*by_option*)	affect the rendition of combined graphs; separate only

Options for mdsshepard

Main

separate displays separate plots of each value of the ID variable. Note that this may be time consuming if the number of distinct ID values is not small.

notransform suppresses the linear transformation of the dissimilarities to match the fitted Euclidean distances.

autoaspect specifies that the aspect ratio be automatically adjusted based on the range of the data to be plotted. By default, mdsshepard uses an aspect ratio of one, producing a square plot.

See the description of the autoaspect option of mdsconfig for more details.

cline_options affect the rendition of the lines connecting the plotted points; see [G] *cline_options*. Note that if you are drawing connected lines, the appearance of the plot depends on the sort order of the data.

marker_options affect the rendition of the markers drawn at the plotted points, including their shape, size, color, and outline; see [G] *marker_options*.

marker_label_options specify if and how the markers are to be labeled; see [G] *marker_label_options*.

Y-Axis, X-Axis, Title, Caption, Overall

twoway_options are any of the options documented in [G] *twoway_options* excluding by(). These include options for titling the graph (see [G] *title_options*) and options for saving the graph to disk (see [G] *saving_option*). See the autoaspect option of mdsconfig for a warning against using options such as xlabel(), xscale(), ylabel(), and yscale().

By

byopts(*by_option*) is documented in [G] *by_option*. This option affects the appearance of the combined graph and is allowed only with the separate option.

Remarks

Remarks are presented under the headings

> Postestimation statistics
> Matching configuration plot and the Shepard diagram
> Predictions

Postestimation statistics

After an MDS analysis, several facilities can help you better understand the analysis and, in particular, to assess the quality of the lower-dimensional Euclidean representation.

▷ Example 1

We illustrate the MDS postestimation facilities using the Morse code digit-similarity dataset; see example 1 in [MV] **mdslong**.

```
. use http://www.stata-press.com/data/r9/morse_long
(Morse data (Rothkopf 1957))
. gen sim = freqsame/100
. mdslong sim, id(digit1 digit2) s2d(standard) noplot
(output omitted)
```

MDS has produced a two-dimensional configuration with Euclidean distances approximating the dissimilarities between the Morse codes for digits. This configuration may be examined using the `estat config` command; see `mdsconfig` if you want to plot the configuration.

```
. estat config
```

Approximating configuration in 2-dimensional Euclidean space

digit1	dim1	dim2
0	0.5690	-0.0162
1	0.4561	0.3384
2	0.0372	0.5854
3	-0.3878	0.4516
4	-0.5800	0.0770
5	-0.5458	0.0196
6	-0.3960	-0.4187
7	-0.0963	-0.5901
8	0.3124	-0.3862
9	0.6312	-0.0608

This configuration is not unique. A translation, a reflection, and an orthonormal rotation of the configuration do not affect the interpoint Euclidean distances. All such transformations are equally reasonable MDS solutions. Thus you should not interpret aspects of these numbers (or of the configuration plot) that are not invariant under these transformations.

We now turn to the three `estat` subcommands that analyze the MDS residuals, i.e., the differences between the dissimilarities and the matching Euclidean distances. There is a catch here. The raw residuals of classical MDS are not well behaved. For instance, the sum of the raw residuals is not zero—often it is not even close. The classical MDS solution does *not* minimize the sum of squares of the raw residuals (Mardia, Kent, and Bibby 1979, 406–408). To create reasonable residuals with classical MDS, the dissimilarities can be transformed to approximate the Euclidean distances, just as in metric MDS. We use a linear transform f, fitted by least squares. This is equivalent to Kruskal's Stress1 loss function. The modified residuals are defined as the differences between the linearly transformed dissimilarities and the matching Euclidean distances.

The three `estat` subcommands summarize the residuals in different ways. `estat stress` displays the Kruskal loss or stress measures for each object and the overall total.

```
. estat stress
```

Stress between adjusted dissimilarities and Euclidean distances

digit1	Kruskal
0	0.1339
1	0.1255
2	0.1972
3	0.2028
4	0.2040
5	0.2733
6	0.1926
7	0.1921
8	0.1715
9	0.1049
Total	0.1848

Second, the quantiles of the residuals are available, both overall and for the subgroup of object pairs in which an object is involved.

. estat quantiles

Quantiles of adjusted residuals

digit1	N	min	p25	q50	q75	max
0	9	-.111732	-.088079	-.028917	.11202	.220399
1	9	-.170063	-.137246	-.041244	.000571	.11202
2	9	-.332717	-.159472	-.072359	.074999	.234866
3	9	-.136251	-.120398	-.072359	.105572	.365833
4	9	-.160797	-.014099	.03845	.208215	.355053
5	9	-.09971	-.035357	.176337	.325043	.365833
6	9	-.137246	-.113564	-.075008	.177448	.325043
7	9	-.332717	-.170063	-.124129	.03845	.176337
8	9	-.186452	-.134831	-.041244	.075766	.220399
9	9	-.160797	-.104403	-.088079	-.064316	-.030032
Total	90	-.332717	-.113564	-.041244	.105572	.365833

The dissimilarities for the Morse code of digit 5 are fitted considerably worse than for all other Morse codes. Digit 5 has the largest Kruskal stress (0.273) and median residual (0.176).

Finally, estat correlations displays the Pearson and Spearman correlations between the (transformed or untransformed) dissimilarities and the Euclidean distances.

. estat correlations

Correlations of dissimilarities and fitted distances

digit1	N	Pearson	Spearman
0	9	0.9510	0.9540
1	9	0.9397	0.7782
2	9	0.7674	0.4017
3	9	0.7922	0.7815
4	9	0.9899	0.9289
5	9	0.9412	0.9121
6	9	0.8226	0.8667
7	9	0.8444	0.4268
8	9	0.8505	0.7000
9	9	0.9954	0.9333
Total	90	0.8602	0.8301

◁

Matching configuration plot and the Shepard diagram

The matching configuration plot and Shepard diagram are easily obtained after an MDS analysis.

▷ Example 2

By default, mds, mdsmat, and mdslong display the MDS matching configuration plot. If you want to exercise control over the graph, you can specify the noplot option of mds, mdsmat, or mdslong and then use the mdsconfig postestimation graph command.

Continuing with the Morse code digit example, we produce a configuration plot with an added title and subtitle.

```
. mdsconfig, title(Morse code digit dissimilarity)
              subtitle(data: Rothkopf 1957)
```

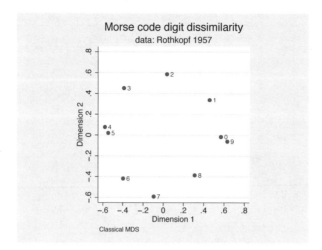

Note that the plot has an aspect ratio of one so that 1 cm on the horizontal dimension equals 1 cm on the vertical dimension. Thus the "straight-line" distance in the plot is really (proportional to) the Euclidean distance between the points in the configuration and hence approximates the dissimilarities between the objects—in this case, the Morse codes for digits.

◁

▷ Example 3

A second popular plot for MDS is the Shepard diagram. This is a plot of the Euclidean distances in the matching configuration against the "observed" dissimilarities. As we explained before, a linear transformation is applied to the dissimilarities to fit the Euclidean distances as close as possible (in the least-squares sense). A Shepard diagram is a plot of the $n(n-1)/2$ transformed dissimilarities against the Euclidean distances.

```
. mdsshepard
```

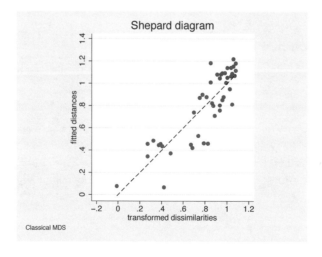

If the Euclidean configuration closely matches the dissimilarities between the objects, all points would be close to the "y = x" line. Deviations indicate lack of fit. To simplify the diagnosis of whether there are specific objects that are poorly represented, Shepard diagrams can be produced for each object separately. Such plots consist of n small plots with $n-1$ points each, namely the transformed dissimilarities and Euclidean distances to all other objects.

```
. mdsshepard, separate
(mdsshepard is producing a separate plot for each obs; this may take a while)
```

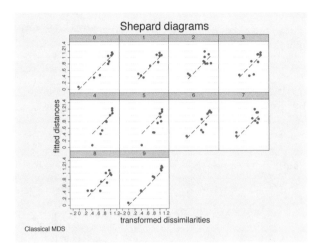

Other examples of `mdsconfig` are found in [MV] **mds**, [MV] **mdslong**, and [MV] **mdsmat**.

Predictions

It is possible to generate variables containing the results from the MDS analysis. MDS operates at two levels: first, at the level of the objects and second, at the level of relations between the objects or pairs of objects. You can generate variables at both of these levels.

The `config` option of `predict` after an MDS requests that the coordinates of the objects in the matching configuration be stored in variables. You may specify as many variables as there are retained dimensions. You may also specify fewer variables. The first variable will store the coordinates from the first dimension. The second variable, if any, will store the coordinates from the second dimension, and so on.

The `pairwise` option specifies that the pairwise statistics are stored in variables. These are the dissimilarities, the fitted distances, and the residuals, i.e., the difference between dissimilarities and the fitted distances. By default, the dissimilarities are linearly transformed to match the fitted distances, but it is possible to skip the transformation step.

There is a complicating issue. With n objects, there are $n(n-1)/2$ pairs of objects. So, to store properties of objects, you need n observations, but to store properties of pairs of objects, you need $n(n-1)/2$ observations. So, where do you store the variables? `predict` after MDS commands has the unique ability to save the predicted variables in a new dataset. Specify the option `saving(`*filename*`)`. Such a dataset will automatically have the appropriate object identification variable or variables.

In some cases, it is also possible to store the variables in the dataset you have in memory: object-level variables in an object-level dataset, and pairwise-level variables in a pairwise-level dataset.

After mds you have a dataset in memory in which the observations are associated with the MDS objects. In this case, you can store object-level variables directly in the dataset in memory. To do so, you just omit the saving() option. In this case, it is not possible to store the pairwise statistics in the dataset in memory. The pairwise statistics have to be stored in a new dataset.

After mdslong, the dataset in memory is in a pairwise form, so the variables predicted with the option pairwise can be stored in the dataset in memory. It is, of course, also possible to store the pairwise variables in a new dataset, the choice is yours. With pairwise data in memory, you cannot store the object-level predicted variables into the data in memory; you need to specify the name of a new dataset.

After mdsmat, you always need to save the predicted variables in a new dataset.

▷ Example 4

Continuing with our Morse code example that used mdslong, the dataset in memory is in long form. Thus we can store the pairwise statistics with the dataset in memory.

```
. predict tdissim eudist resid, pairwise
. list in 1/10
```

	digit1	digit2	freqsame	sim	tdissim	eudist	resid
1.	2	1	62	.62	.3227682	.4862905	-.1635224
2.	3	1	16	.16	.957076	.851504	.1055719
3.	3	2	59	.59	.3732277	.4455871	-.0723594
4.	4	1	6	.06	1.069154	1.068583	.0005709
5.	4	2	23	.23	.8745967	.7995979	.0749989
6.	4	3	38	.38	.6841489	.4209922	.2631567
7.	5	1	12	.12	1.002667	1.051398	-.048731
8.	5	2	8	.08	1.047234	.8123672	.2348665
9.	5	3	27	.27	.8257753	.4599419	.3658335
10.	5	4	56	.56	.4218725	.0668193	.3550532

Since we used mdslong, the object-level statistics have to be saved in a file.

```
. predict d1 d2, config saving(digitdata)
. describe using digitdata
```

```
Contains data                          unit statistics for MDS
                                         (method=classical,dim=2)
  obs:            10                   1 Mar 2005 10:07
  vars:            3
  size:          130
```

variable name	storage type	display format	value label	variable label
digit1	str1	%9s		
d1	float	%9.0g		MDS dimension 1
d2	float	%9.0g		MDS dimension 2

```
Sorted by: digit1
```

The information in these variables was already shown with estat config. The dataset created has variables d1 and d2 with the coordinates of the Morse digits on the two retained dimensions and an identification variable digit1. Use merge to add these variables to the data in memory; see [D] merge.

◁

Saved Results

`estat correlations` saves in `r()`:

Matrices

    `r(R)`   statistics per object; columns with # of obs., Pearson corr., and Spearman corr.

    `r(T)`   overall statistics; # of obs., Pearson corr., and Spearman corr.

`estat quantiles` saves in `r()`:

Macros

    `r(dtype)`   adjusted or raw; dissimilarity transformation

Matrices

    `r(Q)`   statistics per object; columns with # of obs., min., p25, p50, p75, and max.

    `r(T)`   overall statistics; # of obs., min., p25, p50, p75, and max.

`estat stress` saves in `r()`:

Macros

    `r(dtype)`   adjusted or raw; dissimilarity transformation

Matrices

    `r(S)`   Kruskal's stress/loss measure per object

    `r(T)`   1 by 1 matrix with the overall Kruskal stress/loss measure

Methods and Formulas

All postestimation commands listed above are implemented as ado-files.

See [MV] **mdsmat** for information on the methods and formulas for classical multidimensional scaling.

Let D_{ij} be the dissimilarity between objects i and j, $1 \leq i, j \leq n$. We assume $D_{ii} = 0$ and $D_{ij} = D_{ji}$. Let E_{ij} be the Euclidean distance between rows i and j of the matching configuration $\mathbf{Y}$. In classical MDS, $\mathbf{D} - \mathbf{E}$ is not a well-behaved residual matrix. We follow the approach used in metric and nonmetric MDS to transform D_{ij} to "optimally match" E_{ij}, with $\widehat{D}_{ij} = a + b D_{ij}$ where a and b are chosen to minimize the residual sum of squares. This is a simple regression problem and is equivalent to minimizing Kruskal's stress measure (Kruskal 1964; Cox and Cox 2001: 63)

$$\text{Kruskal}(\widehat{\mathbf{D}}, \mathbf{E}) = \left\{ \frac{\sum (E_{ij} - \widehat{D}_{ij})^2}{\sum E_{ij}^2} \right\}^{1/2}$$

with summation over all pairs (i, j). We call the $\widehat{D}_{ij}$ the adjusted or transformed dissimilarities. If the transformation step is skipped by specifying the option `notransform`, we set $\widehat{D}_{ij} = D_{ij}$.

In `estat stress`, the decomposition of Kruskal's stress measure over the objects is displayed. $\text{Kruskal}(\widehat{\mathbf{D}}, \mathbf{E})_i$ is defined analogously with summation over all $j \neq i$.

References

Cox, T. F. and M. A. A. Cox. 2001. *Multidimensional Scaling.* 2nd ed. Boca Raton, FL: Chapman & Hall.

Kruskal, J. B. 1964. Multidimensional scaling by optimizing goodness-of-fit to a nonmetric hypothesis. *Psychometrika* 29: 1–27.

Mardia, K. V., J. T. Kent, and J. M. Bibby. 1979. *Multivariate Analysis.* New York: Academic Press.

Also See

Complementary:	[MV] **mds**, [MV] **mdslong**, [MV] **mdsmat**, [R] **estimates**
Background:	[R] **estat**, [R] **predict**

Title

> **mdslong** — Multidimensional scaling of proximity data in long format

Syntax

mdslong *depvar* $\left[\textit{if}\right]$ $\left[\textit{in}\right]$, id(*var₁ var₂*) $\left[\textit{options}\right]$

options	description
Model	
*id(*var₁ var₂*)	identify comparison pairs (object₁, object₂)
s2d(standard)	convert similarity to dissimilarity: $d_{ij} = \sqrt{s_{ii} + s_{jj} - 2s_{ij}}$
s2d(oneminus)	convert similarity to dissimilarity: $d_{ij} = 1 - s_{ij}$
force	correct problems in proximity information
dimension(#)	configuration dimensions; default is dimension(2)
addconstant	make distance matrix positive definite
Reporting	
neigen(#)	maximum number of eigenvalues to display; default is neigen(10)
config	display table with configuration coordinates
noplot	suppress configuration plot

* id(*var₁ var₂*) is required.

by and statsby are allowed; see [U] **11.1.10 Prefix commands**.

The maximum number of compared objects allowed is the maximum matrix size; see [R] **matsize**.

See [MV] **mds postestimation** for features available after estimation.

Description

mdslong performs classical metric multidimensional scaling (MDS; Torgerson 1952) for two-way proximity data in long format with an explicit measure of similarity or dissimilarity between objects.

For MDS with two-way proximity data in a matrix, see [MV] **mdsmat**. If you are looking for MDS on a dataset, based on dissimilarities between observations over variables, see [MV] **mds**.

Options

> **Model**

id(*var₁ var₂*) is required. The pair of variables *var₁* and *var₂* should uniquely identify comparisons. *var₁* and *var₂* are string or numeric variables that identify the objects to be compared. *var₁* and *var₂* should be of the same datatype; if they are value-labeled, they should be labeled with the same value label. Using value-labeled variables or string variables is generally helpful in identifying the points in plots and tables.

Example data layout for `mdslong proxim, id(i1 i2)`.

proxim	i1	i2
7	1	2
10	1	3
12	1	4
4	2	3
6	2	4
3	3	4

`s2d(standard | oneminus)` specifies how similarities are to be converted into dissimilarities. By default, `mdslong` assumes that you have dissimilarity data. Specifying `s2d()` indicates that your proximity data are similarities.

Dissimilarity data should have zeros on the diagonal (i.e., an object is identical to itself) and non-negative off-diagonal values. Dissimilarities need not satisfy the triangular inequality, $D(i,j)^2 \leq D(i,h)^2 + D(h,j)^2$. Similarity data should have ones on the diagonal (i.e., an object is identical to itself) and have off-diagonal values between zero and one. In either case, proximities should be symmetric. See option `force` if your data violate these assumptions.

The available `s2d()` options, `standard` and `oneminus`, are defined as

$$
\begin{aligned}
\text{standard} \quad & d_{ij} = \sqrt{s_{ii} + s_{jj} - 2s_{ij}} = \sqrt{2(1 - s_{ij})} \\
\text{oneminus} \quad & d_{ij} = 1 - s_{ij}
\end{aligned}
$$

`force` corrects problems with the supplied proximity information. In the long format used by `mdslong`, multiple measurements on (i, j) may be available. Including both (i, j) and (j, i) would be treated as multiple measurements. This is an error, even if the measures are identical. Option `force` uses the mean of the measurements. `force` also resolves problems on the diagonal, i.e., comparisons of objects with themselves; these should have zero dissimilarity or unit similarity. `force` does not resolve incomplete data, i.e., pairs (i, j) for which no measurement is available. Out-of-range values are also not fixed.

`dimension(#)` specifies the dimension of the approximating configuration. `#` defaults to 2 and should not exceed the number of positive eigenvalues of the centered distance matrix.

`addconstant` specifies that if the double-centered distance matrix is not positive semidefinite (psd), a constant should be added to the squared distances to make it psd and, hence, Euclidean.

> Reporting

`neigen(#)` specifies the number of eigenvalues to be included in the table. The default is `neigen(10)`. Specifying `neigen(0)` suppresses the table.

`config` displays the table with the coordinates of the approximating configuration. This table may also be displayed using the postestimation command `estat config`; see [MV] **mds postestimation**.

`noplot` suppresses the graph of the approximating configuration. Note that the graph can still be produced later via `mdsconfig`, which also allows the standard graphics options for fine-tuning the plot; see [MV] **mds postestimation**.

Remarks

Remarks are presented under the headings

Introduction
Proximity data in long format

Introduction

Multidimensional scaling (MDS) is a dimension-reduction and visualization technique. Dissimilarities (for instance, Euclidean distances) between observations in a high-dimensional space are represented in a lower-dimensional space (typically two dimensions) so that the Euclidean distance in the lower-dimensional space approximates the dissimilarities in the higher-dimensional space. See Kruskal and Wish (1978) for a brief nontechnical introduction to MDS. Young and Hamer (1994) and Borg and Groenen (1996) are more advanced textbook-sized treatments.

mdslong performs MDS on data in long format. *depvar* specifies proximity data in either dissimilarity or similarity form. The comparison pairs are identified by two variables specified in the required option id(). Exactly one observation with a nonmissing *depvar* should be included for each pair (i, j). Pairs are unordered; you do not include observations for both (i, j) and (j, i). Observations for comparisons of objects with themselves (i, i) are optional. See option force if your data violate these assumptions.

In some applications, the similarity or dissimilarity of objects is defined by the researcher in terms of variables (attributes) measured on the objects. If you need MDS of this form, you should continue reading in [MV] **mds**.

In many cases, however, proximities—i.e., similarities or dissimilarities—are measured directly. For instance, psychologists studying the similarities or dissimilarities in a set of stimuli—smells, sounds, faces, concepts, etc.—may have subjects rate the dissimilarity of pairs of stimuli. Linguists have subjects rate the similarity or dissimilarity of pairs of dialects. Political scientists have subjects rate the similarity or dissimilarity of political parties or candidates for political office. In other fields, relational data are studied that may be interpreted as proximities in a more abstract sense. For instance, sociologists study interpersonal contact frequencies in groups ("social networks"); these measures are sometimes interpreted in terms of similarities.

A wide variety of MDS methods have been proposed. mdslong performs classical scaling that has its roots in Young and Householder (1938) and Torgerson (1952). Classical scaling requires complete and symmetric dissimilarity interval-level data.

Proximity data in long format

One format for proximity data is called the "long format", with an observation recording the dissimilarity d_{ij} of the "objects" i and j. This requires three variables: one variable to record the dissimilarities and two variables to identify the comparison pair. The MDS command mdslong requires

- Complete data without duplicates: there is exactly one observation for each combination (i, j) or (j, i).

- Optional diagonal: you may, but need not, specify dissimilarities for the reflexive pairs (i, i). If you do, you need not supply values for all (i, i).

▷ Example 1

We illustrate the use of `mdslong` with another popular dataset from the MDS literature. Rothkopf (1957) had 598 subjects listen to pairs of Morse codes for the ten digits and for the 26 letters, recording for each pair of codes the percentage of subjects who declared the codes to be the same. The data on the ten digits are reproduced in Mardia, Kent, and Bibby (1979, 395).

```
. use http://www.stata-press.com/data/r9/morse_long
(Morse data (Rothkopf 1957))
. list in 1/10
```

	digit1	digit2	freqsame
1.	2	1	62
2.	3	1	16
3.	3	2	59
4.	4	1	6
5.	4	2	23
6.	4	3	38
7.	5	1	12
8.	5	2	8
9.	5	3	27
10.	5	4	56

62% of the subjects declare that the Morse codes for 1 and 2 are the same, 16% declare that 1 and 3 are the same, 59% declare 2 and 3 to be the same, etc. We may think that these percentages are similarity measures between the Morse codes: the more similar two Morse codes, the higher the percentage is of subjects who do not distinguish them. The reported percentages suggest, for example, that 1 and 2 are similar to approximately the same extent as 2 and 3, while 1 and 3 are much less similar. This is the kind of relationship you would expect with data that can be adequately represented with MDS.

We transform our data to a zero to one scale.

```
. gen sim = freqsame/100
```

and invoke `mdslong` on `sim`, with `s2d(standard)` specifying that the proximity variable `sim` be interpreted as similarities and converted to dissimilarities using the standard conversion.

```
. mdslong sim, id(digit1 digit2) s2d(standard)
```

Classical metric multidimensional scaling
    similarity variable: sim in long format
        dissimilarity: sqrt(2(1-similarity))

```
                                      Number of obs        =         10
Eigenvalues > 0        =         9    Mardia fit measure 1 =    0.5086
Retained dimensions    =         2    Mardia fit measure 2 =    0.7227
```

Dimension	Eigenvalue	abs(eigenvalue) Percent	Cumul.	(eigenvalue)^2 Percent	Cumul.
1	1.9800226	30.29	30.29	49.47	49.47
2	1.344165	20.57	50.86	22.80	72.27
3	1.063133	16.27	67.13	14.26	86.54
4	.66893922	10.23	77.36	5.65	92.18
5	.60159396	9.20	86.56	4.57	96.75
6	.42722301	6.54	93.10	2.30	99.06
7	.21220785	3.25	96.35	0.57	99.62
8	.1452025	2.22	98.57	0.27	99.89
9	.09351288	1.43	100.00	0.11	100.00

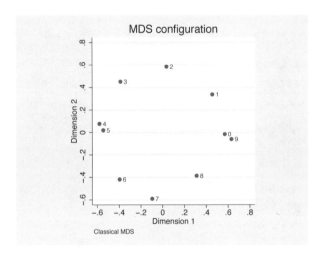

The two-dimensional representation provides a reasonable, but certainly not impressive, fit to the data. The plot itself is interesting though, with the digits being roughly 45 degrees apart, with the exception of the pairs (0,9) and (4,5), which are mapped almost at the same locations. Interpretation is certainly helped if you see the circular structure in the Morse codes.

(Continued on next page)

digit	morse code
1	. - - - -
2	. . - - -
3	. . . - -
4	. . . . -
5	. . . . .
6	- . . . .
7	- - . . .
8	- - - . .
9	- - - - .
0	- - - - -

◁

▷ Example 2

You might have your data in wide instead of long format. The Morse code data in wide format has 10 observations, 10 data variables d1, ..., d9, d0, and one case identifier.

```
. use http://www.stata-press.com/data/r9/morse_wide
(Morse data (Rothkopf 1957))
. describe
Contains data from http://www.stata-press.com/data/r9/morse_wide.dta
  obs:              10                        Morse data (Rothkopf 1957)
  vars:             11                        14 Feb 2005 20:28
  size:            150 (99.9% of memory free)  (_dta has notes)
```

variable name	storage type	display format	value label	variable label
digit	byte	%9.0g		
d1	byte	%9.0g		
d2	byte	%9.0g		
d3	byte	%9.0g		
d4	byte	%9.0g		
d5	byte	%9.0g		
d6	byte	%9.0g		
d7	byte	%9.0g		
d8	byte	%9.0g		
d9	byte	%9.0g		
d0	byte	%9.0g		

```
Sorted by:
. list
```

	digit	d1	d2	d3	d4	d5	d6	d7	d8	d9	d0
1.	1	84	62	16	6	12	12	20	37	57	52
2.	2	62	89	59	23	8	14	25	25	28	18
3.	3	16	59	86	38	27	33	17	16	9	9
4.	4	6	23	38	89	56	34	24	13	7	7
5.	5	12	8	27	56	90	30	18	10	5	5
6.	6	12	14	33	34	30	86	65	22	8	18
7.	7	20	25	17	24	18	65	85	65	31	15
8.	8	37	25	16	13	10	22	65	88	58	39
9.	9	57	28	9	7	5	8	31	58	91	79
10.	0	52	18	9	7	5	18	15	39	79	94

Stata does not provide an MDS command to deal directly with the wide format since it is easy to convert the wide format into the long format using the `reshape` command; see [D] **reshape**.

```
. reshape long d, i(digit) j(other)
(note: j = 0 1 2 3 4 5 6 7 8 9)
Data                                wide   ->   long

Number of obs.                        10   ->     100
Number of variables                   11   ->       3
j variable (10 values)                     ->   other
xij variables:
                           d0 d1 ... d9    ->   d
```

Now our data is in long format, and we can use `mdslong` to obtain a MDS analysis.

```
. gen sim = d/100
. mdslong sim, id(digit other) s2d(standard) noplot
objects should have unit similarity to themselves
r(198);
```

`mdslong` complains. The wide data—and hence also the long data that we now have—also contain the frequencies in which two identical Morse codes were recognized as the same. This is not 100%. Auditive memory is not perfect. This poses a problem for the standard MDS model. We can solve this by ignoring the diagonal observations:

```
. mdslong ... if digit != other ...
```

Alternatively, we may specify the `force` option. The `force` option will take care of a problem that has not yet surfaced, namely that `mdslong` requires a single observation for each pair (i, j). In the long data as now created, we have duplicates.

```
. mdslong sim, id(digit other) s2d(standard) force noplot
Classical metric multidimensional scaling
    similarity variable: sim in long format
        dissimilarity: sqrt(2(1-similarity))
                                     Number of obs       =         10
    Eigenvalues > 0      =       9   Mardia fit measure 1 =     0.5086
    Retained dimensions  =       2   Mardia fit measure 2 =     0.7227
```

Dimension	Eigenvalue	abs(eigenvalue) Percent	abs(eigenvalue) Cumul.	(eigenvalue)^2 Percent	(eigenvalue)^2 Cumul.
1	1.9800226	30.29	30.29	49.47	49.47
2	1.344165	20.57	50.86	22.80	72.27
3	1.063133	16.27	67.13	14.26	86.54
4	.66893922	10.23	77.36	5.65	92.18
5	.60159396	9.20	86.56	4.57	96.75
6	.42722301	6.54	93.10	2.30	99.06
7	.21220785	3.25	96.35	0.57	99.62
8	.1452025	2.22	98.57	0.27	99.89
9	.09351288	1.43	100.00	0.11	100.00

The output produced by `mdslong` in this case is identical to what we saw earlier.

◁

After `mdslong`, all MDS postestimation tools are available. For instance, you may analyze residuals with `estat quantile`, you may produce a Shepard diagram, etc.; see [MV] **mds postestimation**.

Saved Results

mdslong saves in e():

Scalars

e(N)	number of underlying observations
e(p)	number of dimensions in the approximating configuration
e(np)	number of strictly positive eigenvalues
e(addcons)	constant added to squared dissimilarities to force positive semidefiniteness
e(mardia1)	Mardia measure 1
e(mardia2)	Mardia measure 2

Macros

e(cmd)	mdslong
e(method)	classical
e(id)	two ID variable names identifying compared object pairs
e(depvar)	dependent variable (containing dissimilarities)
e(dtype)	similarity or dissimilarity
e(s2d)	standard or oneminus (when e(dtype) is similarity)
e(unique)	1 if eigenvalues are distinct, 0 otherwise
e(properties)	nob noV eigen
e(predict)	mds_p
e(estat_cmd)	mds_estat

Matrices

e(D)	dissimilarity matrix
e(Y)	approximating configuration coordinates
e(Ev)	eigenvalues
e(linearf)	two element vector defining the linear transformation; distance equals first element plus second element times dissimilarity

Functions

e(sample)	marks estimation sample

Methods and Formulas

mdslong is implemented as an ado-file.

See [MV] **mdsmat** for information on the methods and formulas.

References

Borg, I. and P. Groenen. 1996. *Modern Multidimensional Scaling: Theory and Applications.* New York: Springer.

Kruskal, J. B. and M. Wish. 1978. *Multidimensional Scaling.* Newbury Park: Sage.

Lingoes, J. C. 1971. Some boundary conditions for a monotone analysis of symmetric matrices. *Psychometrika* 36: 195–203.

Mardia, K. V., J. T. Kent, and J. M. Bibby. 1979. *Multivariate Analysis.* New York: Academic Press.

Rothkopf, E. Z. 1957. A measure of stimulus similarity and errors in some paired-associate learning tasks. *Journal of Experimental Psychology* 53: 94–101.

Torgerson, W. S. 1952. Multidimensional scaling: I. Theory and method. *Psychometrika* 17: 401–419.

Young, F. W. and R. M. Hamer. 1994. *Theory and Applications of Multidimensional Scaling.* Hillsdale, NJ: Erlbaum Associates.

Young, G. and A. S. Householder. 1938. Discussion of a set of points in terms of their mutual distances. *Psychometrika* 3: 19–22.

Also See

Complementary:	[MV] **mds postestimation**
Related:	[MV] **biplot**, [MV] **ca**, [MV] **factor**, [MV] **mds**, [MV] **mdsmat**, [MV] **pca**
Background:	[U] **11.1.10 Prefix commands**,
	[U] **20 Estimation and postestimation commands**

Title

> **mdsmat** — Multidimensional scaling of proximity data in a matrix

Syntax

$$\texttt{mdsmat} \; \textit{matname} \; \big[\, , \; \textit{options} \,\big]$$

options	description
Model	
shape(<u>f</u>ull)	*matname* is a square symmetric matrix; the default
shape(<u>l</u>ower)	*matname* is a vector with the rowwise lower triangle (with diagonal)
shape(<u>ll</u>ower)	*matname* is a vector with the rowwise strictly lower triangle (no diagonal)
shape(<u>u</u>pper)	*matname* is a vector with the rowwise upper triangle (with diagonal)
shape(<u>uu</u>pper)	*matname* is a vector with the rowwise strictly upper triangle (no diagonal)
<u>names</u>(*namelist*)	object names; required with all but shape(full)
s2d(<u>st</u>andard)	convert similarity to dissimilarity: $d_{ij} = \sqrt{s_{ii} + s_{jj} - 2s_{ij}}$
s2d(<u>one</u>minus)	convert similarity to dissimilarity: $d_{ij} = 1 - s_{ij}$
force	fix problems in proximity information
<u>dim</u>ension(#)	configuration dimensions; default is dimension(2)
<u>add</u>constant	make distance matrix positive definite
Reporting	
<u>neigen</u>(#)	maximum number of eigenvalues to display; default is neigen(10)
config	display table with configuration coordinates
<u>nopl</u>ot	suppress configuration plot

See [MV] **mds postestimation** for features available after estimation.

Description

mdsmat performs classical metric multidimensional scaling (MDS; Torgerson 1952) for two-way proximity data with an explicit measure of similarity or dissimilarity between objects, where the proximities are found in matrix *matname*.

If your proximities are stored as variables in long format, see [MV] **mdslong**. If you are looking for MDS on a dataset based on dissimilarities between observations over variables, see [MV] **mds**.

Options

> ___Model___

shape(*shape*) specifies the storage mode of the existing similarity or dissimilarity matrix *matname*. The following storage modes are allowed:

full specifies that *matname* is a symmetric $n \times n$ matrix.

lower specifies that *matname* is a row or column vector of length $n(n+1)/2$, with the rowwise lower triangle of the similarity or dissimilarity matrix including the diagonal.

$$D_{11}\ D_{21}\ D_{22}\ D_{31}\ D_{32}\ D_{33}\ \ldots\ D_{n1}\ D_{n2}\ \ldots\ D_{nn}$$

llower specifies that *matname* is a row or column vector of length $n(n-1)/2$, with the rowwise lower triangle of the similarity or dissimilarity matrix excluding the diagonal.

$$D_{21}\ D_{31}\ D_{32}\ D_{41}\ D_{42}\ D_{43}\ \ldots\ D_{n1}\ D_{n2}\ \ldots\ D_{n,n-1}$$

upper specifies that *matname* is a row or column vector of length $n(n+1)/2$, with the rowwise upper triangle of the similarity or dissimilarity matrix including the diagonal.

$$D_{11}\ D_{12}\ \ldots\ D_{1n}\ D_{22}\ D_{23}\ \ldots\ D_{2n}\ D_{33}\ D_{34}\ \ldots\ D_{3n}\ \ldots\ D_{nn}$$

uupper specifies that *matname* is a row or column vector of length $n(n-1)/2$, with the rowwise upper triangle of the similarity or dissimilarity matrix excluding the diagonal.

$$D_{12}\ D_{13}\ \ldots\ D_{1n}\ D_{23}\ D_{24}\ \ldots\ D_{2n}\ D_{34}\ D_{35}\ \ldots\ D_{3n}\ \ldots\ D_{n-1,n}$$

names(*namelist*) is required with all but shape(full). The number of names should equal the number of rows (and columns) of the full similarity or dissimilarity matrix and should not contain duplicates.

s2d(standard|oneminus) specifies how similarities are converted into dissimilarities, as expected by mdsmat. By default, mdsmat assumes that you have dissimilarity data. Specifying s2d() indicates that your proximity data are similarities.

Dissimilarity data should have zeros on the diagonal (i.e., an object is identical to itself) and non-negative off diagonal values. Dissimilarities need not satisfy the triangular inequality, $D(i,j)^2 \leq D(i,h)^2 + D(h,j)^2$. Similarity data should have ones on the diagonal (i.e., an object is identical to itself) and have off-diagonal values between zero and one. In either case, proximities should be symmetric. See option force if your data violate these assumptions.

The available s2d() options, standard and oneminus, are defined as

$$
\begin{aligned}
\text{standard} \qquad & d_{ij} = \sqrt{s_{ii} + s_{jj} - 2s_{ij}} = \sqrt{2(1 - s_{ij})} \\
\text{oneminus} \qquad & d_{ij} = 1 - s_{ij}
\end{aligned}
$$

force corrects problems with the supplied proximity information. force specifies that the dissimilarity matrix be symmetrized; the mean of D_{ij} and D_{ji} is used. Also, problems on the diagonal (similarities: $D_{ii} \neq 1$; dissimilarities: $D_{ii} \neq 0$) are fixed. force does not fix missing values or out-of-range values (i.e., $D_{ij} < 0$; or similarities with $D_{ij} > 1$.)

dimension(#) specifies the dimension of the approximating configuration. # defaults to 2 and should not exceed the number of positive eigenvalues of the centered distance matrix.

addconstant specifies that if the double-centered distance matrix is not positive semidefinite (psd), a constant should be added to the squared distances to make it psd and, hence, Euclidean.

`neigen(#)` specifies the number of eigenvalues to be included in the table. The default is `neigen(10)`. Specifying `neigen(0)` suppresses the table.

`config` displays the table with the coordinates of the approximating configuration. This table may also be displayed using the postestimation command `estat config`; see [MV] **mds postestimation**.

`noplot` suppresses the graph of the approximating configuration. Note that the graph can still be produced later via `mdsconfig`, which also allows the standard graphics options for fine-tuning the plot; see [MV] **mds postestimation**.

Remarks

Remarks are presented under the headings

Introduction
Proximity data in a Stata matrix

Introduction

Multidimensional scaling (MDS) is a dimension-reduction and visualization technique. Dissimilarities (for instance, Euclidean distances) between observations in a high-dimensional space are represented in a lower-dimensional space (typically two dimensions) so that the Euclidean distance in the lower-dimensional space approximates the dissimilarities in the higher-dimensional space. See Kruskal and Wish (1978) for a brief nontechnical introduction to MDS. Young and Hamer (1994) and Borg and Groenen (1996) are more advanced textbook-sized treatments.

`mdsmat` performs MDS on a similarity or dissimilarity matrix *matname*. You may enter the matrix as a symmetric square matrix or as a vector (matrix with one row or column) with only the upper or lower triangle; see option `shape()` for details. *matname* should not contain missing values. The diagonal elements should be 0 (dissimilarities) or 1 (similarities). If you provide a square matrix (i.e., `shape(full)`), names of the objects are obtained from the matrix row and column names. The row names should all be distinct, and the column names should equal the row names. Equation names, if any, are ignored. In any of the vectorized shapes, names are specified with option `names()`, and the matrix row and column names are ignored.

See option `force` if your matrix violates these assumptions.

In some applications, the similarity or dissimilarity of objects is defined by the researcher in terms of variables (attributes) measured on the objects. If you need to do MDS of this form, you should continue reading in [MV] **mds**.

In many cases, however, proximities—i.e., similarities or dissimilarities—are measured directly. For instance, psychologists studying the similarities or dissimilarities in a set of stimuli—smells, sounds, faces, concepts, etc.—may have subjects rate the dissimilarity of pairs of stimuli. Linguists have subjects rate the similarity or dissimilarity of pairs of dialects. Political scientists have subjects rate the similarity or dissimilarity of political parties or candidates for political office. In other fields, relational data are studied that may be interpreted as proximities in a more abstract sense. For instance, sociologists study interpersonal contact frequencies in groups ("social networks"); these measures are sometimes interpreted in terms of similarities.

A wide variety of MDS methods have been proposed. `mdsmat` performs classical scaling that has its roots in Young and Householder (1938) and Torgerson (1952). Classical scaling requires complete and symmetric dissimilarity interval-level data.

Proximity data in a Stata matrix

To perform MDS of relational data, you must enter the data in a suitable format. One such format is a Stata matrix. You may want to use this format for analyzing data that you obtain from a printed source.

▷ Example 1

Many texts on multidimensional scaling illustrate how locations can be inferred from a table of geographic distances. We will do this too, using an example of distances in miles between 14 locations in Texas, representing both man-made and natural treasures:

Big Bend	0	523	551	243	322	412	263	596	181	313	553
Corpus Christi	523	0	396	280	705	232	619	226	342	234	30
Dallas	551	396	0	432	643	230	532	243	494	317	426
Del Rio	243	280	432	0	427	209	339	353	62	70	310
El Paso	322	705	643	427	0	528	110	763	365	525	735
Enchanted Rock	412	232	230	209	528	0	398	260	271	69	262
Guadalupe Mnt	263	619	532	339	110	398	0	674	277	280	646
Houston	596	226	243	353	763	260	674	0	415	292	256
Langtry	181	342	494	62	365	271	277	415	0	132	372
Lost Maples	313	234	317	70	525	69	280	292	132	0	264
Padre Island	553	30	426	310	735	262	646	256	372	264	0
Pedernales Falls	434	216	235	231	550	40	420	202	293	115	246
San Antonio	397	141	274	154	564	91	475	199	216	93	171
StataCorp	426	205	151	287	606	148	512	83	318	202	316

Big Bend	434	397	426
Corpus Christi	216	141	205
Dallas	235	274	151
Del Rio	231	154	287
El Paso	550	564	606
Enchanted Rock	40	91	148
Guadalupe Mnt	420	475	512
Houston	202	199	83
Langtry	293	216	318
Lost Maples	115	93	202
Padre Island	246	171	316
Pedernales Falls	0	75	116
San Antonio	75	0	154
StataCorp	116	154	0

You may note the inclusion of StataCorp, located in the twin cities of Bryan/College Station (BCS). To get the data into Stata, we will enter only the strictly upper triangle as a Stata one-dimensional matrix and collect the names in a `global` macro for later use. We are using the strictly upper triangle (i.e., omitting the diagonal) since the diagonal of a dissimilarity matrix contains all zeros—there is no need to enter them.

(Continued on next page)

```
. matrix input D = (
  523  551  243  322  412  263  596  181  313  553  434  397  426
       396  280  705  232  619  226  342  234   30  216  141  205
            432  643  230  532  243  494  317  426  235  274  151
                 427  209  339  353   62   70  310  231  154  287
                      528  110  763  365  525  735  550  564  606
                           398  260  271   69  262   40   91  148
                                674  277  280  646  420  475  512
                                     415  292  256  202  199   83
                                          132  372  293  216  318
                                               264  115   93  202
                                                    246  171  316
                                                          75  116
                                                             154 )
. global names
  Big_Bend       Corpus_Christi  Dallas          Del_Rio
  El_Paso        Enchanted_Rock  Guadalupe_Mnt   Houston
  Langtry        Lost_Maples     Padre_Island    Pedernales_Falls
  San_Antonio    StataCorp
```

The triangular data entry is just typographical and is useful for catching data-entry errors. As far as Stata is concerned, we could have typed all the numbers in one long row. We use `matrix input D =` rather than `matrix define D =` or just `matrix D =` so that we do not have to separate entries with commas.

With the data now in Stata, we may use `mdsmat` to infer the locations in Texas and produce a map:

```
. mdsmat D, names($names) shape(uupper)
Classical metric multidimensional scaling
     dissimilarity matrix: D
```

		Number of obs	=	14
Eigenvalues > 0	=	8	Mardia fit measure 1 =	0.7828
Retained dimensions	=	2	Mardia fit measure 2 =	0.9823

Dimension	Eigenvalue	abs(eigenvalue)		(eigenvalue)^2	
		Percent	Cumul.	Percent	Cumul.
1	691969.62	62.63	62.63	92.45	92.45
2	172983.05	15.66	78.28	5.78	98.23
3	57771.995	5.23	83.51	0.64	98.87
4	38678.916	3.50	87.01	0.29	99.16
5	19262.579	1.74	88.76	0.07	99.23
6	9230.7695	0.84	89.59	0.02	99.25
7	839.70996	0.08	89.67	0.00	99.25
8	44.989372	0.00	89.67	0.00	99.25

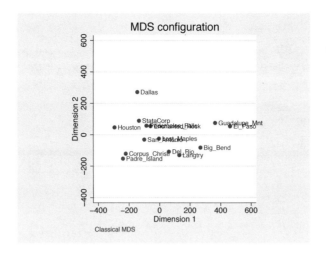

The representation of the distances in two dimensions provides a reasonable, but not great, fit; the percentage of eigenvalues accounted for is 78%.

By default `mdsmat` produces a configuration plot. Enhancements to the configuration plot are possible using the `mdsconfig` postestimation graphics command; see [MV] **mds postestimation**. We present the configuration plot using the `autoaspect` option to obtain better use of the available space while preserving the equivalence of distance in the x- and y-axes. We negate the direction of the x-axis with the `xnegate` option in order to flip the configuration horizontally. We also change the default title and control the placement of labels.

```
. set obs 14
obs was 0, now 14
. generate pos = 3
. replace pos = 4 in 6
(1 real change made)
. replace pos = 2 in 10
(1 real change made)
. mdsconfig, autoaspect xnegate mlabvpos(pos) title(MDS for 14 Texas locations)
```

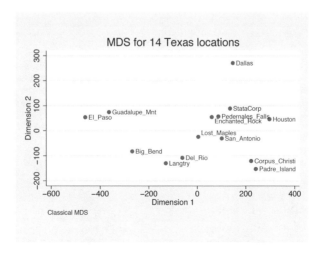

Look at the graph produced by mdsconfig after mdsmat. You will probably recognize a twisted (and slightly distorted) map of Texas. The vertical orientation of the map is not correctly north-south; you would probably want to turn the map some 20 degrees clockwise. Why didn't mdsmat get it right? It could not have concluded the correct rotation from the available distance information. Any orthogonal rotation of the map would produce the same distances. The orientation of the map is *not identified*. Finally, the "location" of the map can not be inferred from the distances. Translating the coordinates does not change the distances. As far as mdsmat is concerned, Texas could be part of China.

◁

After mdsmat, all MDS postestimation tools are available. For instance, you may analyze residuals with estat quantile, you may produce a Shepard diagram, etc.; see [MV] **mds postestimation**.

Saved Results

mdsmat saves in e():

Scalars

e(N)	number of rows or columns (i.e., number of observations)
e(p)	number of dimensions in the approximating configuration
e(np)	number of strictly positive eigenvalues
e(addcons)	constant added to squared dissimilarities to force positive semidefiniteness
e(mardia1)	Mardia measure 1
e(mardia2)	Mardia measure 2

Macros

e(cmd)	mdsmat
e(method)	classical
e(id)	Category
e(dmatrix)	name of analyzed matrix
e(dtype)	similarity or dissimilarity
e(s2d)	standard or oneminus (when e(dtype) is similarity)
e(unique)	1 if eigenvalues are distinct, 0 otherwise
e(properties)	nob noV eigen
e(predict)	mds_p
e(estat_cmd)	mds_estat

Matrices

e(D)	dissimilarity matrix
e(Y)	approximating configuration coordinates
e(Ev)	eigenvalues
e(linearf)	two element vector defining the linear transformation; distance equals first element plus second element times dissimilarity

Methods and Formulas

mdsmat is implemented as an ado-file.

Let $\mathbf{D}$ be an $n \times n$ dissimilarity matrix. The matrix $\mathbf{D}$ is said to be *Euclidean* if there are coordinates $\mathbf{Y}$ so that

$$D_{ij}^2 = (y_{i.} - y_{j.})(y_{i.} - y_{j.})'$$

Let $\mathbf{A} = -\frac{1}{2}\mathbf{D}\cdot\mathbf{D}$, with $\cdot$ being the direct (element-wise) product, and define $\mathbf{B}$ as the double-centered distance matrix

$$\mathbf{B} = \mathbf{HAH} \quad \text{with} \quad \mathbf{H} = \mathbf{I} - \frac{1}{n}\mathbf{1}\mathbf{1}'$$

$\mathbf{D}$ is Euclidean if and only if $\mathbf{B}$ is positive semidefinite. Assume for now that $\mathbf{D}$ is indeed Euclidean. The spectral or eigen decomposition of $\mathbf{B}$ is written as $\mathbf{B} = \mathbf{U}\mathbf{\Lambda}\mathbf{U}'$, with $\mathbf{U}$ the orthonormal matrix of eigenvectors normed to 1, and $\mathbf{\Lambda}$ a diagonal matrix with non-negative values (the eigenvalues of $\mathbf{B}$) in decreasing order. The coordinates $\mathbf{Y}$ are defined in terms of the spectral decomposition $\mathbf{Y} = \mathbf{U}\mathbf{\Lambda}^{1/2}$. Note that these coordinates are centered $\mathbf{Y}'\mathbf{1} = \mathbf{0}$.

The spectral decomposition can also be used to obtain a low-dimensional configuration $\widetilde{\mathbf{Y}}$, $n \times p$, so that the inter-row distances of $\widetilde{\mathbf{Y}}$ approximate $\mathbf{D}$. Mardia, Kent, and Bibby (1979, section 14.4) discuss some characterizations under which the leading p columns of $\mathbf{Y}$ are an optimal choice of $\widetilde{\mathbf{Y}}$. These characterizations also apply to the case when $\mathbf{B}$ is not positive semidefinite, so some of the λ's are negative; we do require that $\lambda_p > 0$.

Various other approaches have been proposed to deal with the case when the matrix $\mathbf{B}$ is not positive semidefinite, i.e., when $\mathbf{B}$ has negative eigenvalues (see Cox and Cox 2001, 45–48). An easy solution is to add a constant to the off-diagonal elements of $\mathbf{D}\cdot\mathbf{D}$ to make $\mathbf{B}$ positive semidefinite. The smallest such constant is $-2\lambda_n$, where λ_n is the smallest eigenvalue of $\mathbf{B}$ (Lingoes 1971). See Cailliez (1983) for a solution to the additive constant problem in terms of the dissimilarities instead of the squared dissimilarities.

If a similarity measure was selected, it is turned into a dissimilarity measure using one of two methods. The *standard* conversion method is

$$\mathrm{dissim}_{ij} = \sqrt{\mathrm{sim}_{ii} + \mathrm{sim}_{jj} - 2\mathrm{sim}_{ij}}$$

With the similarity of an object to itself being 1, this is equivalent to

$$\mathrm{dissim}_{ij} = \sqrt{2(1 - \mathrm{sim}_{ij})}$$

This conversion method has the attractive property that it transforms a positive semidefinite similarity matrix into an Euclidean distance matrix (see Mardia, Kent, and Bibby 1979, 402).

We also offer the *one-minus* method

$$\mathrm{dissim}_{ij} = 1 - \mathrm{sim}_{ij}$$

Goodness-of-fit statistics for a configuration in p dimensions have been proposed by Mardia (1978) in the context of characterizations of optimality properties of the classical solution

$$\mathrm{Mardia}_1 = \frac{\sum_{i=1}^{p}|\lambda_i|}{\sum_{i=1}^{n}|\lambda_i|}$$

and

$$\mathrm{Mardia}_2 = \frac{\sum_{i=1}^{p}\lambda_i^2}{\sum_{i=1}^{n}\lambda_i^2}$$

References

Borg, I. and P. Groenen. 1996. *Modern Multidimensional Scaling: Theory and Applications.* New York: Springer.

Cailliez F. 1983. The analytical solution of the additive constant problem. *Psychometrika* 48: 305–308.

Cox, T. F. and M. A. A. Cox. 2001. *Multidimensional Scaling.* 2nd ed. Boca Raton, FL: Chapman & Hall.

Kruskal, J. B. and M. Wish. 1978. *Multidimensional Scaling.* Newbury Park: Sage.

Lingoes, J. C. 1971. Some boundary conditions for a monotone analysis of symmetric matrices. *Psychometrika* 36: 195–203.

Mardia, K. V. 1978. Some properties of classical multidimensional scaling. *Communications in Statistics—Theory and Methods* A, 7: 1233–1241.

Mardia, K. V., J. T. Kent, and J. M. Bibby. 1979. *Multivariate Analysis.* New York: Academic Press.

Torgerson, W. S. 1952. Multidimensional scaling: I. Theory and method. *Psychometrika* 17: 401–419.

Young, F. W. and R. M. Hamer. 1994. *Theory and Applications of Multidimensional Scaling.* Hillsdale, NJ: Erlbaum Associates.

Young, G. and A. S. Householder. 1938. Discussion of a set of points in terms of their mutual distances. *Psychometrika* 3: 19–22.

Also See

Complementary:	[MV] **mds postestimation**
Related:	[MV] **biplot**, [MV] **ca**, [MV] **factor**, [MV] **mds**, [MV] **mdslong**, [MV] **pca**
Background:	[U] **11.1.10 Prefix commands**,
	[U] **20 Estimation and postestimation commands**

Title

measure_option — Option for similarity and dissimilarity measures

Syntax

command ... , ... <u>mea</u>sure(measure) ...

or

command ... , ... measure ...

measure	description
cont_measure	similarity or dissimilarity measure for continuous data
binary_measure	similarity measure for binary data

cont_measure	description
L2	Euclidean distance (Minkowski with argument 2)
<u>Eucl</u>idean	alias for L2
L(2)	alias for L2
L2squared	squared Euclidean distance
Lpower(2)	alias for L2squared
L1	absolute-value distance (Minkowski with argument 1)
<u>abso</u>lute	alias for L1
<u>cityb</u>lock	alias for L1
<u>manhat</u>tan	alias for L1
L(1)	alias for L1
Lpower(1)	alias for L1
<u>Linf</u>inity	maximum-value distance (Minkowski with infinite argument)
<u>maxi</u>mum	alias for Linfinity
L(#)	Minkowski distance with # arguments
<u>Lpow</u>er(#)	Minkowski distance with # arguments raised to # power
<u>Canb</u>erra	Canberra distance
<u>corre</u>lation	correlation coefficient similarity measure
<u>angu</u>lar	angular separation similarity measure
<u>ang</u>le	alias for angular

(Continued on next page)

357

binary_measure	description
matching	simple matching similarity coefficient
Jaccard	Jaccard binary similarity coefficient
Russell	Russell and Rao similarity coefficient
Hamann	Hamann similarity coefficient
Dice	Dice similarity coefficient
antiDice	antiDice similarity coefficient
Sneath	Sneath and Sokal similarity coefficient
Rogers	Rogers and Tanimoto similarity coefficient
Ochiai	Ochiai similarity coefficient
Yule	Yule similarity coefficient
Anderberg	Anderberg similarity coefficient
Kulczynski	Kulczynski similarity coefficient
Pearson	Pearson's ϕ similarity coefficient
Gower2	similarity coefficient with same denominator as Pearson

Description

Several commands have options that allow you to specify a similarity or dissimilarity measure designated as *measure* in the syntax; see [MV] **cluster**, [MV] **mds**, and [MV] **matrix dissimilarity**. These options are documented here. Most analysis commands (e.g., cluster and mds) transform similarity measures to dissimilarity measures as needed.

Options

Measures are divided into those for continuous data and binary data. *measure* is not case sensitive. Full definitions are presented in *Similarity and dissimilarity measures for continuous data* and *Similarity measures for binary data*.

Most often, the similarity or dissimilarity measure is used to determine the similarity or dissimilarity between observations. However, in some cases, it is the similarity or dissimilarity between variables that is of interest.

Similarity and dissimilarity measures for continuous data

Here are the similarity and dissimilarity measures for continuous data available in Stata. In the following formulas, p represents the number of variables, N is the number of observations, and x_{iv} denotes the value of observation i for variable v.

The formulas are presented in two forms. The first is the formula used when computing the similarity or dissimilarity between observations. The second is the formula used when computing the similarity or dissimilarity between variables.

L2 (aliases <u>Eucl</u>idean and L(2))
requests the Minkowski distance metric with argument 2. When comparing observations i and j, the formula is

$$\left\{\sum_{a=1}^{p}(x_{ia}-x_{ja})^2\right\}^{1/2}$$

and when comparing variables u and v the formula is

$$\left\{\sum_{k=1}^{N}(x_{ku}-x_{kv})^2\right\}^{1/2}$$

L2 is best known as Euclidean distance and is the default dissimilarity measure for mds, matrix dissimilarity, and all the cluster subcommands except for centroidlinkage, median-linkage, and wardslinkage, which default to using L2squared; see [MV] **mds**, [MV] **matrix dissimilarity**, and [MV] **cluster**.

L2squared
requests the square of the Minkowski distance metric with argument 2. When comparing observations i and j, the formula is

$$\sum_{a=1}^{p}(x_{ia}-x_{ja})^2$$

and when comparing variables u and v, the formula is

$$\sum_{k=1}^{N}(x_{ku}-x_{kv})^2$$

L2squared is best known as squared Euclidean distance and is the default dissimilarity measure for the centroidlinkage, medianlinkage, and wardslinkage subcommands of cluster; see [MV] **cluster**.

L1 (aliases <u>absolute</u>, <u>cityblock</u>, <u>manhat</u>tan, and L(1))
requests the Minkowski distance metric with argument 1. When comparing observations i and j, the formula is

$$\sum_{a=1}^{p}|x_{ia}-x_{ja}|$$

and when comparing variables u and v, the formula is

$$\sum_{k=1}^{N}|x_{ku}-x_{kv}|$$

L1 is best known as absolute-value distance.

Linfinity (alias maximum)

requests the Minkowski distance metric with infinite argument. When comparing observations i and j, the formula is

$$\max_{a=1,\dots,p} |x_{ia} - x_{ja}|$$

and when comparing variables u and v, the formula is

$$\max_{k=1,\dots,N} |x_{ku} - x_{kv}|$$

Linfinity is best known as maximum-value distance.

L(#)

requests the Minkowski distance metric with argument #. When comparing observations i and j, the formula is

$$\left(\sum_{a=1}^{p} |x_{ia} - x_{ja}|^{\#} \right)^{1/\#} \qquad \# \geq 1$$

and when comparing variables u and v, the formula is

$$\left(\sum_{k=1}^{N} |x_{ku} - x_{kv}|^{\#} \right)^{1/\#} \qquad \# \geq 1$$

We discourage the use of extremely large values for #. Since the absolute value of the difference is being raised to the value of #, depending on the nature of your data, you could experience numeric overflow or underflow. With a large value of #, the L() option will produce similar results to the Linfinity option. Use the numerically more-stable Linfinity option instead of a large value for # in the L() option.

See Anderberg (1973) for a discussion of the Minkowski metric and its special cases.

Lpower(#)

requests the Minkowski distance metric with argument #, raised to the # power. When comparing observations i and j, the formula is

$$\sum_{a=1}^{p} |x_{ia} - x_{ja}|^{\#} \qquad \# \geq 1$$

and when comparing variables u and v, the formula is

$$\sum_{k=1}^{N} |x_{ku} - x_{kv}|^{\#} \qquad \# \geq 1$$

As with L(#), we discourage the use of extremely large values for #; see the discussion above.

Canberra

requests the following distance metric when comparing observations i and j

$$\sum_{a=1}^{p} \frac{|x_{ia} - x_{ja}|}{|x_{ia}| + |x_{ja}|}$$

and the following distance metric when comparing variables u and v

$$\sum_{k=1}^{N} \frac{|x_{ku} - x_{kv}|}{|x_{ku}| + |x_{kv}|}$$

When comparing observations, the Canberra metric takes values between 0 and p, the number of variables. When comparing variables, the Canberra metric takes values between 0 and N, the number of observations; see Gordon (1999) and Gower (1985). Gordon (1999) explains that the Canberra distance is very sensitive to small changes near zero.

correlation

requests the correlation coefficient similarity measure. When comparing observations i and j, the formula is

$$\frac{\sum_{a=1}^{p} (x_{ia} - \overline{x}_{i.})(x_{ja} - \overline{x}_{j.})}{\left\{ \sum_{a=1}^{p} (x_{ia} - \overline{x}_{i.})^2 \sum_{b=1}^{p} (x_{jb} - \overline{x}_{j.})^2 \right\}^{1/2}}$$

and when comparing variables u and v, the formula is

$$\frac{\sum_{k=1}^{N} (x_{ku} - \overline{x}_{.u})(x_{kv} - \overline{x}_{.v})}{\left\{ \sum_{k=1}^{N} (x_{ku} - \overline{x}_{.u})^2 \sum_{l=1}^{N} (x_{lv} - \overline{x}_{.v})^2 \right\}^{1/2}}$$

where $\overline{x}_{i.} = (\sum_{a=1}^{p} x_{ia})/p$ and $\overline{x}_{.u} = (\sum_{k=1}^{N} x_{ku})/N$.

The correlation similarity measure takes values between -1 and 1. With this measure, the relative direction of the two vectors is important. The correlation similarity measure is related to the angular separation similarity measure (described next). The correlation similarity measure gives the cosine of the angle between the two vectors measured from the mean; see Gordon (1999).

angular (alias angle)

requests the angular separation similarity measure. When comparing observations i and j, the formula is

$$\frac{\sum_{a=1}^{p} x_{ia} x_{ja}}{\left(\sum_{a=1}^{p} x_{ia}^2 \sum_{b=1}^{p} x_{jb}^2 \right)^{1/2}}$$

and when comparing variables u and v, the formula is

$$\frac{\sum_{k=1}^{N} x_{ku} x_{kv}}{\left(\sum_{k=1}^{N} x_{ku}^2 \sum_{l=1}^{N} x_{lv}^2 \right)^{1/2}}$$

The angular separation similarity measure is the cosine of the angle between the two vectors measured from zero and takes values from -1 to 1; see Gordon (1999).

Similarity measures for binary data

Similarity measures for binary data are based on the four values from the cross-tabulation of observation i and j (when comparing observations) or variables u and v (when comparing variables).

When comparing observation i and j, the cross-tabulation is

		obs. j	
		1	0
obs.	1	a	b
i	0	c	d

a is the number of variables where observations i and j both had ones, and d is the number of variables where observations i and j both had zeros. The number of variables where observation i is one and observation j is zero is b, and the number of variables where observation i is zero and observation j is one is c.

When comparing variables u and v, the cross-tabulation is

		var. v	
		1	0
var.	1	a	b
u	0	c	d

a is the number of observations where variables u and v both had ones, and d is the number of observations where variables u and v both had zeros. The number of observations where variable u is one and variable v is zero is b, and the number of observations where variable u is zero and variable v is one is c.

Stata treats nonzero values as one when a binary value is expected. Specifying one of the binary similarity measures imposes this behavior unless some other option overrides it (for instance, the `allbinary` option of `matrix dissimilarity`; see [MV] **matrix dissimilarity**).

Hubálek (1982) gives an extensive list of binary similarity measures. Gower (1985) gives a list of fifteen binary similarity measures, fourteen of which are implemented in Stata. (The excluded measure has many cases where the quantity is undefined, so it was not implemented.) Anderberg (1973) gives an interesting table where many of these measures are compared based on whether the zero–zero matches are included in the numerator, whether these matches are included in the denominator, and how the weighting of matches and mismatches is handled. Hilbe (1992a, 1992b) implemented an early Stata command for computing some of these (as well as other) binary similarity measures.

The formulas for some of these binary similarity measures are undefined when either one or both of the vectors (observations or variables depending on which are being compared) are all zeros (or, in some cases, all ones). Gower (1985) says concerning these cases, "These coefficients are then conventionally assigned some appropriate value, usually zero."

The following binary similarity coefficients are available. Unless stated otherwise, the similarity measures range from 0 to 1.

<u>matching</u>
   requests the simple matching (Zubin 1938, Sokal and Michener 1958) binary similarity coefficient

$$\frac{a+d}{a+b+c+d}$$

   which is the proportion of matches between the two observations or variables.

<u>Jac</u>card
   requests the Jaccard (1901 and 1908) binary similarity coefficient

$$\frac{a}{a+b+c}$$

   which is the proportion of matches when at least one of the vectors had a one. If both vectors are all zeros, this measure is undefined. In this case, Stata declares the answer to be one, meaning perfect agreement. This is a reasonable choice for most applications and will cause an all-zero vector to have similarity of one only with another all-zero vector. In all other cases, an all-zero vector will have Jaccard similarity of zero to the other vector.

   The Jaccard coefficient was discovered earlier by Gilbert (1884).

<u>Russell</u>
   requests the Russell and Rao (1940) binary similarity coefficient

$$\frac{a}{a+b+c+d}$$

Hamann
   requests the Hamann (1961) binary similarity coefficient

$$\frac{(a+d)-(b+c)}{a+b+c+d}$$

   which is the number of agreements minus disagreements divided by the total. The Hamann coefficient ranges from −1, perfect disagreement, to 1, perfect agreement. The Hamann coefficient is equal to twice the simple matching coefficient minus 1.

Dice
   requests the Dice binary similarity coefficient

$$\frac{2a}{2a+b+c}$$

   suggested by Czekanowski (1932), Dice (1945), and Sørensen (1948). The Dice coefficient is similar to the Jaccard similarity coefficient but gives twice the weight to agreements. Like the Jaccard coefficient, the Dice coefficient is declared by Stata to be one if both vectors are all zero, thus avoiding the case where the formula is undefined.

antiDice
>   requests the binary similarity coefficient

$$\frac{a}{a + 2(b + c)}$$

>   which is credited to Anderberg (1973) but was shown earlier by Sokal and Sneath (1963, 129). We did not call this the Anderberg coefficient since there is another coefficient better known by that name; see the `Anderberg` option. The name antiDice is our creation. This coefficient takes the opposite view from the Dice coefficient and gives double weight to disagreements. As with the Jaccard and Dice coefficients, the antiDice coefficient is declared to be one if both vectors are all zeros.

Sneath
>   requests the Sneath and Sokal (1962) binary similarity coefficient

$$\frac{2(a + d)}{2(a + d) + (b + c)}$$

>   which is similar to the simple matching coefficient but gives double weight to matches. Also compare the Sneath and Sokal coefficient with the Dice coefficient, which differs only in whether it includes d.

Rogers
>   requests the Rogers and Tanimoto (1960) binary similarity coefficient

$$\frac{a + d}{(a + d) + 2(b + c)}$$

>   which takes the opposite approach from the Sneath and Sokal coefficient and gives double weight to disagreements. Also compare the Rogers and Tanimoto coefficient with the antiDice coefficient, which differs only in whether it includes d.

Ochiai
>   requests the Ochiai (1957) binary similarity coefficient

$$\frac{a}{\left\{(a + b)(a + c)\right\}^{1/2}}$$

>   The formula for the Ochiai coefficient is undefined when one or both of the vectors being compared are all zeros. If both are all zeros, Stata declares the measure to be one, and if only one of the two vectors is all zeros, the measure is declared to be zero.

>   The Ochiai coefficient was presented earlier by Driver and Kroeber (1932).

Yule
>   requests the Yule (see Yule 1900 and Yule and Kendall 1950) binary similarity coefficient

$$\frac{ad - bc}{ad + bc}$$

which ranges from -1 to 1. The formula for the Yule coefficient is undefined when one or both of the vectors are either all zeros or all ones. Stata declares the measure to be 1 when $b + c = 0$, meaning that there is complete agreement. Stata declares the measure to be -1 when $a + d = 0$, meaning that there is complete disagreement. Otherwise, if $ad - bc = 0$, Stata declares the measure to be 0. These rules, applied before using the Yule formula, avoid the cases where the formula would produce an undefined result.

Anderberg
requests the Anderberg binary similarity coefficient

$$\left(\frac{a}{a+b} + \frac{a}{a+c} + \frac{d}{c+d} + \frac{d}{b+d} \right) \Big/ 4$$

The Anderberg coefficient is undefined when one or both vectors are either all zeros or all ones. This difficulty is overcome by first applying the rule that if both vectors are all ones (or both vectors are all zeros), the similarity measure is declared to be one. Otherwise, if any of the marginal totals $(a + b,\ a + c,\ c + d,\ b + d)$ are zero, then the similarity measure is declared to be zero.

Though this similarity coefficient is best known as the Anderberg coefficient, it appeared earlier in Sokal and Sneath (1963, 130).

Kulczynski
requests the Kulczynski (1927) binary similarity coefficient

$$\left(\frac{a}{a+b} + \frac{a}{a+c} \right) \Big/ 2$$

The formula for this measure is undefined when one or both of the vectors are all zeros. If both vectors are all zeros, Stata declares the similarity measure to be one. If only one of the vectors is all zeros, the similarity measure is declared to be zero.

Pearson
requests Pearson's (1900) ϕ binary similarity coefficient

$$\frac{ad - bc}{\{(a+b)(a+c)(d+b)(d+c)\}^{1/2}}$$

which ranges from -1 to 1. The formula for this coefficient is undefined when one or both of the vectors are either all zeros or all ones. Stata declares the measure to be 1 when $b + c = 0$, meaning that there is complete agreement. Stata declares the measure to be -1 when $a + d = 0$, meaning that there is complete disagreement. Otherwise, if $ad - bc = 0$, Stata declares the measure to be 0. These rules, applied before using Pearson's ϕ coefficient formula, avoid the cases where the formula would produce an undefined result.

Gower2
requests the binary similarity coefficient

$$\frac{ad}{\{(a+b)(a+c)(d+b)(d+c)\}^{1/2}}$$

which is presented by Gower (1985) but appeared earlier in Sokal and Sneath (1963, 130). Stata uses the name Gower2 to avoid confusion with the better-known Gower coefficient (not currently in Stata), which is used to combine continuous and categorical similarity or dissimilarity measures computed on a dataset into one measure.

The formula for this similarity measure is undefined when one or both of the vectors are all zeros or all ones. This is overcome by first applying the rule that if both vectors are all ones (or both vectors are all zeros) then the similarity measure is declared to be one. Otherwise, if $ad = 0$, the similarity measure is declared to be zero.

❏ Technical Note

Binary similarity measures applied to averages

Some cluster-analysis methods (such as Stata's kmeans and kmedians clustering) need to compute the similarity or dissimilarity between observations and group averages or group medians; see [MV] **cluster**. With binary data, a group average is interpreted as a proportion.

A group median for binary data will be zero or one, except when there are an equal number of zeros and ones. In this case, Stata calls the median 0.5, which can also be interpreted as a proportion.

In Stata's cluster kmeans and cluster kmedians commands for the case of comparing a binary observation to a group proportion (see *Partition cluster-analysis methods* in [MV] **cluster**), the values of a, b, c, and d are obtained by assigning the appropriate fraction of the count to these values. In our earlier table showing the relationship of a, b, c, and d in the cross-tabulation of observation i and observation j, we replace observation j by the group-proportions vector. Then when observation i is 1, we add the corresponding proportion to a and add one minus that proportion to b. When observation i is 0, we add the corresponding proportion to c, and add one minus that proportion to d. After the values of a, b, c, and d are computed in this way, the binary similarity measures are computed using the formulas as already described.

❏

References

Anderberg, M. R. 1973. *Cluster Analysis for Applications*. New York: Academic Press.

Czekanowski, J. 1932. "Coefficient of racial likeness" und "durchschnittliche Differenz". *Anthropologischer Anzeiger* 9: 227–249.

Dice, L. R. 1945. Measures of the amount of ecologic association between species. *Ecology* 26: 297–302.

Driver, H. E. and A. L. Kroeber. 1932. Quantitative expression of cultural relationships. *The University of California Publications in American Archaeology and Ethnology* 31: 211–256.

Gilbert, G. K. 1884. Finley's tornado predictions. *American Meteorological Journal* 1: 166–172.

Gordon, A. D. 1999. *Classification*. 2nd ed. Boca Raton, FL: CRC.

Gower, J. C. 1985. Measures of similarity, dissimilarity, and distance. In *Encyclopedia of Statistical Sciences*, Vol. 5, ed. S. Kotz, N. L. Johnson, and C. B. Read, 397–405. New York: Wiley.

Hamann, U. 1961. Merkmalsbestand und Verwandtschaftsbeziehungen der Farinosae. Ein Beitrag zum System der Monokotyledonen. *Willdenowia* 2: 639–768.

Hilbe, J. 1992a. sg9: Similarity coefficients for 2 x 2 binary data. *Stata Technical Bulletin* 9: 14–15. Reprinted in *Stata Technical Bulletin Reprints*, vol. 2, pp. 130–131.

———. 1992b. sg9.1: Additional statistics to similari output. *Stata Technical Bulletin* 10: 22. Reprinted in *Stata Technical Bulletin Reprints*, vol. 2, p. 132.

Hubálek, Z. 1982. Coefficients of association and similarity, based on binary (presence–absence) data: an evaluation. *Biological Reviews* 57: 669–689.

Jaccard, P. 1901. Distribution de la flore alpine dans le Bassin des Dranses et dans quelques régions voisines. *Bulletin de la Société Vaudoise des Sciences Naturelles* 37: 241–272.

——. 1908. Nouvelles recherches sur la distribution florale. *Bulletin de la Société Vaudoise des Sciences Naturelles* 44: 223–270.

Kulczynski, S. 1927. Die Pflanzenassoziationen der Pieninen. [In Polish, German summary.] *Bulletin International de l'Academie Polonaise des Sciences et des Lettres, Classe des Sciences Mathematiques et Naturelles, B (Sciences Naturelles)* 1927 (Suppl. 2): 57–203.

Ochiai, A. 1957. Zoogeographic studies on the soleoid fishes found in Japan and its neighbouring regions. [In Japanese, English summary.] *Bulletin of the Japanese Society of Scientific Fisheries* 22: 526–530.

Pearson, K. 1900. Mathematical contributions to the theory of evolution VII: On the correlation of characters not quantitatively measurable. *Philosophical Transactions of the Royal Society* Series A 195: 1–47.

Rogers, D. J. and T. T. Tanimoto. 1960. A computer program for classifying plants. *Science* 132: 1115–1118.

Russell, P. F. and T. R. Rao. 1940. On habitat and association of species of anopheline larvae in south-eastern Madras. *Journal of the Malaria Institute of India* 3: 153–178.

Sneath, P. H. A. and R. R. Sokal. 1962. Numerical taxonomy. *Nature* 193: 855–860.

Sokal, R. R. and C. D. Michener. 1958. A statistical method for evaluating systematic relationships. *University of Kansas Science Bulletin* 38: 1409–1438.

Sokal, R. R. and P. H. A. Sneath. 1963. *Principles of Numerical Taxonomy*. San Francisco: W. H. Freeman.

Sørensen, T. 1948. A method of establishing groups of equal amplitude in plant sociology based on similarity of species content and its application to analyses of the vegetation on Danish commons. *Kongel. Danske Vidensk. Selsk. Biol. Skr.* 5(4): 1–34.

Yule, G. U. 1900. On the association of attributes in statistics: with illustrations from the material of the Childhood Society, etc. *Philosophical Transactions of the Royal Society, Series A* 194: 257–319. Reprinted in Stuart, A. and M. G. Kendall (eds) 1971. *Statistical papers of George Udny Yule*. London: Griffin, 7–69.

Yule, G. U. and M. G. Kendall. 1950. *An Introduction to the Theory of Statistics*. 14th ed. New York: Hafner.

Zubin, J. 1938. A technique for measuring like-mindedness. *Journal of Abnormal and Social Psychology* 33: 508–516.

Also See

Complementary:	[MV] **cluster averagelinkage**, [MV] **cluster centroidlinkage**, [MV] **cluster completelinkage**, [MV] **cluster kmeans**, [MV] **cluster kmedians**, [MV] **cluster medianlinkage**, [MV] **cluster programming utilities**, [MV] **cluster singlelinkage**, [MV] **cluster wardslinkage**, [MV] **cluster waveragelinkage**, [MV] **mds**, [MV] **mdslong**
Related:	[MV] **matrix dissimilarity**
Background:	[MV] **cluster**

Title

pca — Principal component analysis

Syntax

Principal component analysis of data

pca *varlist* [*if*] [*in*] [*weight*] [, *options*]

Principal component analysis of a correlation or covariance matrix

pcamat *matname*, n(#) [*options pcamat_options*]

options	description
Model 2	
<u>com</u>ponents(#)	retain maximum of # principal components; <u>factors</u>() is a synonym
<u>mineigen</u>(#)	retain eigenvalues larger than #; default is 1e-5
<u>correlation</u>	perform PCA of the correlation matrix; the default
<u>cov</u>ariance	perform PCA of the covariance matrix
vce(<u>none</u>)	do not compute VCE of the eigenvalues and vectors; the default
vce(<u>normal</u>)	compute VCE of the eigenvalues and vectors assuming mult. normality
Reporting	
<u>level</u>(#)	set confidence level; default is level(95)
<u>bl</u>anks(#)	display loadings as blanks when |loadings| < #
novce	suppress display of SEs even though calculated
* <u>means</u>	display summary statistics of variables
Advanced	
tol(#)	advanced option; see *Options* for details
ignore	advanced option; see *Options* for details
† <u>norot</u>ated	display unrotated results, even if rotated results are available (replay only)

* means is not allowed with pcamat.

† norotated is not shown in the dialog box.

368

pcamat_options	description
Model	
shape(<u>f</u>ull)	*matname* is a square symmetric matrix; the default
shape(<u>l</u>ower)	*matname* is a vector with the rowwise lower triangle (with diagonal)
shape(<u>u</u>pper)	*matname* is a vector with the rowwise upper triangle (with diagonal)
<u>names</u>(*namelist*)	variable names; required if *matname* is triangular
* n(#)	number of observations
sds(*matname₂*)	vector with standard deviations of variables
means(*matname₃*)	vector with means of variables

* n() is required for pcamat.

bootstrap, by, jackknife, rolling, statsby, and xi are allowed with pca; see [U] **11.1.10 Prefix commands**.
aweights and fweights are allowed with pca; see [U] **11.1.6 weight**.
See [MV] **pca postestimation** for features available after estimation.

Description

Principal component analysis (PCA) is a statistical technique used for data reduction. The leading eigenvectors from the eigen decomposition of the correlation or covariance matrix of the variables describe a series of uncorrelated linear combinations of the variables that contain most of the variance. In addition to data reduction, the eigenvectors from a PCA are often inspected in order to learn more about the underlying structure of the data.

pca and pcamat display the eigenvalues and eigenvectors from the PCA eigen decomposition. The eigenvectors are returned in orthonormal form, i.e., orthogonal (uncorrelated) and normalized (with unit length, $\mathbf{L'L} = \mathbf{I}$). pcamat provides the correlation or covariance matrix directly. For pca, the correlation or covariance matrix is computed from the variables in *varlist*.

pcamat allows the correlation or covariance matrix $\mathbf{C}$ to be specified as a $k \times k$ symmetric matrix with row and column names set to the variable names, or as a $k(k+1)/2$ long row or column vector containing the lower or upper triangle of $\mathbf{C}$ along with the names() option providing the variable names. See the shape() option for details.

The vce(normal) option of pca and pcamat provides standard errors of the eigenvalues and eigenvectors and aids in interpreting the eigenvectors. See *Remarks* for a discussion of the underlying assumptions.

Scores, residuals, rotations, scree plots, score plots, loading plots, and more are available after pca and pcamat, see [MV] **pca postestimation**.

Options

> Model 2

components(#) and mineigen(#) specify the maximum number of components (eigenvectors or factors) to be retained. components() specifies the number directly, and mineigen() specifies it indirectly, keeping all components with eigenvalues greater than the indicated value. The options can be specified individually, together, or not at all. factors() is a synonym for components().

components(#) sets the maximum number of components (factors) to be retained. pca and pcamat always display the full set of eigenvalues but display eigenvectors only for retained components. Specifying a number larger than the number of variables in *varlist* is equivalent to specifying the number of variables in *varlist* and is the default.

mineigen(#) sets the minimum value of eigenvalues to be retained. The default is 1e-5 or the value of tol() if specified.

Specifying components() and mineigen() affects only the number of components to be displayed and stored in e(); it does not enforce the assumption that the other eigenvalues are 0. In particular, the standard errors reported when vce(normal) is specified do not depend on the number of retained components.

correlation and covariance specify that principal components be calculated for the correlation matrix and covariance matrix, respectively. The default is correlation. Unlike factor analysis, PCA is not scale invariant; the eigenvalues and eigenvectors of a covariance matrix differ from those of the associated correlation matrix. Usually, a PCA of a covariance matrix is only meaningful if the variables are expressed in the same units.

In the case of pcamat, do not confuse the type of the matrix to be analyzed with the type of *matname*. Obviously, if *matname* is a correlation matrix and the option sds() is not specified, it is not possible to perform a PCA of the covariance matrix.

vce(none | normal) specifies whether standard errors are to be computed for the eigenvalues, the eigenvectors, and the (cumulative) percentage of explained variance ("confirmatory PCA"). These standard errors are obtained assuming multivariate normality of the data and are valid only for a PCA of a covariance matrix. Be cautious if applying these to correlation matrices.

Reporting

level(#) specifies the confidence level, as a percentage, for confidence intervals. The default is level(95) or as set by set level; see [U] **20.6 Specifying the width of confidence intervals**. level() is only allowed in combination with vce(normal).

blanks(#) shows blanks for loadings with absolute value smaller than #. This option is ignored when specified with vce(normal).

novce suppresses the display of standard errors, even though they are computed, and displays the PCA results in a matrix/table style. You can specify novce during estimation in combination with vce(normal). More likely, you will want to use novce during replay.

means displays summary statistics of the variables over the estimation sample. This option is not available with pcamat.

Advanced

tol(#) is an advanced, rarely used option and is available only in combination with vce(normal). An eigenvalue, ev_i, is classified as being close to zero if $ev_i < \text{tol} \times \max(ev)$. Two eigenvalues, ev_1 and ev_2, are "close" if $\text{abs}(ev_1 - ev_2) < \text{tol} \times \max(ev)$. The default is tol(1e-5). See option ignore and the technical note below.

ignore is an advanced, rarely used option and is available only in combination with vce(normal). It continues the computation of standard errors and tests, even if some eigenvalues are suspiciously close to zero or suspiciously close to other eigenvalues, violating crucial assumptions of the asymptotic theory used to estimate standard errors and tests. See the *technical note* later in this entry.

The following option is available with pca and pcamat but is not shown in the dialog box:

norotated displays the unrotated principal components, even if rotated components are available. This option may only be specified when replaying results.

Options unique to pcamat

⌐ Model ⌐

shape(*shape_arg*) specifies the shape (storage mode) for the covariance or correlation matrix *matname*. The following modes are supported:

full specifies that the correlation or covariance structure of k variables is stored as a symmetric $k \times k$ matrix. Specifying shape(full) is optional in this case.

lower specifies that the correlation or covariance structure of k variables is stored as a vector with $k(k+1)/2$ elements in rowwise lower-triangular order:

$$C_{11} \; C_{21} \; C_{22} \; C_{31} \; C_{32} \; C_{33} \; \ldots \; C_{k1} \; C_{k2} \; \ldots \; C_{kk}$$

upper specifies that the correlation or covariance structure of k variables is stored as a vector with $k(k+1)/2$ elements in rowwise upper-triangular order:

$$C_{11} \; C_{12} \; C_{13} \; \ldots \; C_{1k} \; C_{22} \; C_{23} \; \ldots C_{2k} \; \ldots \; C_{(k-1k-1)} \; C_{(k-1k)} \; C_{kk}$$

names(*namelist*) specifies a list of k different names, which are used to document output and to label estimation results and are used as variable names by predict. By default, pcamat verifies that the row and column names of *matname* and the column or row names of *matname*$_2$ and *matname*$_3$ from the sds() and means() options are in agreement. Using the names() option turns off this check.

n(*#*) is required and specifies the number of observations.

sds(*matname*$_2$) specifies a $k \times 1$ or $1 \times k$ matrix with the standard deviations of the variables. The row or column names should match the variable names, unless the names() option is specified. sds() may only be specified if *matname* is a correlation matrix.

means(*matname*$_3$) specifies a $k \times 1$ or $1 \times k$ matrix with the means of the variables. The row or column names should match the variable names, unless the names() option is specified. Specify means() if you have variables in your dataset and want to use predict after pcamat.

Remarks

Principal component analysis (PCA) is commonly thought of as a statistical technique for data reduction. It helps you reduce the number of variables in an analysis by describing a series of uncorrelated linear combinations of the variables that contain most of the variance.

PCA originated with the work of Pearson (1901) and Hotelling (1933). For an introduction, see Rabe-Hesketh and Everitt (2004, chapter 14) or van Belle et al. (2004, chapter 14). More advanced treatments are Mardia, Kent, and Bibby (1979, chapter 8), and Rencher (2002, chapter 12). For monograph-sized treatments, including extensive discussions of the relationship between PCA and related approaches, see Jackson (1991) and Jolliffe (2002).

The objective of PCA is to find unit-length linear combinations of the variables with the greatest variance. The first principal component has maximal overall variance. The second principal component has maximal variance among all unit length linear combinations that are uncorrelated to the first principal component, etc. The last principal component has the smallest variance among all unit-length linear combinations of the variables. All principal components combined contain the same information as the original variables, but the important information is partitioned over the components in a particular way: The components are orthogonal, and earlier components contain more information than later components. PCA thus conceived is just a linear transformation of the data. It does not assume that the data satisfy a specific statistical model, though it does require that the data be interval-level data—otherwise taking linear combinations is meaningless.

It is important to stress that PCA is scale dependent. The principal components of a covariance matrix and those of a correlation matrix are different. In applied research, PCA of a covariance matrix is useful only if the variables are expressed in commensurable units.

❑ Technical Note

Principal components have a number of useful properties. Some of these are geometric. Both the principal components and the principal scores are uncorrelated (orthogonal) among each other. The f leading principal components have maximal generalized variance among all f unit-length linear combinations.

Alternatively, it is possible to interpret PCA as a fixed effects factor analysis with homoskedastic residuals

$$ y_{ij} = \mathbf{a}_i' \mathbf{b}_j + e_{ij} \qquad i = 1, \ldots, n \qquad j = 1, \ldots, p $$

where y_{ij} are the elements of the matrix $\mathbf{Y}$, $\mathbf{a}_i$ (scores) and $\mathbf{b}_j$ (loadings) are f-vectors of parameters, and e_{ij} are independent homoskedastic residuals. (In factor analysis, the scores $\mathbf{a}_i$ are random rather than fixed, and the residuals are allowed to be heteroskedastic in j.) It follows that $E(\mathbf{Y})$ is a matrix of rank f, with f typically substantially less than n or p. Thus we may think of PCA as a regression model with a restricted number but unknown independent variables. We may also say that the expected values of the rows (or columns) of $\mathbf{Y}$ are in some unknown f-dimensional space.

For additional information on these properties and for additional characterizations of PCA, see Jackson (1991) and Jolliffe (2002).

❑

▷ Example 1

We consider a dataset of audiometric measurements on 100 males, age 9. The measurements are minimal discernable intensities at 4 different frequencies with the left and right ear (see Jackson 1991, 106). The variable lft1000 refers to the left ear at 1000Hz.

```
. use http://www.stata-press.com/data/r9/audiometric
(Audiometric measures)

. correlate lft* rght*
(obs=100)
```

	lft500	lft1000	lft2000	lft4000	rght500	rght1000	rght2000
lft500	1.0000						
lft1000	0.7775	1.0000					
lft2000	0.4012	0.5366	1.0000				
lft4000	0.2554	0.2749	0.4250	1.0000			
rght500	0.6963	0.5515	0.2391	0.1790	1.0000		
rght1000	0.6416	0.7070	0.4460	0.2632	0.6634	1.0000	
rght2000	0.2372	0.3597	0.7011	0.3165	0.1589	0.4142	1.0000
rght4000	0.2041	0.2169	0.3262	0.7097	0.1321	0.2201	0.3746

	rght4000
rght4000	1.0000

As you may have expected, measurements on the same ear are more highly correlated than measurements on different ears. Also, measurements on different ears at the same frequency are more highly correlated than at different frequencies. Since the variables are in commensurable units, it would make theoretical sense to analyze the covariance matrix of these variables. However, the variances of the measures are very different:

```
. summarize lft* rght*, sep(4)
    Variable |        Obs        Mean    Std. Dev.        Min         Max
-------------+--------------------------------------------------------------
      lft500 |        100        -2.8     6.408643        -10          15
     lft1000 |        100         -.5     7.571211        -10          20
     lft2000 |        100           2     10.94061        -10          45
     lft4000 |        100       21.35     19.61569        -10          70
-------------+--------------------------------------------------------------
     rght500 |        100        -2.6     7.123726        -10          25
    rght1000 |        100         -.7     6.396811        -10          20
    rght2000 |        100         1.6     9.289942        -10          35
    rght4000 |        100       21.35     19.33039        -10          75
```

In an analysis of the covariances, the higher frequency measures would dominate the results. There is no clinical reason for such an effect (see also Jackson 1991). Therefore, we will analyze the correlation matrix.

```
. pca lft* rght*
Principal components/correlation                 Number of obs    =         100
                                                 Number of comp.  =           8
                                                 Trace            =           8
        Rotation: (unrotated = principal)        Rho              =      1.0000

    ------------------------------------------------------------------------
     Component |  Eigenvalue   Difference       Proportion   Cumulative
    -----------+------------------------------------------------------------
         Comp1 |    3.92901      2.31068           0.4911       0.4911
         Comp2 |    1.61832      .642997           0.2023       0.6934
         Comp3 |    .975325      .508543           0.1219       0.8153
         Comp4 |    .466782      .126692           0.0583       0.8737
         Comp5 |     .34009     .0241988           0.0425       0.9162
         Comp6 |    .315891       .11578           0.0395       0.9557
         Comp7 |    .200111     .0456375           0.0250       0.9807
         Comp8 |    .154474            .           0.0193       1.0000
    ------------------------------------------------------------------------

Principal components (eigenvectors)

    ------------------------------------------------------------------------------
       Variable |     Comp1      Comp2      Comp3      Comp4      Comp5      Comp6
    ------------+-----------------------------------------------------------------
         lft500 |    0.4011    -0.3170     0.1582    -0.3278     0.0231     0.4459
        lft1000 |    0.4210    -0.2255    -0.0520    -0.4816    -0.3792    -0.0675
        lft2000 |    0.3664     0.2386    -0.4703    -0.2824     0.4392    -0.0638
        lft4000 |    0.2809     0.4742     0.4295    -0.1611     0.3503    -0.4169
        rght500 |    0.3433    -0.3860     0.2593     0.4876     0.4975     0.1948
       rght1000 |    0.4114    -0.2318    -0.0289     0.3723    -0.3513    -0.6136
       rght2000 |    0.3115     0.3171    -0.5629     0.3914    -0.1108     0.2650
       rght4000 |    0.2542     0.5135     0.4262     0.1591    -0.3960     0.3660
    ------------------------------------------------------------------------------
```

Variable	Comp7	Comp8	Unexplained
lft500	0.3293	-0.5463	0
lft1000	-0.0331	0.6227	0
lft2000	-0.5255	-0.1863	0
lft4000	0.4269	0.0839	0
rght500	-0.1594	0.3425	0
rght1000	-0.0837	-0.3614	0
rght2000	0.4778	0.1466	0
rght4000	-0.4139	-0.0508	0

pca shows two panels. The first panel lists the eigenvalues of the correlation matrix, ordered from largest to smallest. The corresponding eigenvectors are listed in the second panel. These are the principal components and have unit length; the columnwise sum of the squares of the loadings is 1 ($0.4011^2 + 0.4210^2 + \cdots + 0.2542^2 = 1$).

Remark: Literature and software that treat principal components in combination with factor analysis tend to display principal components normed to the associated eigenvalues rather than to 1. This normalization is available in the post estimation command estat loadings; see [MV] **pca postestimation**.

The eigenvalues add up to the sum of the variances of the variables in the analysis—the "total variance" of the variables. Since we are analyzing a correlation matrix, the variables are standardized to have unit variance, so the total variance is 8. The eigenvalues are the variances of the principal components. The first principal component has variance 3.93, explaining 49% (3.93/8) of the total variance. The second principal component has variance 1.61 or 20% (1.61/8) of the total variance. We stress that principal components are uncorrelated. You may want to verify that, for instance,

$$0.4011(-0.3170) + 0.4210(-0.2255) + \cdots + 0.2542(0.5135) = 0$$

As a consequence, we may also say that the first two principal components explain the sum of the variances of the individual components, or $49 + 20 = 69\%$ of the total variance. Had the components been correlated, they would have partly represented the same information, so the information contained in the combination would not have been equal to the sum of the information of the components. All 8 principal components combined explain all variance in all variables; therefore, the unexplained variances listed in the second panel are all zero, and Rho = 1.00 as shown above the first panel.

Over 85% of the variance is contained in the first 4 principal components. We can list just these components with the option components(4).

```
. pca lft* rght*, components(4)
```

Principal components/correlation

Number of obs	=	100
Number of comp.	=	4
Trace	=	8

Rotation: (unrotated = principal)   Rho   =   0.8737

Component	Eigenvalue	Difference	Proportion	Cumulative
Comp1	3.92901	2.31068	0.4911	0.4911
Comp2	1.61832	.642997	0.2023	0.6934
Comp3	.975325	.508543	0.1219	0.8153
Comp4	.466782	.126692	0.0583	0.8737
Comp5	.34009	.0241988	0.0425	0.9162
Comp6	.315891	.11578	0.0395	0.9557
Comp7	.200111	.0456375	0.0250	0.9807
Comp8	.154474	.	0.0193	1.0000

Principal components (eigenvectors)

Variable	Comp1	Comp2	Comp3	Comp4	Unexplained
lft500	0.4011	-0.3170	0.1582	-0.3278	.1308
lft1000	0.4210	-0.2255	-0.0520	-0.4816	.1105
lft2000	0.3664	0.2386	-0.4703	-0.2824	.1275
lft4000	0.2809	0.4742	0.4295	-0.1611	.1342
rght500	0.3433	-0.3860	0.2593	0.4876	.1194
rght1000	0.4114	-0.2318	-0.0289	0.3723	.1825
rght2000	0.3115	0.3171	-0.5629	0.3914	.07537
rght4000	0.2542	0.5135	0.4262	0.1591	.1303

The first panel is not affected. The second panel now lists the first 4 principal components. These 4 components do not contain all information in the data, and therefore some of the variances in the variables are unaccounted for or unexplained. These equal the sums of squares of the loadings in the deleted components, weighted by the associated eigenvalues. The unexplained variances in all variables are of similar order. Note that the average unexplained variance is exactly equal to the overall unexplained variance of 13% $(1 - 0.87)$.

Look more closely at the principal components. The first component has positive loadings of roughly equal size on all variables. It can be interpreted as overall sensitivity of a person's ears. The second principal component has positive loadings on the higher frequencies with both ears, and negative loadings for the lower frequencies. Thus the second principal component distinguishes sensitivity for higher frequencies versus lower frequencies. The third principal component similarly differentiates sensitivity at medium frequencies from sensitivity at other frequencies. Finally, the fourth principal component has negative loadings on the left ear and positive loadings on the right ear; it obviously differentiates the left and right ear.

We stated earlier that the first principal component had similar loadings on all eight variables. This can be tested if we are willing to assume that the data are multivariate normal distributed. For this case, pca can estimate the standard errors and related statistics. To conserve paper, we request only the results of the first two principal components and specify the option vce(normal).

```
. pca l* r*, comp(2) vce(normal)
(with PCA/correlation, SEs and tests are approximate)
```

Principal components/correlation

```
                                              Number of obs    =       100
                                              Number of comp.  =         2
                                              Trace            =         8
                                              Rho              =    0.6934
SEs assume multivariate normality             SE(Rho)          =    0.0273
```

	Coef.	Std. Err.	z	P>|z|	[95% Conf. Interval]	
Eigenvalues						
Comp1	3.929005	.5556453	7.07	0.000	2.839961	5.01805
Comp2	1.618322	.2288653	7.07	0.000	1.169754	2.066889
Comp1						
lft500	.4010948	.0429963	9.33	0.000	.3168236	.485366
lft1000	.4209908	.0359372	11.71	0.000	.3505551	.4914264
lft2000	.3663748	.0463297	7.91	0.000	.2755702	.4571794
lft4000	.2808559	.0626577	4.48	0.000	.1580491	.4036628
rght500	.343251	.0528285	6.50	0.000	.2397091	.446793
rght1000	.4114209	.0374312	10.99	0.000	.3380571	.4847846
rght2000	.3115483	.0551475	5.65	0.000	.2034612	.4196354
rght4000	.2542212	.066068	3.85	0.000	.1247303	.3837121
Comp2						
lft500	-.3169638	.067871	-4.67	0.000	-.4499885	-.1839391
lft1000	-.225464	.0669887	-3.37	0.001	-.3567595	-.0941686
lft2000	.2385933	.1079073	2.21	0.027	.0270989	.4500877
lft4000	.4741545	.0967918	4.90	0.000	.284446	.6638629
rght500	-.3860197	.0803155	-4.81	0.000	-.5434352	-.2286042
rght1000	-.2317725	.0674639	-3.44	0.001	-.3639994	-.0995456
rght2000	.317059	.1215412	2.61	0.009	.0788427	.5552752
rght4000	.5135121	.0951842	5.39	0.000	.3269544	.7000697

```
LR test for independence:      chi2(28)  =    448.21  Prob > chi2 =   0.0000
LR test for    sphericity:     chi2(35)  =    451.11  Prob > chi2 =   0.0000
```

Explained variance by components

Components	Eigenvalue	Proportion	SE_Prop	Cumulative	SE_Cum	Bias
Comp1	3.929005	0.4911	0.0394	0.4911	0.0394	.056663
Comp2	1.618322	0.2023	0.0271	0.6934	0.0273	.015812
Comp3	.9753248	0.1219	0.0178	0.8153	0.0175	-.014322
Comp4	.4667822	0.0583	0.0090	0.8737	0.0127	.007304
Comp5	.34009	0.0425	0.0066	0.9162	0.0092	.026307
Comp6	.3158912	0.0395	0.0062	0.9557	0.0055	-.057717
Comp7	.2001111	0.0250	0.0040	0.9807	0.0031	-.013961
Comp8	.1544736	0.0193	0.0031	1.0000	0.0000	-.020087

Here pca acts like an estimation command. The output is organized in different equations. The first equation contains the eigenvalues. The second equation named, Comp1, is the first principal component, etc. pca reports, for instance, standard errors of the eigenvalues. While testing the values of eigenvalues may, up to now, be rare in applied research, interpretation of results should take stability into consideration. It makes little sense to report the first eigenvalue as 3.929 if you see that the standard error is 0.55.

pca has also reported the standard errors of the principal components. It has also estimated the covariances.

```
. estat vce
(output omitted)
```

Showing the large amount of information contained in the VCE matrix is not useful by itself. The fact that it has been estimated, however, enables us to test properties of the principal components. Does it make good sense to talk about the loadings of the first principal component being of the same size? We use `testparm` with two options; see [R] **test**. `eq(Comp1)` specifies that we are testing coefficients for equation `Comp1`, i.e., the first principal component. `equal` specifies that instead of testing that the coefficients are zero, we want to test that the coefficients are equal to each other—a more sensible hypothesis since principal components are normalized to 1.

```
. testparm lft* rght*, equal eq(Comp1)
 ( 1) - [Comp1]lft500 + [Comp1]lft1000 = 0
 ( 2) - [Comp1]lft500 + [Comp1]lft2000 = 0
 ( 3) - [Comp1]lft500 + [Comp1]lft4000 = 0
 ( 4) - [Comp1]lft500 + [Comp1]rght500 = 0
 ( 5) - [Comp1]lft500 + [Comp1]rght1000 = 0
 ( 6) - [Comp1]lft500 + [Comp1]rght2000 = 0
 ( 7) - [Comp1]lft500 + [Comp1]rght4000 = 0

           chi2(  7) =     7.56
         Prob > chi2 =    0.3729
```

We cannot reject the null hypothesis of equal loadings, so our interpretation of the first component does not seem to be in conflict with the data.

`pca` also displays standard errors of the proportions of variance explained by the leading principal components. Again this information is useful primarily to indicate the strength of formulations of results rather than to test hypotheses about these statistics. The information is also useful to compare studies: if in one study the leading two principal components explain 70% of variance while in a replicating study they explain 80%, are these differences significant given the sampling variation?

Since `pca` is an estimation command just like `regress` or `xtlogit`, you may replay the output by typing just `pca`. If you have used `pca` with the `vce(normal)` option, you may use the option `novce` at estimation or during replay to display the standard PCA output.

```
. pca, novce
Principal components/correlation                 Number of obs    =       100
                                                 Number of comp.  =         2
                                                 Trace            =         8
          Rotation: (unrotated = principal)      Rho              =    0.6934
```

Component	Eigenvalue	Difference	Proportion	Cumulative
Comp1	3.92901	2.31068	0.4911	0.4911
Comp2	1.61832	.642997	0.2023	0.6934
Comp3	.975325	.508543	0.1219	0.8153
Comp4	.466782	.126692	0.0583	0.8737
Comp5	.34009	.0241988	0.0425	0.9162
Comp6	.315891	.11578	0.0395	0.9557
Comp7	.200111	.0456375	0.0250	0.9807
Comp8	.154474	.	0.0193	1.0000

(*Continued on next page*)

Principal components (eigenvectors)

Variable	Comp1	Comp2	Unexplained
lft500	0.4011	-0.3170	.2053
lft1000	0.4210	-0.2255	.2214
lft2000	0.3664	0.2386	.3805
lft4000	0.2809	0.4742	.3262
rght500	0.3433	-0.3860	.2959
rght1000	0.4114	-0.2318	.248
rght2000	0.3115	0.3171	.456
rght4000	0.2542	0.5135	.3193

◁

❏ Technical Note

Inference on the eigenvalues and eigenvectors of a covariance matrix is based on a series of assumptions:

(A1) the variables are multivariate normal distributed, and

(A2) the variance–covariance matrix of the observations has all distinct and strictly positive eigenvalues.

Under assumptions A1 and A2, the eigenvalues and eigenvectors of the sample covariance matrix can be seen as maximum likelihood estimates for the population analogues that are asymptotically (multivariate) normally distributed (Andersen 1963; Jackson 1991). See Tyler (1981) for related results for elliptic distributions. Be cautious in interpreting since the asymptotic variances are rather sensitive to violations of assumption A1 (and A2). Wald tests of hypotheses that are in conflict with assumption A2 (e.g., testing that the first and second eigenvalue are the same) produce incorrect p-values.

Because the statistical theory for a PCA of a correlation matrix is much more complicated, pca and pcamat compute standard errors and tests of a correlation matrix as if it were a covariance matrix. This practice is in line with the application of asymptotic theory in Jackson (1991). This will usually lead to some underestimation of standard errors, but we believe that this problem is smaller than the consequences of deviations from normality.

You may conduct tests for marginal normality of the variables (see [R] **sktest** and [R] **swilk**), but recall that marginal normality does not imply multivariate normality.

. sktest lft* rght*

Skewness/Kurtosis tests for Normality

Variable	Pr(Skewness)	Pr(Kurtosis)	adj chi2(2)	——— joint ——— Prob>chi2
lft500	0.017	0.300	6.34	0.0420
lft1000	0.011	0.813	6.16	0.0460
lft2000	0.000	0.000	29.35	0.0000
lft4000	0.069	0.332	4.36	0.1129
rght500	0.000	0.009	19.45	0.0001
rght1000	0.005	0.089	9.33	0.0094
rght2000	0.000	0.015	16.38	0.0003
rght4000	0.001	0.819	10.32	0.0057

These tests cast some serious doubts on univariate normality of the variables. Multivariate normality is thus not plausible, as well. We advise caution in interpreting the inference results. Time permitting, you may want to turn to bootstrap methods for inference on the principal components and eigenvalues, but you should be aware of some serious identification problems in using the bootstrap here (Milan and Whittaker 1995).

❑

▷ Example 2

We remarked before that the principal components of a correlation matrix are generally different from the principal components of a covariance matrix. pca defaults to performing the PCA of the correlation matrix. To obtain a PCA of the covariance matrix, specify the covariance option.

```
. pca l* r*, comp(4) covariance
Principal components/covariance                Number of obs     =      100
                                               Number of comp.   =        4
                                               Trace             =   1154.5
       Rotation: (unrotated = principal)       Rho               =   0.9396
```

Component	Eigenvalue	Difference	Proportion	Cumulative
Comp1	706.795	527.076	0.6122	0.6122
Comp2	179.719	68.3524	0.1557	0.7679
Comp3	111.366	24.5162	0.0965	0.8643
Comp4	86.8501	57.4842	0.0752	0.9396
Comp5	29.366	9.53428	0.0254	0.9650
Comp6	19.8317	6.67383	0.0172	0.9822
Comp7	13.1578	5.74352	0.0114	0.9936
Comp8	7.41432	.	0.0064	1.0000

Principal components (eigenvectors)

Variable	Comp1	Comp2	Comp3	Comp4	Unexplained
lft500	0.0835	0.2936	-0.0105	0.3837	7.85
lft1000	0.1091	0.3982	0.0111	0.3162	11.71
lft2000	0.2223	0.5578	0.0558	-0.4474	11.13
lft4000	0.6782	-0.1163	-0.7116	-0.0728	.4024
rght500	0.0662	0.2779	-0.0226	0.4951	12.42
rght1000	0.0891	0.3119	0.0268	0.2758	11.14
rght2000	0.1707	0.3745	0.2721	-0.4496	14.71
rght4000	0.6560	-0.3403	0.6441	0.1550	.4087

As expected, the results are less clear. The total variance to be analyzed is 1154.5; this is the sum of the variances of the 8 variables, i.e., the trace of the covariance matrix. The leading principal components now account for a larger fraction of the variance; this is often the case with covariance matrices where the variables have widely different variances. The principal components are somewhat harder to interpret; mainly the loadings are no longer of roughly comparable size.

◁

▷ Example 3

Sometimes you do not have the original data, but only the correlation or covariance matrix. pcamat performs a PCA for such a matrix. To simplify presentation, we use the data on the left ear.

```
. correlate lft*, cov
(obs=100)
                 |   lft500   lft1000   lft2000   lft4000
        ---------+----------------------------------------
          lft500 |  41.0707
         lft1000 |  37.7273   57.3232
         lft2000 |  28.1313   44.4444   119.697
         lft4000 |   32.101   40.8333   91.2121   384.775
```

Suppose that we only have the covariances of the variables. We can enter the covariances into a square Stata matrix.

```
. matrix Cfull  = ( 41.0707, 37.7273, 28.1313,  32.101 \
                    37.7273, 57.3232, 44.4444, 40.8333 \
                    28.1313, 44.4444, 119.697, 91.2121 \
                    32.101 , 40.8333, 91.2121, 384.775 )
```

and invoke pcamat with the options n(100), specifying the number of observations and names()
providing the variable names.

```
. pcamat Cfull, comp(2) n(100) names(lft500 lft1000 lft2000 lft4000)
Principal components/correlation              Number of obs    =       100
                                              Number of comp.  =         2
                                              Trace            =         4
          Rotation: (unrotated = principal)   Rho              =    0.8169
```

Component	Eigenvalue	Difference	Proportion	Cumulative
Comp1	2.37181	1.47588	0.5930	0.5930
Comp2	.895925	.366238	0.2240	0.8169
Comp3	.529687	.327107	0.1324	0.9494
Comp4	.20258	.	0.0506	1.0000

```
Principal components (eigenvectors)
```

Variable	Comp1	Comp2	Unexplained
lft500	0.5384	-0.4319	.1453
lft1000	0.5730	-0.3499	.1116
lft2000	0.4958	0.2955	.3387
lft4000	0.3687	0.7770	.1367

Alternatively, we could have defined the rownames and colnames of Cfull as the variable names.

```
. matrix rownames Cfull = lft500 lft1000 lft2000 lft4000
. matrix colnames Cfull = lft500 lft1000 lft2000 lft4000
```

The option names() on pcamat would then have been superfluous.

In providing the full symmetric matrix, we have had to enter the off-diagonal elements twice, once for the lower triangle and once for the upper triangle of the matrix. For this small example, it is not much of a burden, but with a larger covariance or correlation matrix it would be nice to be able to avoid excess typing. pcamat allows you to provide the covariance or correlation matrix with just the upper or lower triangular elements including the diagonal. (Thus for correlations you have to enter the 1's for the diagonal.) We enter the lower triangle of our covariance matrix row by row up to and including the diagonal as a one-row Stata matrix.

```
. matrix Clow = ( 41.0707, 37.7273, 57.3232, 28.1313, 44.4444,
                  119.697, 32.101,  40.8333, 91.2121, 384.775 )
```

The matrix Clow has 1 row and 10 columns. We prefer to enter these numbers in the following way:

```
. matrix Clow = ( 41.0707,
                  37.7273, 57.3232,
                  28.1313, 44.4444, 119.697,
                  32.101,  40.8333, 91.2121, 384.775 )
```

so that it is easier to see the structure. When using the lower or upper triangle stored in a row or column vector, it is not possible to define the variable names as row or column names of the matrix; the option names() is required. Moreover, we have to specify the option shape(lower) to inform pcamat that the vector contains the lower triangle, not the upper triangle.

```
. pcamat Clow, comp(2) shape(lower) n(100) names(lft500 lft1000 lft2000 lft4000)
  (output omitted )
```

◁

Saved Results

pca and pcamat without the vce(normal) option save in e():

Scalars

e(N)	number of observations
e(f)	number of retained components
e(rho)	fraction of explained variance
e(trace)	trace of e(C)
e(lndet)	ln of the determinant of e(C)
e(cond)	condition number of e(C)

Macros

e(cmd)	pca (even in the case of pcamat)
e(title)	Principal components/correlation or covariance
e(predict)	pca_p
e(rotate_cmd)	pca_rotate
e(estat_cmd)	pca_estat
e(Ctype)	correlation or covariance
e(wtype)	weight type
e(wexp)	weight expression
e(properties)	nob noV eigen

Matrices

e(C)	$p \times p$ correlation or covariance matrix
e(means)	$1 \times p$ matrix of means
e(sds)	$1 \times p$ matrix of standard deviations
e(Ev)	$1 \times p$ matrix of eigenvalues (sorted)
e(L)	$p \times f$ matrix of eigenvectors = components
e(Psi)	$1 \times p$ matrix of unexplained correlation/covariance

Functions

e(sample)	marks estimation sample

pca and pcamat with the vce(normal) option save the above and additionally save:

Scalars

e(v_rho)	variance of e(rho)
e(chi2_i)	χ^2 statistic for test of independence
e(df_i)	degrees of freedom for test of independence
e(p_i)	significance of test of independence
e(chi2_s)	χ^2 statistic for test of sphericity
e(df_s)	degrees of freedom for test of sphericity
e(p_s)	significance of test of sphericity

Macros

e(vce)	multivariate normality
e(properties)	b V

Matrices

e(b)	$1 \times p+fp$ coefficient vector (all eigenvalues and retained eigenvectors)
e(V)	variance–covariance matrix of the estimates e(b)
e(Ev_bias)	$1 \times p$ matrix: bias of eigenvalues
e(Ev_stats)	$p \times 5$ matrix with statistics on explained variance

Methods and Formulas

pca and pcamat are implemented as ado-files.

Let C be the $p \times p$ correlation or covariance matrix to be analyzed. The spectral or eigen decomposition of C is

$$\mathbf{C} = \mathbf{V}\mathbf{\Lambda}\mathbf{V}' = \sum_{i=1}^{p} \lambda_i \mathbf{v}_i \mathbf{v}_i'$$

$$\mathbf{v}_i'\mathbf{v}_j = \delta_{ij} \qquad \text{(i.e., orthonormality)}$$

$$\lambda_1 \geq \lambda_2 \geq \ldots \geq \lambda_p \geq 0$$

The eigenvectors $\mathbf{v}_i$ are also known as the principal components. Note that the direction (sign) of principal components is not defined. pca returns principal components signed so that $\mathbf{1}'\mathbf{v}_i > 0$. In PCA, "total variance" equals $\text{trace}(\mathbf{C}) = \sum \lambda_j$.

Inference on eigenvalues and eigenvectors

The asymptotic distribution of the eigenvectors $\widehat{\mathbf{v}}_i$ and eigenvalues $\widehat{\lambda}_i$ of a covariance matrix $\mathbf{S}$ for a sample from a multivariate normal distribution $N(\mu, \mathbf{\Sigma})$ was derived by Girshick (1939); for additional results, see also Anderson (1963) and Jackson (1991). Higher-order expansions are discussed in Lawley (1956). See Tyler (1981) for related results for elliptic distributions. The theory of the exact distribution is rather complicated (Muirhead 1982, chapter 9) and hard to implement. Assuming that eigenvalues of $\mathbf{\Sigma}$ are distinct and strictly positive, the eigenvalues and eigenvectors of $\mathbf{S}$ are jointly asymptotically multivariate normal distributed with the following moments (up to order n^{-3}):

$$\mathrm{E}(\widehat{\lambda}_i) = \lambda_i \left\{ 1 + \frac{1}{n} \sum_{j \neq i}^{k} \left(\frac{\lambda_j}{\lambda_i - \lambda_j} \right) \right\} + O(n^{-3})$$

$$\mathrm{Var}(\widehat{\lambda}_i) = \frac{2\lambda_i^2}{n} \left\{ 1 - \frac{1}{n} \sum_{j \neq i}^{k} \left(\frac{\lambda_j}{\lambda_i - \lambda_j} \right)^2 \right\} + O(n^{-3})$$

$$\mathrm{Cov}(\widehat{\lambda}_i, \widehat{\lambda}_j) = \frac{2}{n^2} \left(\frac{\lambda_i \lambda_j}{\lambda_i - \lambda_j} \right)^2 + O(n^{-3})$$

$$\mathrm{Var}(\widehat{\mathbf{v}}_i) = \frac{1}{n} \sum_{j \neq i}^{k} \frac{\lambda_i \lambda_j}{(\lambda_i - \lambda_j)^2} \mathbf{v}_j \mathbf{v}_j'$$

$$\mathrm{Cov}(\widehat{\mathbf{v}}_i, \widehat{\mathbf{v}}_j) = -\frac{1}{n} \frac{\lambda_i \lambda_j}{(\lambda_i - \lambda_j)^2} \mathbf{v}_i \mathbf{v}_j'$$

For the asymptotic theory of the cumulative proportion of variance explained, see Kshirsagar (1972, 454).

Additional general tests for multivariate normal distributions

The likelihood-ratio χ^2 test of independence (Basilevsky 1994, 187) is

$$\chi^2 = -\left(n - \frac{2p + 5}{6} \right) \ln\{\det(\mathbf{C})\}$$

with $p(p - 1)/2$ degrees of freedom.

The likelihood-ratio χ^2 test of sphericity (Basilevsky 1994, 192) is

$$\chi^2 = -\left(n - \frac{2p^2 + p + 2}{6p} \right) \left[\ln\{\det(\widetilde{\mathbf{\Lambda}})\} - p \ln \left\{ \frac{\mathrm{trace}(\widetilde{\mathbf{\Lambda}})}{p} \right\} \right]$$

with $(p + 2)(p - 1)/2$ degrees of freedom and with $\widetilde{\mathbf{\Lambda}}$ the eigenvalues of the correlation matrix.

References

Anderson, T. W. 1963. Asymptotic theory for principal component analysis. *Annals of Mathematical Statistics* 34: 122–148.

Basilevsky, A. T. 1994. *Statistical Factor Analysis and Related Methods: Theory and Applications*. New York: Wiley.

Belle, G. van, L. D. Fisher, P. J. Heagerty, and T. Lumley. 2004. *Biostatistics: A Methodology for the Health Sciences*. 2nd ed. New York: Wiley.

Girshick, M. A. 1939. On the sampling theory of the roots of determinantal equations. *Annals of Mathematical Statistics* 10: 203–224.

Hotelling, H. 1933. Analysis of a complex of statistical variables into principal components. *Journal of Educational Psychology* 24: 417–441, 498–520.

Jackson, J. E. 1991. *A User's Guide to Principal Components*. New York: Wiley.

Jolliffe, I. T. 2002. *Principal Component Analysis*. 2nd ed. New York: Springer.

Kshirsagar, A. M. 1972. *Multivariate Analysis*. New York: Marcel Dekker.

Lawley, D. N. 1956. Tests of significance for the latent roots of covariance and correlation matrices. *Biometrika* 43: 128–136.

Mardia, K. V., J. T. Kent, and J. M. Bibby. 1979. *Multivariate Analysis*. New York: Academic Press.

Milan, L. M. and J. Whittaker. 1995. Application of the parametric bootstrap to models that incorporate a singular value decomposition. *Journal of the Royal Statistical Society, Series C* 44: 31–49.

Muirhead R. J. 1982. *Aspects of Multivariate Statistical Theory*. New York: Wiley.

Pearson, K. 1901. On lines and planes of closest fit to systems of points in space. *Philosophical Magazine, Series 6* 2: 559–572.

Rabe-Hesketh, S. and B. S. Everitt. 2004. *A Handbook of Statistical Analysis using Stata*. 3rd ed. Boca Raton, FL: Chapman & Hall/CRC.

Rencher, A. C. 2002. *Methods of Multivariate Analysis*. 2nd ed. New York: Wiley.

Tyler, D. E. 1981. Asymptotic inference for eigenvectors. *Annals of Statistics* 9: 725–736.

Weesie, J. 1997. smv7: Inference on principal components. *Stata Technical Bulletin* 37: 22–23. Reprinted in *Stata Technical Bulletin Reprints*, vol. 7, pp. 229–231.

Also See

Complementary:	[MV] **pca postestimation**,
	[R] **tetrachoric**
Related:	[MV] **biplot**, [MV] **canon**, [MV] **factor**,
	[D] **corr2data**, [R] **alpha**
Background:	[U] **11.1.10 Prefix commands**,
	[U] **20 Estimation and postestimation commands**

Title

> **pca postestimation** — Postestimation tools for pca and pcamat

Description

The following postestimation commands are of special interest after pca and pcamat:

command	description
estat anti	anti-image correlation and covariance matrices
estat kmo	Kaiser–Meyer–Olkin measure of sampling adequacy
estat loadings	component-loading matrix in one of several normalizations
estat residuals	matrix of correlation or covariance residuals
estat rotatecompare	compare rotated and unrotated components
estat smc	squared multiple correlations between each variable and the rest
* estat summarize	display summary statistics over the estimation sample
loadingplot	plot component loadings
rotate	rotate component loadings
scoreplot	plot score variables
screeplot	plot eigenvalues

* estat summarize is not available after pcamat.

For information about loadingplot and scoreplot, see [MV] **scoreplot**; for information about screeplot, see [MV] **screeplot**; and for all other commands, see below.

In addition, the following standard postestimation commands are available:

command	description
† estat	examine the VCE matrix
estimates	cataloging estimation results
* lincom	point estimates, standard errors, testing, and inference for linear combinations of coefficients
* nlcom	point estimates, standard errors, testing, and inference for nonlinear combinations of coefficients
predict	score variables, predictions, and residuals
* predictnl	point estimates, standard errors, testing, and inference for generalized predictions
* test	Wald tests for simple and composite linear hypotheses
* testnl	Wald tests of nonlinear hypotheses

† estat is available after pca and pcamat with the vce(normal) option.

* lincom, nlcom, predictnl, test, and testnl are available only after pca with the vce(normal) option.

See the corresponding entries in the *Stata Base Reference Manual* for details.

Special-interest postestimation commands

estat anti displays the anti-image correlation and anti-image covariance matrices. These are minus the partial covariance and minus the partial correlation of all pairs of variables, holding all other variables constant.

estat kmo displays the Kaiser–Meyer–Olkin measure of sampling adequacy. KMO takes values between 0 and 1, with small values indicating that overall the variables have too little in common to warrant a factor analysis. Historically, the following labels are given to values of KMO (Kaiser 1974):

0.00 to 0.49	unacceptable
0.50 to 0.59	miserable
0.60 to 0.69	mediocre
0.70 to 0.79	middling
0.80 to 0.89	meritorious
0.90 to 1.00	marvelous

estat loadings displays the component-loading matrix in one of several normalizations of the columns (eigenvectors).

estat residuals displays the difference between the observed correlation or covariance matrix and the fitted (reproduced) matrix using the retained factors.

estat rotatecompare displays the unrotated (principal) components next to the most recent rotated components.

estat smc displays the squared multiple correlations between each variable and all other variables. SMC is a theoretical lower bound for communality, and thus an upper bound for the unexplained variance.

estat summarize displays summary statistics of the variables in the principal component analysis over the estimation sample. This subcommand is not available after pcamat.

Syntax for predict

$$predict\ [type]\ newvarlist\ [if]\ [in]\ [,\ statistic\ options]$$

statistic	# of vars.	description (k = # of orig. vars.; f = # of components)
Main		
score	$1, \ldots, f$	scores based on the components; the default
fit	k	fitted values using the retained components
residual	k	raw residuals from the fit using the retained components
q	1	residual sum of squares

options	description
Main	
norotated	use unrotated results, even when rotated results are available
center	base scores on centered variables
notable	suppress table of scoring coefficients
format(%fmt)	format for displaying the scoring coefficients

Options for predict

Note on pcamat: predict requires that variables with the correct names be available in memory. Apart from centered scores, means() should have been specified with pcamat. If you used pcamat because you only have access to the correlation or covariance matrix, you cannot use predict.

⌐ Main ⌐

score calculates the scores for components 1, ..., #, where # the number of variables in *newvarlist*.

fit calculates the fitted values, using the retained components, for each of the variables. The number of variables in *newvarlist* should equal the number of variables in the *varlist* of pca.

residual calculates for each of the variables the raw residuals (residual = observed − fitted), with the fitted values computed using the retained components.

q calculates the Rao-statistics (i.e., the sum of squares of the omitted components) weighted by the respective eigenvalues. This equals the residual sum of squares between the original variables and the fitted values.

norotated uses unrotated results, even when rotated results are available.

center bases scores on centered variables. This option is only relevant for a PCA of a covariance matrix, in which the scores are based on uncentered variables by default. Scores for a PCA of a correlation matrix are always based on the standardized variables.

notable suppresses the table of scoring coefficients.

format(*%fmt*) specifies the display format for scoring coefficients.

Syntax for estat

Display the anti-image correlation and covariance matrices

   estat anti [, nocorr nocov format(*%fmt*)]

Display the Kaiser–Meyer–Olkin measure of sampling adequacy

   estat kmo [, novar format(*%fmt*)]

Display the component-loading matrix

   estat loadings [, cnorm(unit | eigen | inveigen) format(*%fmt*)]

Display the differences in matrices

   estat residuals [, obs fitted format(*%fmt*)]

Display the unrotated and rotated components

   estat rotatecompare [, format(*%fmt*)]

Display the squared multiple correlations

   estat smc [, format(*%fmt*)]

Display the summary statistics

> estat <u>su</u>mmarize [, <u>label</u> <u>nohea</u>der <u>noweights</u>]

Options for estat

┌─── Main └──

nocorr, an option used with estat anti, suppresses the display of the anti-image correlation matrix.

nocov, an option used with estat anti, suppresses the display of the anti-image covariance matrix.

format(% *fmt*) specifies the display format. The defaults differ between the subcommands.

novar, an option used with estat kmo, suppresses the Kaiser–Meyer–Olkin measures of sampling adequacy for the variables in the principal component analysis, displaying the overall KMO measure only.

cnorm(unit | eigen | inveigen), an option used with estat loadings, selects the normalization of the eigenvectors, the columns of the principal-component loading matrix. The following normalizations are available

unit	ssq(column) = 1; the default
eigen	ssq(column) = eigenvalue
inveigen	ssq(column) = 1/eigenvalue

with ssq(column) the sum of squares of the elements in a column, and eigenvalue the eigenvalue associated with the column (eigenvector).

obs, an option used with estat residuals, displays the observed correlation or covariance matrix for which the PCA was performed.

fitted, an option used with estat residuals, displays the fitted (reconstructed) correlation or covariance matrix based on the retained components.

label, noheader, and noweights are the same as for the generic estat summarize command; see [R] **estat**.

Remarks

After computing the principal components and the associated eigenvalues, you have additional issues to resolve. How many components do you want to retain? How well is the correlation or covariance matrix approximated by the retained components? How can you interpret the principal components? Is it possible to improve the interpretability by rotating the retained principal components? And, when these issues have been settled, the component scores are probably needed for subsequent research.

The remainder of this entry describes the specific tools available for these purposes. Remarks are presented under the headings

> *Postestimation statistics*
> *Plots of eigenvalues, component loadings, and scores*
> *Rotating the components*
> *How rotate interacts with pca*
> *Predicting the component scores*

In addition to these specific postestimation tools, general tools are available as well. Recall that `pca` is an estimation command, so it is possible to manage a series of PCA analyses with the `estimates` command; see [R] **estimates**. If you have specified the `vce(normal)` option, `pca` has saved the coefficients `e(b)` and the associated variance–covariance matrix `e(V)`, and you can use standard Stata commands to test hypotheses about the principal components and eigenvalues ("confirmatory principal component analysis"), for instance, with the `test`, `lincom`, and `testnl` commands. We caution you to only test hypotheses that do not violate the assumptions of the theory underlying the derivation of the covariance matrix. In particular, all eigenvalues are assumed to be different and strictly positive. Thus it makes no sense to use `test` to test the hypothesis that the smallest four eigenvalues are equal (let alone that they are equal to zero.)

Postestimation statistics

`pca` displays the principal components in unit normalization; the sum of squares of the principal loadings equals 1. This parallels the standard conventions in mathematics concerning eigenvectors. Some texts and some software use a different normalization. Some texts multiply the eigenvectors by the square root of the eigenvalues. In this normalization, the sum of the squared loadings equals the variance explained by that component. `estat loadings` can display the loadings in this normalization.

```
. use audiometric
(Audiometric measures)
. pca l* r*, comp(4)
 (output omitted )
. estat loadings, cnorm(eigen)
```
```
Principal component loadings (unrotated)
    component normalization: sum of squares(column) = eigenvalue
```

	Comp1	Comp2	Comp3	Comp4
lft500	.795	-.4032	.1562	-.2239
lft1000	.8345	-.2868	-.05132	-.3291
lft2000	.7262	.3035	-.4645	-.193
lft4000	.5567	.6032	.4242	-.1101
rght500	.6804	-.4911	.2561	.3331
rght1000	.8155	-.2948	-.0285	.2544
rght2000	.6175	.4033	-.5559	.2674
rght4000	.5039	.6533	.4209	.1087

How close the retained principal components approximate the correlation matrix can be seen from the fitted (reconstructed) correlation matrix and from the residuals, i.e., the difference between the observed and fitted correlations.

```
. estat residual, fit format(%7.3f)
```
```
Fitted correlation matrix
```

Variable	lft500	lft1000	lft2000	lft4000	rght500	rg~1000
lft500	0.869					
lft1000	0.845	0.890				
lft2000	0.426	0.606	0.872			
lft4000	0.290	0.306	0.412	0.866		
rght500	0.704	0.586	0.162	0.155	0.881	
rght1000	0.706	0.683	0.467	0.236	0.777	0.818
rght2000	0.182	0.340	0.778	0.322	0.169	0.469
rght4000	0.179	0.176	0.348	0.841	0.166	0.234

Variable	rg~2000	rg~4000
rght2000	0.925	
rght4000	0.370	0.870

Residual correlation matrix

Variable	lft500	lft1000	lft2000	lft4000	rght500	rg~1000
lft500	0.131					
lft1000	-0.067	0.110				
lft2000	-0.024	-0.070	0.128			
lft4000	-0.035	-0.031	0.013	0.134		
rght500	-0.008	-0.034	0.077	0.024	0.119	
rght1000	-0.064	0.024	-0.021	0.027	-0.114	0.182
rght2000	0.056	0.020	-0.076	-0.005	-0.010	-0.054
rght4000	0.025	0.041	-0.022	-0.131	-0.034	-0.014

Variable	rg~2000	rg~4000
rght2000	0.075	
rght4000	0.005	0.130

All off diagonal residuals are small, except perhaps the two measurements at the highest frequency.

estat also provides some of the standard methods for studying correlation matrices to assess whether the variables have strong linear relations with each other. In a sense, these methods could be seen as pre-estimation rather than as postestimation methods. The first method is the inspection of the squared multiple correlation (the regression R^2) of each variable on all other variables.

. estat smc

Squared multiple correlations of variables with all other variables

Variable	smc
lft500	0.7113
lft1000	0.7167
lft2000	0.6229
lft4000	0.5597
rght500	0.5893
rght1000	0.6441
rght2000	0.5611
rght4000	0.5409

The SMC measures help identify variables that can not be explained well from the other variables. For such variables, you should re-evaluate whether they should be included in the analysis. In our examples, none of the SMCs are so small as to warrant exclusion. Two additional statistics are offered. First, we can inspect the anti-image correlation and covariance matrices, i.e., the negative of correlations (covariances) of the variables partialing out all other variables. If many of these correlations or covariances are "high", the relationships between some of the variables have little to do with the other variables, indicating that it will not be possible to obtain a low-dimensional reduction of the data.

```
. estat anti, nocov format(%7.3f)
```

Anti-image correlation coefficients —— partialing out all other variables

Variables	lft500	lft1000	lft2000	lft4000	rght500	rg~1000
lft500	1.000					
lft1000	-0.561	1.000				
lft2000	-0.051	-0.267	1.000			
lft4000	-0.014	0.026	-0.285	1.000		
rght500	-0.466	0.131	0.064	-0.017	1.000	
rght1000	0.023	-0.389	0.043	-0.042	-0.441	1.000
rght2000	0.085	0.068	-0.617	0.161	0.067	-0.248
rght4000	-0.047	-0.002	0.150	-0.675	0.019	0.023

Variables	rg~2000	rg~4000
rght2000	1.000	
rght4000	-0.266	1.000

The Kaiser–Meyer–Olkin measure of sampling adequacy compares the correlations and the partial correlations between variables. If the partial correlations are relatively high compared to the correlations, the KMO measure is small, and a low-dimensional representation of the data is not possible.

```
. estat kmo
```

Kaiser-Meyer-Olkin measure of sampling adequacy

Variable	kmo
lft500	0.7701
lft1000	0.7767
lft2000	0.7242
lft4000	0.6449
rght500	0.7562
rght1000	0.8168
rght2000	0.6673
rght4000	0.6214
Overall	0.7328

Using the Kaiser (1974) characterization of KMO values

0.00 to 0.49	unacceptable
0.50 to 0.59	miserable
0.60 to 0.69	mediocre
0.70 to 0.79	middling
0.80 to 0.89	meritorious
0.90 to 1.00	marvelous

we declare our KMO value, 0.73, middling.

Plots of eigenvalues, component loadings, and scores

After computing the principal components, we likely wish to determine how many components to keep. In factor analysis the question of the "true" number of factors is a complicated one. With PCA, it is a little more straightforward. We may set a percentage of variance we wish to account for, say 90%, and retain just enough components to account for at least that much of the variance. Usually you will want to weigh the costs associated with using additional components in subsequent analyses against the benefits of the additional variance they account for. The relative magnitudes of the eigenvalues indicate the amount of variance they account for. A useful tool for visualizing the eigenvalues relative to one another, so that you can decide the number of components to retain, is the scree plot proposed by Cattell (1966); see [MV] **screeplot**.

```
. screeplot, mean
```

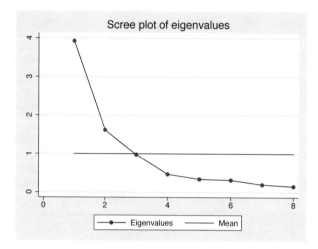

Since we are analyzing a correlation matrix, the mean eigenvalue is 1. We wish to retain the components associated with the high part of the scree plot and drop the components associated with the lower flat part of the scree plot. The boundary between high and low is not very clear here, but we would choose two or three components, although the fourth component had the nice interpretation of the left versus the right ear; see [MV] **pca**.

A problem in interpreting the scree plot is that no guidance is given with respect to its stability under sampling. How different could the plot be with different samples? The approximate variance of an eigenvalue $\widehat{\lambda}$ of a covariance matrix for multivariate normal distributed data is $2\lambda^2/n$. From this we can derive confidence intervals for the eigenvalues. These scree plot confidence intervals aid in the selection of important components.

```
. screeplot, ci
(caution is advised in interpreting an asymptotic theory-based confidence
 interval of eigenvalues of a correlation matrix)
```

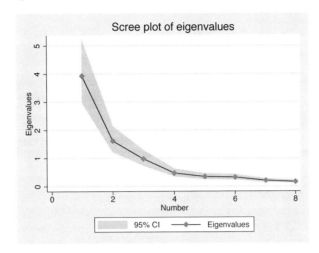

Despite our appreciation of the underlying interpretability of the fourth component, the evidence still points to retaining two or three principal components.

Plotting the components is sometimes useful in interpreting a PCA. We may look at the components from the perspective of the columns (variables) or the rows (observations). The associated plots are produced by the commands loadingplot (variables) and scoreplot (observations).

By default, the first 2 components are used to produce the loading plot.

```
. loadingplot
```

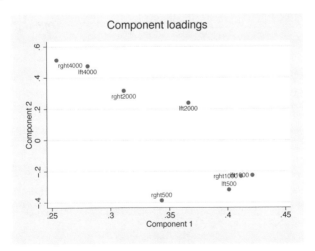

You may request more components, in which case each possible pair of requested components will be graphed. You can choose between a matrix or combined graph layout for the multiple graphs. Here we show the combined layout.

. loadingplot, comp(3) combined

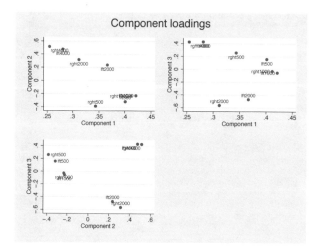

Score plots approach the display of principal components from the perspective of the observations. scoreplot and loadingplot have most of their options in common; see [MV] **scoreplot**. Unlike loadingplot, which automatically uses the variable names as marker labels, with scoreplot you use the mlabel() graph option to provide meaningful marker labels. Score plots are especially helpful if the observations are well-known objects, such as countries, firms, or brands. The score plot may help you visualize the principal components with your background knowledge of these objects. Score plots are sometimes useful for the detection of outliers; see Jackson (1991).

❏ Technical Note

In [MV] **pca** we noted that PCA may also be interpreted as fixed effects factor analysis; in that interpretation, the selection of the number of components to be retained is of comparable complexity as in factor analysis.

❏

Rotating the components

Rotating principal components is a disputed issue and one in which reasonable people may disagree. pca computes the principal components. Rotating the solution destroys some of the properties of principal components. In particular, the first rotated component no longer has maximal variance, the second rotated component no longer has maximal variance among those linear combinations uncorrelated to the first component, etc. If preserving the maximal variance property is very important to your interpretations, do not rotate.

On the other hand, when we rotate, say, the leading three principal components, the total variance explained by the three rotated components is exactly equal to the variance explained by the three principal components. If you applied an orthogonal rotation, the rotated components are still uncorrelated. The only thing that has changed is that the explanation is distributed differently among the three rotated components. If the rotated components have a clearer interpretation, you may actually prefer to use them in your subsequent work.

After pca, a wide variety of rotations are available; see [MV] **rotate**. The default method of rotation is varimax, rotating the principal components to maximize the sum over the columns of the within-column variances.

```
. rotate
```

Principal components/correlation

				Number of obs	=	100
				Number of comp.	=	4
				Trace	=	8
Rotation: orthogonal varimax (Horst off)				Rho	=	0.8737

Component	Variance	Difference	Proportion	Cumulative
Comp1	2.11361	.400444	0.2642	0.2642
Comp2	1.71316	.118053	0.2141	0.4783
Comp3	1.59511	.0275517	0.1994	0.6777
Comp4	1.56756	.	0.1959	0.8737

Rotated components

Variable	Comp1	Comp4	Comp3	Comp2	Unexplained
lft500	0.5756	0.0265	-0.1733	0.1781	.1308
lft1000	0.6789	-0.0289	-0.0227	-0.0223	.1105
lft2000	0.3933	0.0213	0.5119	-0.2737	.1275
lft4000	0.1231	0.6987	-0.0547	-0.0885	.1342
rght500	-0.0005	0.0158	-0.0380	0.7551	.1194
rght1000	0.0948	-0.0248	0.2289	0.5481	.1825
rght2000	-0.1173	-0.0021	0.8047	0.0795	.07537
rght4000	-0.1232	0.7134	0.0550	0.0899	.1303

Component rotation matrix

	Comp1	Comp4	Comp3	Comp2
Comp1	0.6663	0.3784	0.4390	0.4692
Comp2	-0.3055	0.6998	0.4012	-0.5059
Comp3	-0.0657	0.6059	-0.7365	0.2936
Comp4	-0.6770	-0.0022	0.3224	0.6616

Note that pca now labels one of the columns of the first table as "Variance" instead of "Eigenvalue"; the rotated components have been ordered in decreasing order of variance. The variance explained by the 4 rotated components equals 87.37%, which is identical to the explained variance by the 4 leading principal components. But while the principal components have rather dispersed eigenvalues, the 4 rotated components all explain about the same fraction of the variance.

You may also choose to rotate only a few of the retained principal components. In contrast to most methods of factor analysis, the principal components are not affected by the number of retained components. However, the first two rotated components are different if you are rotating all four components or only the leading two or three principal components.

(Continued on next page)

```
. rotate, comp(3)
```

Principal components/correlation

			Number of obs	=	100
			Number of comp.	=	4
			Trace	=	8
Rotation: orthogonal varimax (Horst off)			Rho	=	0.8737

Component	Variance	Difference	Proportion	Cumulative
Comp1	2.99422	1.16842	0.3743	0.3743
Comp2	1.8258	.123163	0.2282	0.6025
Comp3	1.70264	1.23585	0.2128	0.8153
Comp4	.466782	.	0.0583	0.8737

Rotated components

Variable	Comp1	Comp2	Comp3	Comp4	Unexplained
lft500	0.5326	-0.0457	0.0246	-0.3278	.1308
lft1000	0.4512	0.1618	-0.0320	-0.4816	.1105
lft2000	0.0484	0.6401	0.0174	-0.2824	.1275
lft4000	0.0247	0.0011	0.6983	-0.1611	.1342
rght500	0.5490	-0.1799	0.0163	0.4876	.1194
rght1000	0.4521	0.1368	-0.0259	0.3723	.1825
rght2000	-0.0596	0.7148	-0.0047	0.3914	.07537
rght4000	-0.0200	0.0059	0.7138	0.1591	.1303

Component rotation matrix

	Comp1	Comp2	Comp3	Comp4
Comp1	0.7790	0.5033	0.3738	0.0000
Comp2	-0.5932	0.3987	0.6994	0.0000
Comp3	0.2030	-0.7666	0.6092	-0.0000
Comp4	-0.0000	0.0000	0.0000	1.0000

The three component varimax-rotated solution differs from the leading three components from the four component varimax-rotated solution. The fourth component is not affected by a rotation among the leading three component—it is still the fourth principal component.

So, how interpretable are rotated components? We believe that for this example the original components had a much clearer interpretation than the rotated components. Notice how the clear symmetry in the treatment of left and right ears has been broken.

To add further to an already controversial method, we may use oblique rotation methods. An example is the oblique oblimin method.

```
. rotate, oblimin oblique
```

Principal components/correlation

			Number of obs	=	100
			Number of comp.	=	4
			Trace	=	8
Rotation: oblique oblimin (Horst off)			Rho	=	0.8737

Component	Variance	Proportion	Rotated comp. are correlated
Comp1	2.21066	0.2763	
Comp2	1.71164	0.2140	
Comp3	1.69708	0.2121	
Comp4	1.62592	0.2032	

Rotated components

Variable	Comp2	Comp4	Comp1	Comp3	Unexplained
lft500	0.5834	0.0259	0.1994	-0.1649	.1308
lft1000	0.6797	-0.0292	0.0055	-0.0157	.1105
lft2000	0.3840	0.0216	-0.2489	0.5127	.1275
lft4000	0.1199	0.6988	-0.0857	-0.0545	.1342
rght500	0.0261	0.0146	0.7561	-0.0283	.1194
rght1000	0.1140	-0.0257	0.5575	0.2370	.1825
rght2000	-0.1158	-0.0022	0.0892	0.8048	.07537
rght4000	-0.1209	0.7134	0.0848	0.0549	.1303

Component rotation matrix

	Comp2	Comp4	Comp1	Comp3
Comp1	0.6836	0.3773	0.5053	0.4523
Comp2	-0.3250	0.7008	-0.5137	0.3916
Comp3	-0.0550	0.6054	0.2774	-0.7337
Comp4	-0.6557	-0.0029	0.6408	0.3238

You may observe that the oblique rotation methods do not change the variance that is unexplained by the components. But this time, the rotated components are no longer uncorrelated. This makes measuring the importance of the rotated components more ambiguous, a problem that is very similar to ambiguities in interpreting importance of correlated independent variables. In this oblique case, the sum of the variances of the rotated components equals 90.5% (0.2763 + 0.2140 + 0.2121 + 0.2032) of the total variance. This is larger than the 87.37% of variance explained by the 4 principal components. The oblique rotated components partly explain the same variance, and this shared variance is entering multiple times into the total.

How rotate interacts with pca

rotate stores the rotated component loadings and associated statistics in e(), the estimation storage area, along with the regular pca estimation results. Replaying pca will display the rotated results again.

Other postestimation statistics also use the rotated results whenever this is meaningful. For instance, loadingplot would display the rotated loadings. These postestimation commands have an option norotated that specifies that the unrotated results, i.e., the principal components, be used. Thus

```
. pca, norotated
  (output omitted )
```

displays the standard pca output for the unrotated (principal) solution, and

```
. loadingplot, norotated
  (output omitted )
```

produces the loading plot for the unrotated (principal) solution.

If you execute rotate again, the new rotate results are stored with the pca estimation, replacing the previous rotate results. Thus pca knows about at most one rotation.

To compare rotated and unrotated results, it is of course possible to replay the rotated results (pca) and unrotated results (pca, norotate) consecutively. You would especially seek to compare the loadings. Such a comparison is easier if the loadings are displayed in parallel. This feature is provided with the estat command rotatecompare.

```
. estat rotatecompare
```

Rotation matrix — oblique oblimin (Horst off)

Variables	Comp2	Comp4	Comp1	Comp3
Comp1	0.6836	0.3773	0.5053	0.4523
Comp2	-0.3250	0.7008	-0.5137	0.3916
Comp3	-0.0550	0.6054	0.2774	-0.7337
Comp4	-0.6557	-0.0029	0.6408	0.3238

Rotated component loadings

Variables	Comp2	Comp4	Comp1	Comp3
lft500	0.5834	0.0259	0.1994	-0.1649
lft1000	0.6797	-0.0292	0.0055	-0.0157
lft2000	0.3840	0.0216	-0.2489	0.5127
lft4000	0.1199	0.6988	-0.0857	-0.0545
rght500	0.0261	0.0146	0.7561	-0.0283
rght1000	0.1140	-0.0257	0.5575	0.2370
rght2000	-0.1158	-0.0022	0.0892	0.8048
rght4000	-0.1209	0.7134	0.0848	0.0549

Unrotated component loadings

Variables	Comp1	Comp2	Comp3	Comp4
lft500	0.4011	-0.3170	0.1582	-0.3278
lft1000	0.4210	-0.2255	-0.0520	-0.4816
lft2000	0.3664	0.2386	-0.4703	-0.2824
lft4000	0.2809	0.4742	0.4295	-0.1611
rght500	0.3433	-0.3860	0.2593	0.4876
rght1000	0.4114	-0.2318	-0.0289	0.3723
rght2000	0.3115	0.3171	-0.5629	0.3914
rght4000	0.2542	0.5135	0.4262	0.1591

Finally, sometimes you may want to remove rotation results permanently, e.g., you decide to continue with the unrotated (principal) solution. Since all postestimation commands operate on the rotated solution by default, you would have to add the option norotated over and over again. Instead, you can remove the rotated solution with the command

```
. rotate, clear
```

❏ Technical Note

pca results may be stored and restored with estimates, just like other estimation results. If you have stored PCA estimation results without rotated results, and later rotate the solution, the rotated results are not automatically stored as well. The pca would need to be stored again.

❏

Predicting the component scores

After deciding on the number of components and, possibly, the rotation of the components, you may want to estimate the component scores for all respondents. To estimate only the first component scores, which here is called pc1:

```
. predict pc1
(score assumed)
(3 components skipped)

Scoring coefficients
    sum of squares(column-loading) = 1
```

Variable	Comp1	Comp2	Comp3	Comp4
lft500	0.4011	-0.3170	0.1582	-0.3278
lft1000	0.4210	-0.2255	-0.0520	-0.4816
lft2000	0.3664	0.2386	-0.4703	-0.2824
lft4000	0.2809	0.4742	0.4295	-0.1611
rght500	0.3433	-0.3860	0.2593	0.4876
rght1000	0.4114	-0.2318	-0.0289	0.3723
rght2000	0.3115	0.3171	-0.5629	0.3914
rght4000	0.2542	0.5135	0.4262	0.1591

The table is informing you that pc1 could be obtain as a weighted sum of standardized variables,

```
. egen std_lft500   = std(lft500)
. egen std_lft1000  = std(lft1000)
  .
  .
  .
. egen std_rght4000 = std(rght4000)
. gen pc1 = 0.4011*std_lft500 + 0.4210*std_lft500 + ... + 0.2542*std_rght4000
```

(egen's std() function converts a variable to its standardized form (mean 0, variance 1); see [D] **egen**.) The principal component scores are in standardized units after a PCA of a correlation matrix and in the original units after a PCA of a covariance matrix.

It is possible to predict other statistics, as well. For instance, the fitted values of the eight variables by the first four principal components are obtained as

```
. predict f_1-f_8, fit
```

The predicted values are in the units of the original variables, with the means substituted back in. If we had retained all eight components, the fitted values would have been identical to the observations.

❏ Technical Note

The fitted values are meaningful in the interpretation of PCA as rank-restricted multivariate regression. The component scores are the "x-variables"; the component loadings are the regression coefficients. If the PCA was computed for a correlation matrix, you could think of the regression as being in standardized units. The fitted values are transformed from the standardized units back to the original units.   ❏

❏ Technical Note

You may have observed that the scoring coefficients were equal to the component loadings. This holds true for the principal components in unit normalization and for the orthogonal rotations thereof; it does not hold for oblique rotations.

❏

Saved Results

Let p be the number of variables and f the number of factors.

predict, in addition to generating variables, also saves in r():

> Matrices
>
> r(scoef) $p \times f$ matrix of scoring coefficients

estat anti saves in r():

> Matrices
>
> r(acov) $p \times p$ anti-image covariance matrix
>
> r(acorr) $p \times p$ anti-image correlation matrix

estat kmo saves in r():

> Scalars
>
> r(kmo)   the Kaiser–Meyer–Olkin measure of sampling adequacy
>
> Matrices
>
> r(kmow)   column vector of KMO measures for each variable

estat loadings saves in r():

> Macros
>
> r(cnorm)   component normalization: eigen, inveigen, or unit
>
> Matrices
>
> r(A) $p \times f$ matrix of normalized component loadings

estat residuals saves in r():

> Matrices
>
> r(fit) $p \times p$ matrix of fitted values
>
> r(residual) $p \times p$ matrix of residuals

estat smc saves in r():

> Matrices
>
> r(smc)   vector of squared multiple correlations of variables with all other variables

See [R] **estat** for the returned results of estat summarize and estat vce (available when vce(normal) is specified with pca or pcamat).

rotate after `pca` and `pcamat` add to the existing `e()`:

Scalars

`e(r_f)`	number of components in rotated solution
`e(r_fmin)`	rotation criterion value

Macros

`e(r_class)`	orthogonal or oblique
`e(r_criterion)`	rotation criterion
`e(r_ctitle)`	title for rotation
`e(r_normalization)`	horst or none

Matrices

`e(r_L)`	rotated loadings
`e(r_T)`	rotation
`e(r_Ev)`	explained variance by rotated components

Note that the components in the rotated solution are in decreasing order of `e(r_Ev)`.

Methods and Formulas

All postestimation commands listed above are implemented as ado-files.

`estat anti` computes and displays the anti-image covariance matrix $\mathbf{C}$ and the anti-image correlation matrix $\mathbf{A}$

$$\mathbf{C} = \{\text{diag}(\mathbf{R})\}^{-1/2}\, \mathbf{R}\, \{\text{diag}(\mathbf{R})\}^{-1/2}$$

$$\mathbf{A} = \{\text{diag}(\mathbf{R})\}^{-1}\, \mathbf{R}\, \{\text{diag}(\mathbf{R})\}^{-1}$$

where $\mathbf{R}$ is the inverse of the correlation matrix.

`estat kmo` computes the "Kaiser–Meyer–Olkin measure of sampling adequacy" (KMO) and is defined as

$$KMO = \frac{\sum_{\mathcal{S}} r_{ij}^2}{\sum_{\mathcal{S}} (a_{ij}^2 + r_{ij}^2)}$$

where $\mathcal{S} = (i, j; i \neq j)$; r_{ij} is the correlation of variables i and j; and a_{ij} is the anti-image correlation. The variable-wise measure KMO_i is defined analogously as

$$KMO_i = \frac{\sum_{\mathcal{S}} r_{ij}^2}{\sum_{\mathcal{S}} (a_{ij}^2 + r_{ij}^2)}$$

where $\mathcal{S} = (j; i \neq j)$.

`estat loadings` displays the component loadings in different normalizations (see Jackson 1991, 16–18; he labels them as $\mathbf{U}$, $\mathbf{V}$, and $\mathbf{W}$ vectors). Let $\mathbf{C} = \mathbf{L}\mathbf{\Lambda}\mathbf{L}'$ be the spectral or eigen decomposition of the analyzed correlation or covariance matrix $\mathbf{C}$, with $\mathbf{L}$ the orthonormal eigenvectors of $\mathbf{C}$, and $\mathbf{\Lambda}$ a diagonal matrix of eigenvalues. The principal components $\mathbf{A}$, i.e., the eigenvectors $\mathbf{L}$, are displayed in one of the following normalizations:

cnorm(unit)	$\mathbf{A} = \mathbf{L}$	and so $\mathbf{A}'\mathbf{A} = \mathbf{I}$
normal(eigen)	$\mathbf{A} = \mathbf{L}\mathbf{\Lambda}^{1/2}$	and so $\mathbf{A}'\mathbf{A} = \mathbf{\Lambda}$
normal(inveigen)	$\mathbf{A} = \mathbf{L}\mathbf{\Lambda}^{-1/2}$	and so $\mathbf{A}'\mathbf{A} = \mathbf{\Lambda}^{-1}$

Note that normalization of the component loadings affects the normalization of the component scores.

The standard errors of the components are only available in unit normalization, i.e., as normalized eigenvectors.

`estat residuals` computes the fitted values $\mathbf{F}$ for the analyzed correlation or covariance matrix $\mathbf{C}$ as $\mathbf{F} = \mathbf{L}\mathbf{\Lambda}\mathbf{L}'$ over the retained components, with $\mathbf{L}$ being the retained components in unit normalization and $\mathbf{\Lambda}$ being the associated eigenvalues. The residuals are simply $\mathbf{C} - \mathbf{F}$.

`estat smc` displays the squared multiple correlation coefficients SMC_i of each variable on the other variables in the analysis. These are conveniently computed from the inverse $\mathbf{R}$ of the correlation matrix $\mathbf{C}$,

$$SMC_i = 1 - \mathbf{R}_{ii}^{-1}$$

See [MV] **rotate** and [MV] **rotatemat** for details concerning the rotation methods and algorithms used.

The variance of the rotated loadings $\mathbf{L}_r$ is computed as $\mathbf{L}_{r'}\mathbf{C}\mathbf{L}_r$.

To understand `predict` after `pca` and `pcamat`, think of PCA as a fixed-effect factor analysis with homoskedastic residuals

$$\mathbf{Z} = \mathbf{A}\mathbf{L}' + \mathbf{E}$$

$\mathbf{L}$ contains the loadings, and $\mathbf{A}$ contains the scores. $\mathbf{Z}$ is the centered variables for a PCA of a covariance matrix and standardized variables for a PCA of a correlation matrix. $\mathbf{A}$ is estimated by OLS regression of $\mathbf{Y}$ on $\mathbf{L}$

$$\widehat{\mathbf{A}} = \mathbf{Z}\mathbf{B} \qquad \mathbf{B} = \mathbf{L}(\mathbf{L}'\mathbf{L})^{-}$$

The columns of $\mathbf{A}$ are called the scores. The matrix $\mathbf{B}$ contains the scoring coefficients. The PCA fitted values for $\mathbf{Z}$ are defined as the fitted values from this regression, or in matrix terms

$$\widehat{\mathbf{Z}} = \mathbf{Z}\mathbf{P_L} = \mathbf{Y}\mathbf{L}(\mathbf{L}'\mathbf{L})^{-}\mathbf{L}'$$

with $\mathbf{P_L}$ the orthogonal projection on (the rowspace of) $\mathbf{L}$.

Note that this formulation allows orthogonal as well as oblique loadings $\mathbf{L}$ as well as loadings in different normalizations.

The above formulation is in transformed units. `predict` transforms the fitted values back to the original units. The component scores are left in transformed units, with one exception. After a PCA of covariances, means are substituted back in unless the option `centered` is specified. The residuals are returned in the original units. The residual sum of squares (over the variables) and the normalized versions are in transformed units.

References

Cattell, R. B. 1966. The scree test for the number of factors. *Multivariate Behavioral Research* 1: 245–276.

Jackson, J. E. 1991. *A User's Guide to Principal Components*. New York: Wiley.

Kaiser, H. F. 1974. An index of factor simplicity. *Psychometrika* 39: 31–36.

For additional references see [MV] **pca**.

Also See

Complementary:	[MV] **pca**, [MV] **rotate**, [MV] **scoreplot**, [MV] **screeplot**,
	[R] **estimates**, [R] **lincom**, [R] **nlcom**,
	[R] **predictnl**, [R] **test**, [R] **testnl**
Background:	[R] **estat**, [R] **predict**

Title

procrustes — Procrustes transformation

Syntax

procrustes (*varlist$_y$*) (*varlist$_x$*) [*if*] [*in*] [*weight*] [, *options*]

options	description
Model	
transform(<u>or</u>thogonal)	orthogonal rotation and reflection transformation; the default
transform(<u>ob</u>lique)	oblique rotation transformation
transform(<u>un</u>restricted)	unrestricted transformation
<u>noco</u>nstant	suppress the constant
<u>norho</u>	suppress the dilation factor ρ (set $\rho = 1$)
force	allow overlap and duplicates in *varlist$_x$* and *varlist$_y$* (advanced)
Reporting	
<u>nofit</u>	suppress table of fit statistics by target variable

bootstrap, by, jackknife, and statsby may be used with procrustes; see [U] **11.1.10 Prefix commands**.
aweights and fweights are allowed; see [U] **11.1.6 weight**.
See [MV] **procrustes postestimation** for features available after estimation.

Description

procrustes performs the Procrustean analysis, one of the standard methods of multidimensional scaling. For given "target" variables *varlist$_y$* and "source" variables *varlist$_x$*, the goal is to transform the source **X** to be as close as possible to the target **Y**. The permitted transformations are any combination of dilation (uniform scaling), rotation and reflection (i.e., orthogonal or oblique transformations), and translation. Closeness is measured by the residual sum of squares. procrustes deals with complete cases only.

procrustes assumes equal weights or scaling for the dimensions. It would be inappropriate, for example, to have the first variable measured in grams ranging from 5000 to 8000, the second variable measured in dollars ranging from 3 to 12, and the third variable measured in meters ranging from 100 to 280. In such cases, you would want to operate on standardized variables.

Options

> **Model**

transform(*transform*) specifies the transformation method. The following transformation methods are allowed:

  orthogonal specifies that the linear transformation matrix **A** should be orthogonal, $\mathbf{A'A} = \mathbf{AA'} = \mathbf{I}$. This is the default.

  oblique specifies that the linear transformation matrix **A** should be oblique, $\text{diag}(\mathbf{AA'}) = \mathbf{1}$.

unrestricted applies no restrictions to **A**, making the `procrustes` transformation equivalent to multivariate regression with uncorrelated errors; see [R] **mvreg**.

noconstant specifies that the translation component **c** is fixed at **0** (the 0-vector).

norho specifies that the dilation (scaling) constant ρ is fixed at 1. This option is not relevant with `transform(unrestricted)`; in this case, ρ is always fixed at 1.

force, an advanced option, allows overlap and duplicates in the target variables *varlist*$_y$ and source variables *varlist*$_x$.

⌐ Reporting ⌐

nofit suppresses the table of fit statistics per target variable. This option may be specified during estimation and upon replay.

Remarks

Remarks are presented under the headings

Introduction to Procrustes methods
Orthogonal Procrustes analysis
Is an orthogonal Procrustes analysis symmetric?
Other transformations

Introduction to Procrustes methods

The name *Procrustes analysis* was applied to optimal matching of configurations by Hurley and Cattell (1962) and refers to Greek mythology. The following account follows Cox and Cox (2001, 123). Travelers from Eleusis to Athens were kindly invited by Damastes to spend the night at his place. Damastes, however, practiced a queer kind of hospitality. If guests would not fit the bed, Damastes would either stretch them to make them fit, or chop off extremities if they were too long. Therefore, he was given the nickname Procrustes—ancient Greek for "the stretcher". Theseus, a warrior, finally gave Procrustes some of his own medicine.

Procrustes methods have been applied in many areas. Gower and Dijksterhuis (2004) mention applications in psychometrics (e.g., the matching of factor loading matrices), image analysis, market research, molecular biology, biometric identification, and shape analysis.

Formally, `procrustes` solves the minimization problem

$$\text{Minimize} \mid \mathbf{Y} - (\mathbf{1}\mathbf{c}' + \rho\,\mathbf{X}\,\mathbf{A}) \mid$$

where **c** is a row vector representing the translation, ρ is the scalar "dilation factor", **A** is the rotation and reflection matrix (orthogonal, oblique, or unrestricted), and $|.|$ denotes the L2 norm.

Some of the early work on Procrustes analysis was done by Mosier (1939), Green (1952), Hurley and Cattell (1962), and Browne (1967); see Gower and Dijksterhuis (2004).

Orthogonal Procrustes analysis

▷ Example 1

We illustrate `procrustes` using John Speed's historical 1610 map of the Worcestershire region in England, engraved and printed by Jodocus Hondius in Amsterdam in 1611–1612.

We analyze the accuracy of this map. Cox and Cox (2001) present data on the locations of 20 towns and villages on this old map, as well as the locations on a modern map from the Landranger Series of Ordnance Survey Maps. The locations were measured relative to the lower-left corner of the maps. We list this small dataset, specifying the `noobs` option to prevent wrapping and `sep(0)` to suppress internal horizontal lines.

```
. use speed_survey
(Data on Speed's Worcestershire map (1610))
. list name lname speed_x speed_y survey_x survey_y, sep(0) noobs
```

name	lname	speed_x	speed_y	survey_x	survey_y
Alve	Alvechurch	192	211	1027	725
Arro	Arrow	217	155	1083	565
Astl	Astley	88	180	787	677
Beck	Beckford	193	66	976	358
Beng	Bengeworth	220	99	1045	435
Crad	Cradley	79	93	736	471
Droi	Droitwich	136	171	893	633
Ecki	Eckington	169	81	922	414
Eves	Evesham	211	105	1037	437
Hall	Hallow	113	142	828	579
Hanb	Hanbury	162	180	944	637
Inkb	Inkberrow	188	156	1016	573
Kemp	Kempsey	128	108	848	490
Kidd	Kidderminster	104	220	826	762
Mart	Martley	78	145	756	598
Stud	Studley	212	185	1074	632
Tewk	Tewkesbury	163	40	891	324
UpSn	UpperSnodsbury	163	138	943	544
Upto	Upton	138	71	852	403
Worc	Worcester	125	132	850	545

You will probably conclude immediately that the scales of the two maps differ, and that the coordinates are expressed with respect to different reference points; the lower-left corners of the maps correspond to different physical locations. Another distinction will not be so obvious—at least not by looking at these numbers: the orientations of the maps may well differ. We display as scatter plots Speed's data (`speed_x`, `speed_y`) and the modern survey data (`survey_x`, `survey_y`).

(Continued on next page)

```
. scatter speed_y speed_x, mlabel(name)
    ytitle("") xtitle("") yscale(off) xscale(off) ylabel(,nogrid)
    title(Historic map of 20 towns and villages in Worcestershire)
    subtitle((Speed 1610))
```

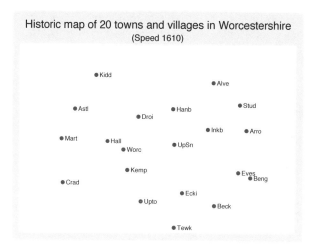

```
. scatter survey_y survey_x, mlabel(name)
    ytitle("") xtitle("") yscale(off) xscale(off) ylabel(,nogrid)
    title(Modern map of 20 towns and villages in Worcestershire)
    subtitle((Landranger series of Ordnance Survey Maps))
```

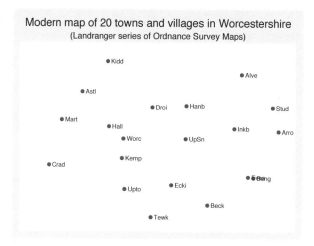

To gauge the accuracy of the historic map, we have to take into account the differences in scale, location, and orientation. Since the area depicted on the map is relatively small, we think it is justified to ignore the curvature of the earth and approximate distances along the globe with Euclidean distances, and apply a Procrustes analysis

$$\text{survey\_map} = \text{transformation}(\text{speed\_map}) + \text{residual}$$

choosing the transformation (from among the allowed transformations) to minimize the residual in terms of the residual sum of squares. The transformation should allow for, in mathematical terms, translation, uniform scaling, and two-dimensional orthogonal rotation. The uniform scaling factor is often described as the dilation factor, a positive scalar. The transformation from source to target configuration can be written as

$$\begin{pmatrix} \text{survey\_x} & \text{survey\_y} \end{pmatrix} = \begin{pmatrix} c_x & c_y \end{pmatrix} + \rho \begin{pmatrix} \text{speed\_x} & \text{speed\_y} \end{pmatrix} \begin{pmatrix} a_{11} & a_{12} \\ a_{21} & a_{22} \end{pmatrix} + \begin{pmatrix} \text{res\_x} & \text{res\_y} \end{pmatrix}$$

or simply as

$$\text{survey\_map} = \text{translation} + \text{dilation} \times \text{speed\_map} \times \text{rotation} + \text{residual}$$

The matix

$$\mathbf{A} = \begin{pmatrix} a_{11} & a_{12} \\ a_{21} & a_{22} \end{pmatrix}$$

should satisfy the constraint that it represents an orthogonal rotation—it should maintain the lengths of vectors and the angles between vectors. We estimate the translation $\begin{pmatrix} c_x & c_y \end{pmatrix}$, dilation factor ρ, and the rotation matrix $\mathbf{A}$ with the `procrustes` command.

```
. procrustes (survey_x survey_y) (speed_x speed_y)
Procrustes analysis (orthogonal)        Number of observations =         20
                                        Model df (df_m)        =          4
                                        Residual df (df_r)     =         36
                                        SS(target)             =     495070
                                        RSS(target)            =   1973.384
                                        RMSE = root(RSS/df_r)  =   7.403797
                                        Procrustes = RSS/SS    =     0.0040
```

Translation c

	survey_x	survey_y
_cons	503.8667	293.9878

Rotation & reflection matrix A (orthogonal)

	survey_x	survey_y
speed_x	.9841521	-.1773266
speed_y	.1773266	.9841521

Dilation factor

```
            rho =      2.3556
```

Fit statistics by target variable

Statistics	survey_x	survey_y
SS	216310.2	278759.8
RSS	1081.36	892.0242
RMSE	7.750841	7.039666
Procrustes	.0049991	.0032
Corr_y_yhat	.9976669	.9985076

We can read the elements of the transformation from the output: the translation from the Speed map onto the survey map is (504, 294). The scale of the survey and Speed maps differ by a factor of 2.36. The orientations of the maps also differ somewhat; if the maps had been oriented the same, we would have expected the rotation to be an identity matrix. Note that $.984^2 + .177^2 = 1$, subject to rounding error—indeed the rotation is "norm-preserving". A counterclockwise rotation in a plane over θ can be written as

$$
\begin{pmatrix}
\cos(\theta) & \sin(\theta) \\
-\sin(\theta) & \cos(\theta)
\end{pmatrix}
$$

See appendix B in Gower and Dijksterhuis (2004). In this case, $\cos(\theta) = 0.984$, so the difference in orientation of the two maps is $\theta = 10.2$ degrees.

The other output produced by `procrustes` may be more familiar. `procrustes` estimated four parameters: the angle of rotation, two translation parameters, and the dilation factor ρ. SS(target) is the centered sum of squares of the survey data, and is meaningful mostly in relation to the residual sum of squares RSS(target). The Procrustes statistic, defined as RSS/SS, measures the size of the residuals relative to the variation in the target variables; it is equivalent to $1 - R^2$ in a regression-analysis context. The smaller the Procrustes statistic, the closer the correspondence of Speed's map to the survey map. The number in this case, 0.004, is very small indeed (by what standard, though?). Another way of looking at fit is via the square root of the mean squared residual, RMSE, a measure for the average size of residuals.

The last output table describes how well the transformed Speed coordinates match the survey coordinates, separately for the horizontal (x) and the vertical (y) coordinates. In this case, we do not see disturbing differences between the coordinates. By definition, the overall Procrustes statistic and the overall-RMSE are averages of the coordinate statistics. Since Procrustes analysis treats (weights) both coordinates the same and independently, analogous to the sphericity assumption in multivariate regression or MANOVA analysis, the comparable statistics for the different coordinates is reassuring.

This example is continued in [MV] **procrustes postestimation**, demonstrating how to generate fitted values and residual sum of squares using `predict`, how to produce a graph showing the target overlaid with the transformed source values using `procoverlay`, and how to produce various summaries and comparisons using `estat`.

◁

A Procrustes analysis fits the transformation subject to the constraint that $\mathbf{A}$ is orthogonal; for other constraints, see below. In two dimensions, there are actually two types of orthogonal matrices: rotations and reflections. Think of left and right hands. A rotation transforms a left hand into a left hand, never into a right hand; rotation preserves orientation. A reflection changes a left hand into a right hand; reflections invert orientation. In algebraic terms, an orthogonal matrix $\mathbf{A}$ satisfies $\det(\mathbf{A}) = \pm 1$. $\mathbf{A}$ is a rotation if $\det(\mathbf{A}) = 1$, and $\mathbf{A}$ is a reflection if $\det(\mathbf{A}) = -1$. In more than two dimensions, the classification of orthogonal transformations is more complicated.

▷ Example 2

In example 1, we treated the location, dilation, and orientation as estimable aspects of the transformation. It is possible to omit the location and dilation aspects—though, admittedly, from a casual inspection as well as the substantial understanding of the data, it is clear that these aspects are crucial. For instance, we may omit the dilation factor—i.e., assume $\rho = 1$—with the `norho` option.

```
. procrustes (survey_x survey_y) (speed_x speed_y), norho
Procrustes analysis (orthogonal)          Number of observations =        20
                                          Model df (df_m)        =         3
                                          Residual df (df_r)     =        37
                                          SS(target)             =    495070
                                          RSS(target)            =  165278.1
                                          RMSE = root(RSS/df_r)  =  66.83544
                                          Procrustes = RSS/SS    =    0.3338
```

Translation c

	survey_x	survey_y
_cons	741.4458	435.6215

Rotation & reflection matrix A (orthogonal)

	survey_x	survey_y
speed_x	.9841521	-.1773266
speed_y	.1773266	.9841521

Dilation factor

```
        rho =    1.0000
```

Fit statistics by target variable

Statistics	survey_x	survey_y
SS	216310.2	278759.8
RSS	70385.78	94892.36
RMSE	61.68174	71.61925
Procrustes	.3253928	.340409
Corr_y_yhat	.9976669	.9985076

As expected, the optimal transformation without dilation implies a much weaker relation between the Speed and Survey maps; the Procrustes statistic has increased from 0.0040 to 0.3338. We conclude that we cannot adequately describe the correspondence between the maps if we ignore differences in scale

◁

Is an orthogonal Procrustes analysis symmetric?

In examples 1 and 2, we transformed the Speed map to optimally match the modern Survey map. We could also have reversed the procedure, i.e., transform the Survey map to match the Speed map.

(Continued on next page)

▷ Example 3

Here we change the order of the Speed and Survey map in our call to `procrustes` from example 1.

```
. procrustes (speed_x speed_y) (survey_x survey_y)
```

Procrustes analysis (orthogonal)		
Number of observations	=	20
Model df (df_m)	=	4
Residual df (df_r)	=	36
SS(target)	=	88862.75
RSS(target)	=	354.2132
RMSE = root(RSS/df_r)	=	3.136759
Procrustes = RSS/SS	=	0.0040

Translation c

	speed_x	speed_y
_cons	-187.0142	-159.5801

Rotation & reflection matrix A (orthogonal)

	speed_x	speed_y
survey_x	.9841521	.1773266
survey_y	-.1773266	.9841521

Dilation factor

rho =   0.4228

Fit statistics by target variable

Statistics	speed_x	speed_y
SS	41544.95	47317.8
RSS	218.3815	135.8317
RMSE	3.483146	2.747036
Procrustes	.0052565	.0028706
Corr_y_yhat	.9975074	.9986641

The implied transformations are similar but not identical. For instance, the product of estimated scale factors is $2.3556 \times 0.4228 = 0.9959$, which is close to 1 but not identical to 1—this is not due to roundoff error. Why do the results differ? Think about the analogy with regression analysis. The regression of Y on X and the regression of X on Y generally imply different relationships between the variables. In geometric terms, one minimizes the sum of squares of the "vertical" distances between the data point and the regression line, the other minimizes the "horizontal" distances. The implied regression lines are the same if the variance in X and Y are the same. Even if this does not hold, the proportion of explained variance R^2 in both regressions is the same. In Procrustes analysis an analogous relationship holds between the analyses "Speed = transformed(Survey) + E" and "Survey = transformed(Speed) + E". Both analyses yield the same Procrustes statistic. The implied analyses are equivalent (i.e., the implied transformation in one analysis is the mathematical inverse of the transformation in the other analysis) only if the Speed and Survey data are scaled so that the trace of the associated covariance matrices is the same.

◁

Other transformations

A Procrustes analysis can also be applied with other classes of transformations. Browne (1967) analyzed Procrustes analyses with oblique rotations. Cramer (1974) and ten Berge and Nevels (1977) identified and solved some problems in Browne's solution (but still ignore the problem that the derived oblique rotations are not necessarily orientation preserving). procrustes supports oblique transformations. procrustes also allows dilation; see *Methods and Formulas*.

▷ Example 4

Even though the orthogonal Procrustes analysis of example 1 demonstrated a close match between the two configurations assuming an orthogonal transformation, we now investigate what happens with an oblique transformation.

```
. procrustes (survey_x survey_y) (speed_x speed_y), trans(oblique)
```

Procrustes analysis (oblique)

Number of observations =	20
Model df (df_m) =	5
Residual df (df_r) =	35
SS(target) =	495070
RSS(target) =	1967.854
RMSE = root(RSS/df_r) =	7.498294
Procrustes = RSS/SS =	0.0040

Translation c

	survey_x	survey_y
_cons	503.0093	292.4346

Rotation & reflection matrix A (oblique)

	survey_x	survey_y
speed_x	.9835969	-.1737553
speed_y	.1803803	.9847889

Dilation factor

rho = 2.3562

Fit statistics by target variable

Statistics	survey_x	survey_y
SS	216310.2	278759.8
RSS	1080.677	887.1769
RMSE	7.858307	7.1201
Procrustes	.004996	.0031826
Corr_y_yhat	.9976685	.9985163

We see that the optimal oblique transformation is almost orthogonal; the columns of the oblique rotation and reflection matrix are almost perpendicular. The dilation factor and translation vector hardly differ from the orthogonal case shown in example 1. Finally, we see that the residual sum of squares decreased very little, namely from 1973.4 to 1967.9.

◁

Procrustes analysis can be interpreted as multivariate regression $\mathbf{Y} = \mathbf{c} + \mathbf{xB} + \mathbf{e}$ in which some nonlinear restriction is applied to the coefficients $\mathbf{B}$. Procrustes analysis assumes $\mathbf{B} = \rho\mathbf{A}$ with $\mathbf{A}$

assumed to be orthogonal or $\mathbf{A}$ assumed to be oblique. The intercepts of the multivariate regression are, of course, the translation of the Procrustean transform. In contrast to multivariate regression, it is assumed that the distribution of the residuals $\mathbf{e}$ is spherical, that is, all that is assumed is that $\mathrm{var}(\mathbf{e}) = \sigma^2 \mathbf{I}$. This assumption affects standard errors, not the estimated coefficients. Multivariate regression serves as a useful baseline to gauge the extent to which the Procrustean analysis is appropriate. procrustes supports the `transform(unrestricted)` option and displays the fitted model in a format comparable with Procrustes analysis.

▷ Example 5

We demonstrate with Speed's map data.

```
. procrustes (survey_x survey_y) (speed_x speed_y), trans(unrestricted)
```

Procrustes analysis (unrestricted)

Number of observations =	20
Model df (df_m) =	6
Residual df (df_r) =	34
SS(target) =	495070
RSS(target) =	1833.435
RMSE = root(RSS/df_r) =	7.343334
Procrustes = RSS/SS =	0.0037

Translation c

	survey_x	survey_y
_cons	510.8028	288.243

Rotation & reflection matrix A (unrestricted)

	survey_x	survey_y
speed_x	2.27584	-.4129564
speed_y	.4147244	2.355725

Fit statistics by target variable

Statistics	survey_x	survey_y
SS	216310.2	278759.8
RSS	1007.14	826.2953
RMSE	7.696981	6.971772
Procrustes	.004656	.0029642
Corr_y_yhat	.9976693	.9985168

Since we already saw that there is almost no room to improve on the orthogonal Procrustes transform with this particular dataset, it will not come as a big surprise that dropping the restrictions on the coefficients hardly improves the fit. For instance, the residual sum of squares further decreases from 1967.9 in the oblique case to 1833.4 in the unrestricted case, with only a very small reduction in the value of the Procrustes statistic.

◁

Saved Results

`procrustes` saves in `e()`:

Scalars
e(N)	number of observations
e(rho)	dilation factor
e(P)	Procrustes statistic
e(ss)	total sum of squares, summed over all y variables
e(rss)	residual sum of squares, summed over all y variables
e(rmse)	root mean squared error
e(urmse)	root mean squared error (unadjusted for # of estimated parameters)
e(df_m)	model degrees of freedom
e(df_r)	residual degrees of freedom

Macros
e(cmd)	procrustes
e(ylist)	y variables (target variables)
e(xlist)	x variables (source variables)
e(transform)	orthogonal, oblique, or unrestricted
e(wtype)	weight type
e(wexp)	weight expression
e(properties)	nob noV
e(estat_cmd)	program used to implement estat
e(predict)	program used to implement predict

Matrices
e(c)	translation vector
e(A)	orthogonal transformation matrix
e(ystats)	matrix containing fit statistics

Functions
e(sample)	marks estimation sample

Methods and Formulas

`procrustes` is implemented as an ado-file.

A Procrustes analysis is accomplished by solving a matrix minimization problem

$$\text{Minimize } | \mathbf{Y} - (\mathbf{1c}' + \rho\, \mathbf{X}\, \mathbf{A}) |$$

with respect to $\mathbf{A}$, $\mathbf{c}$, and ρ. $\mathbf{A}$ is a matrix representing a linear transformation, $\rho > 0$ is a scalar called the "dilation factor", $\mathbf{c}$ is a translation (row-)vector, and $|.|$ is the Frobenius (or L2) norm. Three classes of transformations are available in `procrustes`: orthogonal, oblique, and unrestricted. The orthogonal class consists of all orthonormal matrices $\mathbf{A}$, i.e., all square matrices that satisfy $\mathbf{A}'\mathbf{A} = \mathbf{I}$, representing orthogonal norm-preserving rotations and reflections. The oblique class comprises all normal matrices $\mathbf{A}$, characterized by $\text{diag}(\mathbf{A}'\mathbf{A}) = \mathbf{1}$. Oblique transformations preserve the length of vectors, but not the angles between vectors—orthogonal vectors will generally not remain orthogonal under oblique transformations. Finally, the unrestricted class consists of all conformable regular matrices $\mathbf{A}$.

Define $\widetilde{\mathbf{Y}}$ and $\widetilde{\mathbf{X}}$ as the centered $\mathbf{Y}$ and $\mathbf{X}$, respectively, if a constant $\mathbf{c}$ is included in the analysis and as the uncentered $\mathbf{Y}$ and $\mathbf{X}$ otherwise.

The derivation of the optimal $\mathbf{A}$ obviously differs for the three classes of transformations.

Orthogonal transformations

The solution for the orthonormal case can be expressed in terms of the singular value decomposition of $\widetilde{\mathbf{Y}}'\widetilde{\mathbf{X}}$,

$$\widetilde{\mathbf{Y}}'\widetilde{\mathbf{X}} = \mathbf{U}\mathbf{\Lambda}\mathbf{V}'$$

where $\mathbf{U}'\mathbf{U} = \mathbf{V}'\mathbf{V} = \mathbf{I}$. Then

$$\widehat{\mathbf{A}} = \mathbf{V}\mathbf{U}'$$

$\widehat{\mathbf{A}}$ is the same whether or not scaling is required, i.e., whether ρ is a free parameter or a fixed parameter. When ρ is a free parameter, the optimal ρ is

$$\widehat{\rho} = \frac{\text{trace}(\widehat{\mathbf{A}}\widetilde{\mathbf{Y}}'\widetilde{\mathbf{X}})}{\text{trace}(\widetilde{\mathbf{X}}'\widetilde{\mathbf{X}})}$$

See ten Berge (1977) for a modern and elementary derivation; see Mardia, Kent, and Bibby (1979) for a derivation using matrix differential calculus.

Oblique transformations

Improving on earlier studies by Browne (1967) and Cramer (1974), ten Berge and Nevels (1977) provide a full algorithm to compute the optimal oblique rotation without dilation, i.e., with uniform scaling $\rho = 1$. In contrast to the orthogonal case, the optimal oblique rotation $\widehat{\mathbf{A}}$ depends on ρ. To the best of our knowledge, this case has not been treated in the literature (ten Berge 2004). However, an "alternating least squares" extension of the ten Berge and Nevels (1977) algorithm deals with this case.

For each iteration, step (a) follows ten Berge and Nevels (1977) for calculating $\widetilde{\mathbf{Y}}$ and $\widehat{\rho}\widetilde{\mathbf{X}}$. In step (b) of an iteration, ρ is optimized, keeping $\widehat{\mathbf{A}}$ fixed, with solution

$$\widehat{\rho} = \frac{\text{trace}(\widehat{\mathbf{A}}\widetilde{\mathbf{Y}}'\widetilde{\mathbf{X}})}{\text{trace}(\widetilde{\mathbf{X}}'\widetilde{\mathbf{X}}\widehat{\mathbf{A}}\widehat{\mathbf{A}}')}$$

Iteration continues while the relative decrease in the residual sum of squares is sufficiently large. This algorithm is ensured to yield a local optimum of the residual sum of squares as the RSS decreases both when updating the rotation $\mathbf{A}$ and when updating the dilation factor ρ. Beware that the algorithm is not guaranteed to find the global minimum.

Unrestricted transformations

In the unrestricted solution, the dilation factor ρ is fixed at 1. The computation of the Procrustes transformation is obviously equivalent to the least-squares solution of multivariate regression

$$\widehat{\mathbf{A}} = (\widetilde{\mathbf{X}}'\widetilde{\mathbf{X}})^{-1}\widetilde{\mathbf{X}}'\widetilde{\mathbf{Y}}$$

Given $\widehat{\mathbf{A}}$ and $\widehat{\rho}$, the optimal translation $\widehat{\mathbf{c}}$ can be written as

$$\widehat{\mathbf{c}} = \mathbf{Y}'\mathbf{1} - \widehat{\rho}\widehat{\mathbf{A}}\mathbf{X}$$

If the constant is suppressed, $\mathbf{c}$ is simply set to $\mathbf{0}$.

Reported statistics

procrustes computes and displays the following statistics for each target variable separately and globally by adding the appropriate sums of squares over all target variables. The predicted values $\widehat{\mathbf{Y}}$ for $\mathbf{Y}$ are defined as

$$\widehat{\mathbf{Y}} = \mathbf{1}\widehat{\mathbf{c}}' + \widehat{\rho}\mathbf{X}\widehat{\mathbf{A}}$$

The Procrustes statistic, P, is a scaled version of the squared distance of $\mathbf{Y}$:

$$P = \mathrm{RSS/SS}$$

where

$$\mathrm{RSS} = \mathrm{trace}((\mathbf{Y} - \widehat{\mathbf{Y}})(\mathbf{Y} - \widehat{\mathbf{Y}})')$$

$$\mathrm{SS} = \mathrm{trace}(\widetilde{\mathbf{Y}}'\widetilde{\mathbf{Y}})$$

Note that $0 \leq P \leq 1$, and a small value of P means that $\mathbf{Y}$ is close to the transformed value of $\mathbf{X}$, that is, the $\mathbf{X}$ and $\mathbf{Y}$ configurations are similar. In the literature, this statistic is often denoted by R^2. It is easy to confuse this with the R^2 statistic in a regression context, which is actually $1 - P$.

A measure for the size of the residuals is the root mean squared error,

$$\mathrm{RMSE} = \sqrt{\mathrm{RSS/df_r}}$$

Here $\mathrm{df_r}$ are $Nn_y - \mathrm{df_m}$, with $\mathrm{df_m} = n_y n_x + n_y + 1 - k$, and with N the number of observations, n_y and n_x the number of target variables and source variables respectively, and k, the number of restrictions, defined as

orthogonal: $k = n_x(n_x - 1)/2$
oblique: $k = n_y$
unrestricted: $k = 1$

procrustes computes $\mathrm{RMSE}(j)$ for target variable y_j as

$$\mathrm{RMSE}(j) = \sqrt{\mathrm{RSS}(j)/(\mathrm{df_r}/n_y)}$$

Finally, procrustes computes the Pearson correlation between y_j and $\widehat{y}_j$. For the unrestricted transformation, this is just the square root of the explained variance $1 - P(j)$ where $P(j) = \mathrm{RSS}(j)/\mathrm{SS}$. For the orthogonal and oblique transformation, this relationship does not hold.

References

Browne, M. W. 1967. On oblique Procrustes rotation. *Psychometrika* 32: 125–132.

Cox, T. F. and M. A. A. Cox. 2001. *Multidimensional Scaling*. 2nd ed. Boca Raton, FL: Chapman & Hall.

Cramer, E. M. 1974. On Browne's solution for oblique Procrustes rotation. *Psychometrika* 39: 159–163.

Gower, J. C. and G. B. Dijksterhuis. 2004. *Procrustes Problems*. Oxford: Oxford University Press.

Green, B. F. 1952. The orthogonal approximation of an oblique structure in factor analysis. *Psychometrika* 17: 429–440.

Hurley, J. R. and R. B. Cattell. 1962. The Procrustes program: producing direct rotation to test a hypothesized factor structure. *Behavioral Science* 7: 258–262.

Mardia, K. V., J. T. Kent, and J. M. Bibby. 1979. *Multivariate Analysis*. New York: Academic Press.

Mosier, C. I. 1939. Determining a simple structure when loadings for certain tests are known. *Psychometrika* 4: 149–162.

ten Berge, J. M. F. 1977. Orthogonal Procrustes rotation for two or more matrices. *Psychometrika* 42: 267–276.

——. 2004. Private communication.

ten Berge, J. M. F. and K. Nevels. 1977. A general solution to Mosier's oblique Procrustes problem. *Psychometrika* 42: 593–600.

Also See

Complementary:	[MV] **procrustes postestimation**; [MV] **ca**, [MV] **pca**, [MV] **rotate**
Related:	[R] **mvreg**
Background:	[U] **11.1.10 Prefix commands**,
	[U] **20 Estimation and postestimation commands**

Title

procrustes postestimation — Postestimation tools for procrustes

Description

The following postestimation commands are of special interest after `procrustes`:

command	description
estat compare	fit statistics for orthogonal, oblique, and unrestricted transformations
estat mvreg	display multivariate regression resembling unrestricted transformation
estat summarize	display summary statistics over the estimation sample
procoverlay	produce a Procrustes overlay graph

For information about these commands, see below.

In addition, the following postestimation commands are available:

command	description
*estimates	cataloging estimation results
predict	compute fitted values and residuals

* All `estimates` subcommands except `table` and `stats` are available; see [R] **estimates**.

See the corresponding entries in the *Stata Base Reference Manual* for details.

Special-interest postestimation commands

`estat compare` displays a table with fit statistics of the three transformations provided by `procrustes`: `orthogonal`, `oblique`, and `unrestricted`. The two additional procrustes analyses are performed on the same sample as the original procrustes analysis and with the same options. F-tests comparing the models are provided.

`estat mvreg` produces the `mvreg` (see [R] **mvreg**) output related to the unrestricted Procrustes analysis (the `transform(unrestricted)` option of `procrustes`).

`estat summarize` displays summary statistics over the estimation sample of the target and source variables (*varlist$_y$* and *varlist$_x$*).

`procoverlay` displays a plot of the target variables overlaid with the fitted values derived from the source variables. If there are more than two target variables, multiple plots are shown in one graph.

Syntax for predict

predict $\big[$ *type* $\big]$ *newvarlist* $\big[$ *if* $\big]$ $\big[$ *in* $\big]$, $\big[$ *statistic* $\big]$

statistic	description
Main	
<u>fitted</u>	fitted values $\mathbf{1} \, \mathbf{c}' + \rho \, \mathbf{X} \, \mathbf{A}$; the default (specify #$_y$ vars)
<u>res</u>iduals	unstandardized residuals (specify #$_y$ vars)
q	residual sum of squares over the target variables (specify one var)

These statistics are available both in and out of sample; type predict ... if e(sample) ... if wanted only for the estimation sample.

Options for predict

┌─── **Main** └───

fitted, the default, computes fitted values, i.e., the least-squares approximations of the target (*varlist$_y$*) variables. You must specify the same number of new variables as there are target variables.

residuals computes the raw (unstandardized) residuals for each of the target (*varlist$_y$*) variables. You must specify the same number of new variables as there are target variables.

q computes the residual sum of squares over all variables, i.e., the squared Euclidean distance between the target and transformed source points. Specify one new variable.

Syntax for estat

Table of fit statistics

estat <u>com</u>pare $\big[$, <u>det</u>ail $\big]$

Comparison of mvreg *and* procrustes *output*

estat <u>mv</u>reg $\big[$, *mvreg_options* $\big]$

Display summary statistics

estat <u>su</u>mmarize $\big[$, <u>label</u> <u>nohea</u>der <u>noweights</u> $\big]$

Options for estat

detail, an option with estat compare, displays the standard procrustes output for the two additional transformations.

mvreg_options, allowed with estat mvreg, are any of the options allowed by mvreg; see [R] **mvreg**. Note that the constant is already suppressed if the Procrustes analysis suppressed it.

label, noheader, and noweights are the same as for the generic estat summarize command; see [R] **estat**.

Syntax for procoverlay

procoverlay [*if*] [*in*] [*procoverlay_options*]

procoverlay_options	description
Main	
<u>auto</u>aspect	adjust aspect ratio based on the data; default aspect ratio is 1
<u>target</u>opts(*target_opts*)	affect the rendition of the target
<u>source</u>opts(*source_opts*)	affect the rendition of the source
Y-Axis, X-Axis, Title, Caption, Legend, Overall	
twoway_options	any options other than by() documented in [G] ***twoway_options***
By	
<u>byopts</u>(*by_option*)	affect the rendition of combined graphs

target_opts	description
Main	
<u>nolabel</u>	removes the default observation label from the target
marker_options	change look of markers (color, size, etc.)
marker_label_options	change look or position of marker labels

source_opts	description
Main	
<u>nolabel</u>	removes the default observation label from the source
marker_options	change look of markers (color, size, etc.)
marker_label_options	change look or position of marker labels

Options for procoverlay

⌐ Main ⌐

autoaspect specifies that the aspect ratio be automatically adjusted based on the range of the data to be plotted. This option can make some procoverlay plots more readable. By default, procoverlay uses an aspect ratio of one, producing a square plot.

As an alternative to autoaspect, the *twoway_option* aspectratio() can be used to override the default aspect ratio. procoverlay accepts the aspectratio() option as a suggestion only and will override it when necessary to produce plots with balanced axes, i.e., where distance on the x-axis equals distance on the y-axis.

twoway_options, such as xlabel(), xscale(), ylabel(), and yscale(), should be used with caution. These options are accepted but may have unintended side effects on the aspect ratio.

targetopts(*target_options*) affect the rendition of the target plot. The following *target_options* are allowed:

nolabel removes the default target observation label from the graph.

marker_options affect the rendition of markers drawn at the plotted points, including their shape, size, color, and outline; see [G] ***marker_options***

marker_label_options specify if and how the markers are to be labeled; see
[G] ***marker_label_options***.

sourceopts(*source_options*) affect the rendition of the source plot. The following *source_options*
are allowed

nolabel removes the default source observation label from the graph.

marker_options affect the rendition of markers drawn at the plotted points, including their shape,
size, color, and outline; see [G] ***marker_options***

marker_label_options specify if and how the markers are to be labeled; see
[G] ***marker_label_options***.

_____⌐ Y-Axis, X-Axis, Title, Caption, Legend, Overall ⌐_____

twoway_options are any of the options documented in [G] ***twoway_options***, excluding by(). These
include options for titling the graph (see [G] ***title_options***) and options for saving the graph to
disk (see [G] ***saving_option***).

See autoaspect above for a warning against using options such as xlabel(), xscale(),
ylabel(), and yscale().

_____⌐ By ⌐_____

byopts(*by_option*) is documented in [G] ***by_option***. This option affects the appearance of the
combined graph and is ignored, unless there are more than two target variables specified in
procrustes.

Remarks

The examples in [MV] **procrustes** demonstrated a Procrustes transformation of a historical map,
produced by John Speed in 1610, to a modern map. Here we demonstrate the use of procrustes
postestimation tools in assessing the accuracy of Speed's map. Example 1 of [MV] **procrustes**
performed the following analysis:

```
. use speed_survey
. procrustes (survey_x survey_y) (speed_x speed_y)
  (output omitted)
```

See example 1 of [MV] **procrustes**. The following examples are based on this procrustes analysis.

▷ Example 1

Did John Speed get the coordinates of the towns right—up to the location, scale, and orientation
of his map relative to the modern map? In example 1 of [MV] **procrustes**, we demonstrated how
the optimal transformation from the historical coordinates to the modern (true) coordinates can be
estimated by procrustes.

It is possible to "predict" the configuration of 20 cities on Speed's historical map, optimally
transformed (rotated, dilated, and translated) to approximate the true configuration. predict with the
fitted option expects the same number of variables as the number of target (dependent) variables
(survey_x and survey_y).

```
. predict fitted_x fitted_y
(fitted assumed)
```

We omitted the `fitted` option since it is the default.

It is often useful to also compute the (squared) distance between the true location and the transformed location of the historical map. This can be seen as a quality measure—the larger the value, the more Speed erred in the location of the respective town.

```
. predict q, q
```

We now list the target data (`survey_x` and `survey_y`, the values from the modern map), the fitted values (`fitted_x` and `fitted_y`, produced by `predict`), and the squared distance between them (q, produced by `predict` with the q option).

```
. list name survey_x survey_y fitted_x fitted_y q, sep(0) noobs
```

name	survey_x	survey_y	fitted_x	fitted_y	q
Alve	1027	725	1037.117	702.9464	588.7149
Arro	1083	565	1071.682	562.6791	133.4802
Astl	787	677	783.0652	674.5216	21.62482
Beck	976	358	978.8665	366.3761	78.37637
Beng	1045	435	1055.245	431.6015	116.51
Crad	736	471	725.8594	476.5895	134.075
Droi	893	633	890.5839	633.6066	6.205747
Ecki	922	414	929.4932	411.1757	64.12465
Eves	1037	437	1036.887	449.2707	150.5827
Hall	828	579	825.1494	575.9836	17.22464
Hanb	944	637	954.6189	643.6107	156.4629
Inkb	1016	573	1004.869	577.1111	140.7917
Kemp	848	490	845.7215	490.8959	5.994327
Kidd	826	762	836.8665	760.5699	120.1264
Mart	756	598	745.2623	597.5585	115.4937
Stud	1074	632	1072.622	634.3164	7.264294
Tewk	891	324	898.4571	318.632	84.42448
UpSn	943	544	939.3932	545.8247	16.33858
Upto	852	403	853.449	400.9419	6.335171
Worc	850	545	848.7917	547.7881	9.233305

We see that Speed especially erred in the location of Alvechurch—it is off by no less than $\sqrt{588} = 24$ miles, while the average error is about 8 miles. In a serious analysis of this dataset, we would check the data on Alvechurch, and, if we found it to be in order, consider whether we should actually drop Alvechurch from the analysis. In this illustration, we ignore this potential problem.

◁

▷ Example 2

While the numerical information convinces us that Speed's map is generally accurate, a plot will convey this message more convincingly. `procoverlay` produces a plot that contains the target (survey) coordinates and the Procrustes-transformed historical coordinates. We could just type

```
. procoverlay
```

However, we decide to set a number of options to produce a presentation-quality graph. The suboption `mlabel()` of `target()` (or of `source()` adds labels, identifying the towns. Since the target and source points are so close, there can be no confusing how they are matched. Displaying the labels twice in the plot is not helpful for this dataset. Therefore, we choose to label the target points, but not the source points using the `nolabel` suboption of `source()`. We preserve the equivalence of the x- and y-scale while using as much of the graphing region as possible with the `autoaspect` option.

The span suboption of title() allows the long title to extend beyond the graph region if needed. We override the default legend using the legend() option.

```
. procoverlay, target(mlabel(name)) source(nolabel) autoaspect
      title(Historic map of 20 towns and villages in Worcestershire, span)
      subtitle(overlaid with actual positions)
      legend(label(1 historic map) label(2 actual position))
```

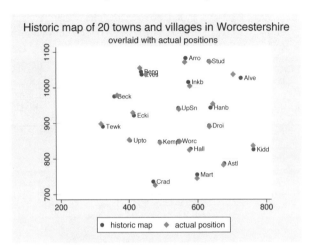

◁

▷ Example 3

estat offers three specific facilities after procrustes. These can all be seen as convenience tools that accomplish simple analyses, ensuring that the same variables and the same observations are used as in the Procrustes analysis.

The variables involved in the Procrustes analysis can be summarized over the estimation sample, for instance, in order to gauge differences in scales and location of the target and source variables.

```
. estat summarize
```

Estimation sample procrustes			Number of obs =	20
Variable	Mean	Std. Dev.	Min	Max
target				
survey_x	916.7	106.6993	736	1083
survey_y	540.1	121.1262	324	762
source				
speed_x	153.95	46.76084	78	220
speed_y	133.9	49.90401	40	220

From the summarization, it is obvious that the two maps have different origins and scale.

As pointed out in [MV] **procrustes**, orthogonal and oblique Procrustes analyses can be thought of as special cases of multivariate regression (see [R] **mvreg**), subject to nonlinear restrictions on the coefficient matrix. Comparing the Procrustes statistics and the transformations for each of the three classes of transformations is helpful in selecting a transformation. The compare subcommand of estat provides summary information for the optimal transformations in each of the three classes.

```
. estat compare
```
Summary statistics for three transformations

	Procrustes	df_m	df_r	rmse
orthogonal	0.0040	4	36	7.403797
oblique	0.0040	5	35	7.498294
unrestricted	0.0037	6	34	7.343334

(F tests comparing the models suppressed)

The Procrustes statistic is ensured to decrease (not increase) from orthogonal to oblique to unrestricted since the associated classes of transformations are getting less restrictive. The model degrees of freedom (df_m) of the three transformation classes are the dimension of the classes, i.e., the number of "free parameters". For instance, with orthogonal transformations between two source and two target variables, there is one degree of freedom for the rotation (representing the rotation angle), two degrees of freedom for the translation, and one degree of freedom for dilation (uniform scaling), that is, four in total. The residual degrees of freedom (RSS) are the number of observations (number of target variables times the number of observations) minus the model degrees of freedom. The root mean squared error RMSE, defined as

$$\text{RMSE} = \sqrt{\frac{\text{RSS}}{\text{df}_r}}$$

does not, unlike the Procrustes statistic, surely become smaller with the less restrictive models. In this example, in fact, the RMSE of the orthogonal transformation is smaller than that of the oblique transformation. This indicates that the additional degree of freedom allowing for skew rotations does not produce a closer fit. In this example, we see little reason to relax orthogonal transformations; very little is gained in terms of the Procrustes statistic (an "illness-of-fit" measure) or the RMSE. In this interpretation, we used our intuition to guide us whether a difference in fit is substantively and statistically meaningful—formal significance tests are not provided.

Finally, the unrestricted transformation can be estimated with procrustes ..., transform(unrestricted). This analysis is related to a multivariate regression with the target variables as the dependent variables and the source variables as the independent variables. While the unrestricted Procrustes analysis assumes spherical (uncorrelated homoskedastic) residuals, this restrictive assumption is not made in multivariate regression as estimated by the mvreg command. The comparable multivariate regression over the same estimation sample can be viewed simply by typing

(*Continued on next page*)

```
. estat mvreg
```

Multivariate regression, similar to "procrustes ..., transform(unrestricted)"

Equation	Obs	Parms	RMSE	"R-sq"	F	P
survey_x	20	3	7.696981	0.9953	1817.102	0.0000
survey_y	20	3	6.971772	0.9970	2859.068	0.0000

	Coef.	Std. Err.	t	P>|t|	[95% Conf. Interval]	
survey_x						
speed_x	2.27584	.0379369	59.99	0.000	2.1958	2.35588
speed_y	.4147244	.0355475	11.67	0.000	.3397257	.489723
_cons	510.8028	8.065519	63.33	0.000	493.7861	527.8196
survey_y						
speed_x	-.4129564	.0343625	-12.02	0.000	-.485455	-.3404579
speed_y	2.355725	.0321982	73.16	0.000	2.287793	2.423658
_cons	288.243	7.305587	39.46	0.000	272.8296	303.6564

Note that this analysis is seen as postestimation after a Procrustes analysis, so it does not change the "last estimation results". We may still replay `procrustes` and use other `procrustes` postestimation commands.

◁

Saved Results

`estat compare` after `procrustes` saves in `r()`:

> Matrices
> > `r(cstat)`   Procrustes statistics, degrees of freedom, and RMSEs
> > `r(fstat)` F statistics, degrees of freedom, and p values

`estat mvreg` does not return results.

`estat summarize` after `procrustes` saves in `r()`:

> Matrices
> > `r(stats)`   means, standard deviations, minimums, and maximums

Methods and Formulas

All postestimation commands listed above are implemented as ado-files.

The predicted values for the jth variable are defined as

$$\widehat{y}_j = \widehat{c}_j + \widehat{\rho}\, \mathbf{X}\, \widehat{\mathbf{A}}[.,j]$$

The residual for y_j is simply $y_j - \widehat{y}_j$. The "rowwise" quality q of the approximation is defined as the residual sum of squares:

$$q = \sum_j (y_j - \widehat{y}_j)^2$$

The entries of the summary table produced by `estat compare` are described in *Methods and Formulas* of [MV] **procrustes**. The F tests produced by `estat compare` are similar to standard nested model tests in linear models.

References

See the references in [MV] **procrustes**.

Also See

Complementary:	[MV] **procrustes**,
	[R] **estimates**
Related:	[R] **mvreg**
Background:	[R] **estat**, [R] **predict**

Title

> **rotate** — Orthogonal and oblique rotations after factor and pca

Syntax

rotate [, *options*]

rotate, clear

options	description
Main	
orthogonal	restrict to orthogonal rotations; the default, except with promax()
oblique	allow oblique rotations
rotation_methods	rotation criterion
normalize	rotate Horst normalized matrix
horst	synonym for normalize
factors(#)	rotate # factors or components; default is to rotate all
components(#)	synonym for factors()
Reporting	
blanks(#)	display loadings as blanks when |loadings| < #; default is blanks(0)
detail	show rotatemat output; seldom used
format(%*fmt*)	display format for matrices; default is format(%9.5f)
noloading	suppress display of rotated loadings
norotation	suppress display of rotation matrix
Opt options	
optimize_options	control the maximization process; seldom used

(Continued on next page)

rotation_methods	description
*varimax	varimax (orthogonal only); the default
vgpf	varimax via the GPF algorithm (orthogonal only)
quartimax	quartimax (orthogonal only)
equamax	equamax (orthogonal only)
parsimax	parsimax (orthogonal only)
entropy	minimum entropy (orthogonal only)
tandem1	Comrey's tandem 1 principle (orthogonal only)
tandem2	Comrey's tandem 2 principle (orthogonal only)
*promax$\left[(\#)\right]$	promax power # (implies oblique); default is promax(3)
oblimin$\left[(\#)\right]$	oblimin with $\gamma = \#$; default is oblimin(0)
cf(#)	Crawford–Ferguson family with $\kappa = \#$, $0 \leq \# \leq 1$
bentler	Bentler's invariant pattern simplicity
oblimax	oblimax
quartimin	quartimin
target(Tg)	rotate toward matrix Tg
partial(Tg W)	rotate toward matrix Tg, weighted by matrix W

* varimax and promax ignore all *optimize_options*.

Description

rotate performs a rotation of the loading matrix after factor, factormat, pca, or pcamat; see [MV] **factor** and [MV] **pca**. Numerous rotation criteria (such as varimax, oblimin, etc.) are available that can be applied with respect to the orthogonal and/or oblique class of rotations. rotate stores in e(r_*name*). For instance, e(r_L) will contain the rotated loadings.

rotate, clear removes the rotation results from the estimation results.

If you want to rotate a given matrix, see [MV] **rotatemat**. Actually, rotate is implemented using rotatemat.

If you want a Procrustes rotation, rotating variables optimally toward other variables, see [MV] **procrustes**.

Options

    ⌐ Main ⌐

orthogonal specifies that an orthogonal rotation be applied. This is the default.

See *Rotation criteria* below for details on the *rotation_methods* available with orthogonal.

oblique specifies that an oblique rotation be applied. This often yields more interpretable factors with a simpler structure than those obtained with an orthogonal rotation. In many applications (e.g., after factor and pca) the factors before rotation are orthogonal (uncorrelated), while the oblique rotated factors are correlated.

See *Rotation criteria* below for details on the *rotation_methods* available with oblique.

clear specifies that rotation results be cleared (removed) from the last estimation command. clear may not be combined with any other option.

rotate stores its results within the e() results of pca and factor, overwriting any previous rotation results. Postestimation commands such as predict operate on the last rotated results, if any, instead of the unrotated results, and allow you to specify norotated to use the unrotated results. The clear option of rotate allows you to remove the rotation results from e(), thus freeing you from having to specify norotated for the postestimation commands.

normalize, and synonym horst, request that the rotation be applied to the Horst (1965) normalization of the matrix $\mathbf{A}$, so that the rowwise sums of squares equal 1. Horst normalization applies to the rotated columns only (see factors() and components() options below).

factors(#), and synonym components(#), specifies the number of factors or components (columns of the loading matrix) to be rotated, counted "from the left", i.e., with the lowest column index. The other columns are left unrotated. All columns are rotated by default.

_____ Reporting _____

blanks(#) shows blanks for loadings with absolute value smaller than #.

detail displays the rotatemat output; it is seldom used.

format(%fmt) specifies the display format for matrices. The default is format(%9.5f).

noloading suppresses the display of the rotated loadings.

norotation suppresses the display of the optimal rotation matrix.

_____ Opt options _____

optimize_options are seldom used; see [MV] **rotatemat**.

Rotation criteria

In the descriptions below, the matrix to be rotated is denoted as $\mathbf{A}$, p denotes the number of rows of $\mathbf{A}$, and f denotes the number of columns of $\mathbf{A}$ (factors or components). If $\mathbf{A}$ is a loading matrix from factor or pca, p is the number of variables, and f is the number of factors or components.

Criteria suitable only for orthogonal rotations

varimax and vgpf apply the orthogonal varimax rotation (Kaiser 1958). varimax maximizes the variance of the squared loadings within factors (columns of $\mathbf{A}$). It is equivalent to cf(1/p) and to oblimin(1). varimax, the most popular rotation, is implemented with a dedicated fast algorithm and ignores all *optimize_options*. Specify vgpf to switch to the general GPF algorithm used for the other criteria.

quartimax uses the quartimax criterion (Harman 1976). quartimax maximizes the variance of the squared loadings within the variables (rows of $\mathbf{A}$). For orthogonal rotations, quartimax is equivalent to cf(0) and to oblimax.

equamax specifies the orthogonal equamax rotation. equamax maximizes a weighted sum of the varimax and quartimax criteria, reflecting a concern for simple structure within variables (rows of $\mathbf{A}$) as well as within factors (columns of $\mathbf{A}$). equamax is equivalent to oblimin(p/2) and cf(#), where $\# = f/(2p)$.

parsimax specifies the orthogonal parsimax rotation. parsimax is equivalent to cf(#), where $\# = (f-1)/(p+f-2)$.

entropy applies the minimum entropy rotation criterion (Jennrich 2004).

tandem1 specifies that the first principle of Comrey's tandem be applied. According to Comrey (1967), this principle should be used to judge which "small" factors should be dropped.

tandem2 specifies that the second principle of Comrey's tandem be applied. According to Comrey (1967), tandem2 should be used for "polishing".

Criteria suitable only for oblique rotations

promax$\bigl[(\#)\bigr]$ specifies the oblique promax rotation. The optional argument specifies the promax power. Not specifying the argument is equivalent to specifying promax(3). Values smaller than 4 are recommended, but the choice is yours. Larger promax powers simplify the loadings (generate numbers closer to zero and one) but at the cost of additional correlation between factors. Choosing a value is a matter of trial-and-error, but most sources find values in excess of 4 undesirable in practice. The power must be greater than 1 but is not restricted to integers.

Promax rotation is an oblique rotation method that was developed before the "analytical methods" (based on criterion optimization) became computationally feasible. Promax rotation is comprised of an oblique Procrustean rotation of the original loadings $\mathbf{A}$ toward the element-wise #-power of the orthogonal varimax rotation of $\mathbf{A}$.

Criteria suitable for orthogonal and oblique rotations

oblimin$\bigl[(\#)\bigr]$ specifies that the oblimin criterion with $\gamma = \#$ be used. When restricted to orthogonal transformations, the oblimin() family is equivalent to the orthomax criterion function. Special cases of oblimin() include

γ	special cases
0	quartimax / quartimin
1/2	biquartimax / biquartimin
1	varimax / covarimin
$p/2$	equamax

$p =$ number of rows of $\mathbf{A}$

γ defaults to zero. Jennrich (1979) recommends $\gamma \leq 0$ for oblique rotations. For $\gamma > 0$, it is possible that optimal oblique rotations do not exist; the iterative procedure used to compute the solution will wander off to a degenerate solution.

cf(#) specifies that a criterion from the Crawford–Ferguson (1970) family be used with $\kappa = \#$. cf(κ) can be seen as $(1 - \kappa)\text{cf}_1(\mathbf{A}) + (\kappa)\text{cf}_2(\mathbf{A})$, where $\text{cf}_1(\mathbf{A})$ is a measure of row parsimony, and $\text{cf}_2(\mathbf{A})$ is a measure of column parsimony. $\text{cf}_1(\mathbf{A})$ attains its greatest lower bound when no row of $\mathbf{A}$ has more than one nonzero element, while $\text{cf}_2(\mathbf{A})$ reaches zero if no column of $\mathbf{A}$ has more than one nonzero element.

(Continued on next page)

For orthogonal rotations, the Crawford–Ferguson family is equivalent to the oblimin() family. For orthogonal rotations, special cases include the following:

κ	special cases
0	quartimax / quartimin
$1/p$	varimax / covarimin
$f/(2p)$	equamax
$(f-1)/(p+f-2)$	parsimax
1	factor parsimony

p = number of rows of $\mathbf{A}$
f = number of columns of $\mathbf{A}$

bentler specifies that the "invariant pattern simplicity" criterion (Bentler 1977) be used.

oblimax specifies the oblimax criterion. oblimax maximizes the number of high and low loadings. oblimax is equivalent to quartimax for orthogonal rotations.

quartimin specifies that the quartimin criterion be used. For orthogonal rotations, quartimin is equivalent to quartimax.

target(Tg) specifies that $\mathbf{A}$ be rotated as near as possible to the conformable matrix Tg. Nearness is expressed by the Frobenius matrix norm.

partial(Tg W) specifies that $\mathbf{A}$ be rotated as near as possible to the conformable matrix Tg. Nearness is expressed by a weighted (by W) Frobenius matrix norm. W should be non-negative and usually is zero–one valued, with ones identifying the target values to be reproduced as closely as possible by the factor loadings, while zeros identify loadings to remain unrestricted.

Remarks

Remarks are presented under the headings

Orthogonal rotations
Oblique rotations
Additional types of rotation

In this entry, we focus primarily on the rotation of factor loading matrices in factor analysis. rotate may also be used after pca, with exactly the same syntax. We advise caution in the interpretation of rotated loadings in principal component analysis since some of the optimality properties of principal components are not preserved under rotation. See [MV] **pca postestimation** for additional discussion of this point.

Orthogonal rotations

The interpretation of a factor analytical solution is not always easy—an understatement, many will agree. This is partly due to the standard way in which the inherent indeterminancy of factor analysis is resolved. Orthogonal transformations of the common factors and the associated factor loadings are possible without affecting the reconstructed (fitted) correlation matrix and preserving the property that common factors are uncorrelated. This gives considerable freedom in selecting an orthogonal rotation to facilitate the interpretation of the factor loadings. Thurstone (1935) offered criteria for a "simple structure" required for a psychologically meaningful factor solution. These informal criteria for interpretation were subsequently formalized into formal rotation criteria, e.g., Harman (1976) and Gorsuch (1983).

▷ Example 1

We illustrate `rotate` using a factor analysis of the correlation matrix of eight physical variables (height, arm span, length of forearm, length of lower leg, weight, bitrochanteric diameter, chest girth, and chest width) of 305 girls.

```
. matlist R, border format(%7.3f)
```

	height	arm_s~n	fore_~m	lower~g	weight	bitrod
height	1.000					
arm_span	0.846	1.000				
fore_arm	0.805	0.881	1.000			
lower_leg	0.859	0.826	0.801	1.000		
weight	0.473	0.376	0.380	0.436	1.000	
bitrod	0.398	0.326	0.319	0.329	0.762	1.000
ch_girth	0.301	0.277	0.237	0.327	0.730	0.583
ch_width	0.382	0.415	0.345	0.365	0.629	0.577

	ch_gi~h	ch_wi~h
ch_girth	1.000	
ch_width	0.539	1.000

We extract two common factors using the iterated principal factors method. See the description of `factormat` in [MV] **factor** for details on running a factor analysis on a Stata matrix rather than on a dataset.

```
. factormat R, n(305) fac(2) ipf
(obs=305)
```

Factor analysis/correlation   Number of obs   =   305
     Method: iterated principal factors   Retained factors =   2
     Rotation: (unrotated)   Number of params =   15

Factor	Eigenvalue	Difference	Proportion	Cumulative
Factor1	4.44901	2.93878	0.7466	0.7466
Factor2	1.51023	1.40850	0.2534	1.0000
Factor3	0.10173	0.04705	0.0171	1.0171
Factor4	0.05468	0.03944	0.0092	1.0263
Factor5	0.01524	0.05228	0.0026	1.0288
Factor6	-0.03703	0.02321	-0.0062	1.0226
Factor7	-0.06025	0.01415	-0.0101	1.0125
Factor8	-0.07440	.	-0.0125	1.0000

LR test: independent vs. saturated: chi2(28) = 2092.68 Prob>chi2 = 0.0000

Factor loadings (pattern matrix) and unique variances

Variable	Factor1	Factor2	Uniqueness
height	0.8560	-0.3244	0.1620
arm_span	0.8482	-0.4115	0.1112
fore_arm	0.8082	-0.4090	0.1795
lower_leg	0.8309	-0.3424	0.1923
weight	0.7503	0.5712	0.1108
bitrod	0.6307	0.4922	0.3600
ch_girth	0.5687	0.5096	0.4169
ch_width	0.6074	0.3507	0.5081

The default factor solution is rather poor from the perspective of a "simple structure", namely that variables should have high loadings on few (one) factors, and factors should ideally have only low and high values. A plot of the loadings is illuminating.

```
. loadingplot, xlab(0(.2)1) ylab(-.4(.2).6) aspect(1) yline(0) xline(0)
```

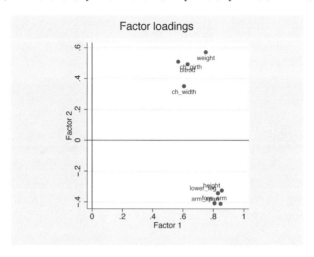

There are two groups of variables. We would like to see one group of variables close to one axis and the other group of variables close to the other axis. Turning the plot by about 45 degrees counter clockwise should make this possible and offer a much "simpler" structure. This is what the `rotate` command accomplishes.

```
. rotate
```

Factor analysis/correlation   Number of obs   =   305
    Method: iterated principal factors   Retained factors =   2
    Rotation: orthogonal varimax (Horst off)   Number of params =   15

Factor	Variance	Difference	Proportion	Cumulative
Factor1	3.39957	0.83989	0.5705	0.5705
Factor2	2.55968	.	0.4295	1.0000

LR test: independent vs. saturated: chi2(28) = 2092.68 Prob>chi2 = 0.0000

Rotated factor loadings (pattern matrix) and unique variances

Variable	Factor1	Factor2	Uniqueness
height	0.8802	0.2514	0.1620
arm_span	0.9260	0.1770	0.1112
fore_arm	0.8924	0.1550	0.1795
lower_leg	0.8708	0.2220	0.1923
weight	0.2603	0.9064	0.1108
bitrod	0.2116	0.7715	0.3600
ch_girth	0.1515	0.7484	0.4169
ch_width	0.2774	0.6442	0.5081

```
Factor rotation matrix
```

	Factor1	Factor2
Factor1	0.8018	0.5976
Factor2	-0.5976	0.8018

See [MV] **factor** for the interpretation of the first panel. Here we will focus on the second and third panel. The rotated factor loadings satisfy

$$\text{Factor1}_{\text{rotated}} = 0.8018 \times \text{Factor1}_{\text{unrotated}} - 0.5976 \times \text{Factor2}_{\text{unrotated}}$$

$$\text{Factor2}_{\text{rotated}} = 0.5976 \times \text{Factor1}_{\text{unrotated}} + 0.8018 \times \text{Factor2}_{\text{unrotated}}$$

The uniqueness—the variance of the specific factors—is not affected, since we are only changing the coordinates in common factor space. The purpose of rotation is to make factor loadings easier to interpret. The first factor loads high on the first four variables and low on the last four variables; for the second factor, the roles are reversed. This is really a simple structure according to Thurstone's criteria. This is very clear in the plot of the factor loadings.

```
. loadingplot, xlab(0(.2)1) ylab(0(.2)1) aspect(1) yline(0) xline(0)
```

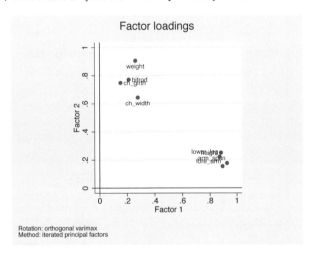

`rotate` provides a fair number of different rotations. You may make your intention more clear by typing the command as

```
. rotate, orthogonal varimax
  (output omitted )
```

`rotate` defaults to orthogonal (angle and length preserving) rotations of the axes; thus `orthogonal` may be omitted. The default rotation method is `varimax`, probably the most popular method. We warn that the varimax rotation is not appropriate if you expect a general factor contributing to all variables (see also Gorsuch 1983, chapter 9). In such a case you could, for instance, consider a quartimax rotation.

◁

▷ Example 2

`rotate` has performed what is known as "raw varimax", rotating the axes to maximize the sum of the variance of the squared loadings in the columns—the variance in a column is large if it is comprised of small and large (in the absolute sense) values. In rotating the axes, rows with large initial loadings—i.e., with high communalities—have more influence than rows with only small values. Kaiser suggested that in the computation of the optimal rotation, all rows should have the same weight. This is usually known as the Horst normalization (Horst 1965). The option `horst` applies this normalization method for rotation.

```
. rotate, horst
```

Factor analysis/correlation   Number of obs   =   305
    Method: iterated principal factors   Retained factors =   2
    Rotation: orthogonal varimax (Horst on) Number of params =   15

Factor	Variance	Difference	Proportion	Cumulative
Factor1	3.31500	0.67075	0.5563	0.5563
Factor2	2.64425	.	0.4437	1.0000

LR test: independent vs. saturated: chi2(28) = 2092.68 Prob>chi2 = 0.0000

Rotated factor loadings (pattern matrix) and unique variances

Variable	Factor1	Factor2	Uniqueness
height	0.8724	0.2775	0.1620
arm_span	0.9203	0.2045	0.1112
fore_arm	0.8874	0.1815	0.1795
lower_leg	0.8638	0.2478	0.1923
weight	0.2332	0.9137	0.1108
bitrod	0.1885	0.7775	0.3600
ch_girth	0.1292	0.7526	0.4169
ch_width	0.2581	0.6522	0.5081

Factor rotation matrix

	Factor1	Factor2
Factor1	0.7837	0.6212
Factor2	-0.6212	0.7837

In this case, the raw and normalized varimax rotated loadings are not very different.

◁

In the first example, `loadingplot` after `rotate` showed the rotated loadings, not the unrotated loadings. How can this be? Remember that Stata estimation commands store their results in e(), which we can list using `ereturn list`.

```
. ereturn list
scalars:
                 e(f) =  2
                 e(N) =  305
              e(df_m) =  15
              e(df_r) =  13
            e(chi2_i) =  2092.68137837692
              e(df_i) =  28
               e(p_i) =  0
             e(evsum) =  5.959224129627425
               e(r_f) =  2
macros:
       e(r_normalization) :  "horst"
               e(r_class) :  "orthogonal"
           e(r_criterion) :  "varimax"
             e(r_ctitle) :  "varimax"
                 e(cmd) :  "factor"
          e(properties) :  "nob noV eigen"
               e(title) :  "Factor analysis"
             e(predict) :  "factor_p"
          e(estat_cmd) :  "factor_estat"
         e(rotate_cmd) :  "factor_rotate"
             e(factors) :  "factors(2)"
             e(mtitle) :  "iterated principal factors"
             e(method) :  "ipf"
         e(matrixname) :  "R"
matrices:
                e(r_Ev) :  1 x 2
               e(r_Phi) :  2 x 2
                e(r_T) :  2 x 2
                e(r_L) :  8 x 2
                 e(C) :  8 x 8
               e(Phi) :  2 x 2
                 e(L) :  8 x 2
               e(Psi) :  1 x 8
                e(Ev) :  1 x 8
```

When you replay an estimation command, it simply knows where to look, so that it can redisplay the output. rotate does something that few other postestimation commands are allowed to do: It adds information to the estimation results computed by factor or pca. But in order to avoid confusion, it writes in e() fields with the prefix r_. For instance the matrix e(r_L) contains the rotated loadings.

If you replay factor after rotate, factor will display the rotated results. And this is what all factor and pca postestimation commands do. For instance, if you predict after rotate, predict will use the rotated results. Of course, it is still possible to operate on the unrotated results. factor, norotated replays the unrotated results. predict with the norotated option computes the factor scores for the unrotated results.

rotate only stores information about the most recent rotation, overwriting any information from the previous rotation. If you need the previous results again, run rotate with the respective options again; you do not need to run factor again. It is also possible to use estimates store to store estimation results for different rotations, which you may later restore and replay at will. See [R] **estimates** for details.

If you no longer need the rotation results, you may type

```
. rotate, clear
```

to clean up the rotation result and return the `factor` results back to their pristine state (as if `rotate` had never been called).

▷ Example 3

rotate provides many more orthogonal rotations. Previously we stated that the varimax rotation can be thought of as the rotation that maximizes the varimax criterion, namely the variance of the squared loadings summed over the columns. A column of loadings with a high variance tends to contain a series of large values and a series of low values, achieving the simplicity aim of factor analytic interpretation. The other types of rotation simply maximize other concepts of simplicity. For instance, the `quartimax` rotation aims at rowwise simplicity—preferably, the loadings within variables fall into a grouping of a few large ones and a few small ones, using again the variance in squared loadings as the criterion to be maximized.

```
. rotate, quartimax horst
```

Factor analysis/correlation			Number of obs	=	305
Method: iterated principal factors			Retained factors =		2
Rotation: orthogonal quartimax (Horst on)			Number of params =		15

Factor	Variance	Difference	Proportion	Cumulative
Factor1	3.32371	0.68818	0.5577	0.5577
Factor2	2.63553	.	0.4423	1.0000

LR test: independent vs. saturated: chi2(28) = 2092.68 Prob>chi2 = 0.0000

Rotated factor loadings (pattern matrix) and unique variances

Variable	Factor1	Factor2	Uniqueness
height	0.8732	0.2749	0.1620
arm_span	0.9210	0.2017	0.1112
fore_arm	0.8880	0.1788	0.1795
lower_leg	0.8646	0.2452	0.1923
weight	0.2360	0.9130	0.1108
bitrod	0.1909	0.7769	0.3600
ch_girth	0.1315	0.7522	0.4169
ch_width	0.2601	0.6514	0.5081

Factor rotation matrix

	Factor1	Factor2
Factor1	0.7855	0.6188
Factor2	-0.6188	0.7855

In this case, the quartimax and the varimax rotated results are rather similar. This need not be the case—varimax focuses on simplicity within columns (factors) and quartimax within rows (variables). It is possible to compromise, rotating to strive for a weighted sum of row simplicity and column simplicity. This is known as the orthogonal oblimin criterion; in the orthogonal case, `oblimin()` is equivalent to the Crawford–Ferguson (option `cf()`) family and also to the orthomax family. These are parameterized families of criteria with, for instance, the following special cases:

oblimin(0)	quartimax rotation
oblimin(0.5)	biquartimax rotation
oblimin(1)	varimax rotation

```
. rotate, oblimin(0.5) horst
```

Factor analysis/correlation	Number of obs	=	305
Method: iterated principal factors	Retained factors	=	2
Rotation: orthogonal oblimin (Horst on)	Number of params	=	15

Factor	Variance	Difference	Proportion	Cumulative
Factor1	3.31854	0.67783	0.5569	0.5569
Factor2	2.64071	.	0.4431	1.0000

LR test: independent vs. saturated: chi2(28) = 2092.68 Prob>chi2 = 0.0000

Rotated factor loadings (pattern matrix) and unique variances

Variable	Factor1	Factor2	Uniqueness
height	0.8727	0.2764	0.1620
arm_span	0.9206	0.2033	0.1112
fore_arm	0.8877	0.1804	0.1795
lower_leg	0.8642	0.2468	0.1923
weight	0.2343	0.9134	0.1108
bitrod	0.1895	0.7772	0.3600
ch_girth	0.1301	0.7525	0.4169
ch_width	0.2589	0.6518	0.5081

Factor rotation matrix

	Factor1	Factor2
Factor1	0.7844	0.6202
Factor2	-0.6202	0.7844

Since the varimax and orthomax rotation are relatively close, it will be of little surprise that the factor loadings resulting from an optimal rotation of a compromise criterion are close as well.

The orthogonal quartimax rotation may be obtained in different ways, namely directly, or by the appropriate member of the oblimin() or cf() families:

```
. rotate, quartimax
(output omitted )
. rotate, oblimin(0)
(output omitted )
. rotate, oblimin(0)
. rotate, cf(0)
(output omitted )
```

◁

❑ Technical Note

The orthogonal varimax rotation also belongs to the oblimin and Crawford–Ferguson families.

```
. rotate, varimax
(output omitted )
. rotate, oblimin(1)
(output omitted )
. rotate, cf(0.125)
(output omitted )
```

(the $0.125 = 1/8$ above is 1 divided by the number of variables). All three produce the orthogonal varimax rotation. (There is actually a fourth way, namely `rotate, vgpf`.) There is, however, a subtle difference in algorithms used. The varimax rotation as specified by the `varimax` option (which is also the default) is computed by the classic algorithm of cycling through rotations of two factors at a time. The other ways use the general "gradient projection" algorithm proposed by Jennrich; see [MV] **rotatemat** for additional information.

❑

Oblique rotations

In addition to orthogonal rotations, oblique rotations are also available.

▷ Example 4

The rotation methods that we have discussed so far are all orthogonal: the angles between the axes are unchanged, so the rotated factors are uncorrelated.

Returning to our original factor analysis

```
. factormat R, n(305) fac(2) ipf
  (output omitted )
```

we examine the correlation matrix of the common factors:

```
. estat common
```

Correlation matrix of the common factors

Factors	Factor1	Factor2
Factor1	1	
Factor2	0	1

and see that they are uncorrelated.

The indeterminancy in the factor analytic model, however, allows us to consider additional transformations of the common factors, namely oblique rotations. These are rotations of the axes that preserve the norms of the rows of the loadings, but not the angles between the axes or the angles between the rows. There are advantages and disadvantages of oblique rotations. See, for instance, Gorsuch (1983, chapter 9). In many substantive theories, there seems little reason to impose the restriction that the common factors be uncorrelated. The additional freedom in choosing the axes generally leads to more easily interpretable factors, sometimes to a great extent. However, while most researchers are willing to accept mildly correlated factors, they would prefer to use fewer of such factors.

`rotate` provides an extensive menu of oblique rotations; with a few exceptions, criteria suitable for orthogonal rotations are also suitable for oblique rotation. Again oblique rotation can be conceived of as maximizing some "simplicity" criterion. We illustrate with the oblimin oblique rotation.

```
. rotate, oblimin oblique horst
```

Factor analysis/correlation   Number of obs   =   305
   Method: iterated principal factors   Retained factors =   2
   Rotation: oblique oblimin (Horst on)   Number of params =   15

Factor	Variance	Proportion	Rotated factors are correlated
Factor1	3.95010	0.6629	
Factor2	3.35832	0.5635	

LR: independence vs saturated: Chi2(28) = 2092.68, Prob > chi2 = 0.0000

Rotated factor loadings (pattern matrix) and unique variances

Variable	Factor1	Factor2	Uniqueness
height	0.8831	0.0648	0.1620
arm_span	0.9560	-0.0288	0.1112
fore_arm	0.9262	-0.0450	0.1795
lower_leg	0.8819	0.0344	0.1923
weight	0.0047	0.9408	0.1108
bitrod	-0.0069	0.8032	0.3600
ch_girth	-0.0653	0.7923	0.4169
ch_width	0.1042	0.6462	0.5081

Factor rotation matrix

	Factor1	Factor2
Factor1	0.9112	0.7930
Factor2	-0.4120	0.6092

The oblique rotation yields a much "simpler" structure in the Thurstone (1935) sense than the orthogonal rotations. This time, the common factors are moderately correlated.

```
. estat common
```

Correlation matrix of the Oblimin(0) rotated common factors

Factors	Factor1	Factor2
Factor1	1	
Factor2	.4716	1

◁

❏ Technical Note

The numerical maximization of a simplicity criterion with respect to the class of orthogonal or oblique rotations proceeds in a stepwise method, making small improvements from an initial guess, until no more small improvements are possible. Such a procedure is not guaranteed to converge to the global optimum; but to a local optimum instead. In practice, we experience very few such problems. To some extent, this is because we have a very reasonable starting value using the unrotated factors or loadings. As a safeguard, Stata starts the improvement from multiple initial positions chosen at random from the classes of orthonormal and normal rotation matrices. If the maximization procedure converges to the same criterion value at each trial, we may be reasonably certain that we have found the global optimum. Let us illustrate.

```
. set seed 123
. rotate, oblimin oblique horst protect(10)
Trial   1 : min criterion    .0181657
Trial   2 : min criterion   46234.38
Trial   3 : min criterion    .0181657
Trial   4 : min criterion    .0181657
Trial   5 : min criterion    .0181657
Trial   6 : min criterion    .0181657
Trial   7 : min criterion    .0181657
Trial   8 : min criterion   1769.989
Trial   9 : min criterion   250205.6
Trial  10 : min criterion    .0181657
```

Factor analysis/correlation			Number of obs =	305
Method: iterated principal factors			Retained factors =	2
Rotation: oblique oblimin (Horst on)			Number of params =	15

Factor	Variance	Proportion	Rotated factors are correlated
Factor1	3.95010	0.6629	
Factor2	3.35832	0.5635	

LR test: independent vs. saturated: chi2(28) = 2092.68 Prob>chi2 = 0.0000

Rotated factor loadings (pattern matrix) and unique variances

Variable	Factor1	Factor2	Uniqueness
height	0.8831	0.0648	0.1620
arm_span	0.9560	-0.0288	0.1112
fore_arm	0.9262	-0.0450	0.1795
lower_leg	0.8819	0.0344	0.1923
weight	0.0047	0.9408	0.1108
bitrod	-0.0069	0.8032	0.3600
ch_girth	-0.0653	0.7923	0.4169
ch_width	0.1042	0.6462	0.5081

Factor rotation matrix

	Factor1	Factor2
Factor1	0.9112	0.7930
Factor2	-0.4120	0.6092

In this case, three of the random trials converged to distinct rotations from the rest. Specifying options log and trace would demonstrate that in these cases, the initial configurations were so far off that no improvements could be found. In a real application, we would likely rerun rotate with a higher number of trials, say protect(50), for additional reassurance.

❑

❑ Technical Note

There is an additional but almost trivial source of nonuniqueness. All simplicity criteria supported by rotate and rotatemat are invariant with respect to permutations of the rows and of the columns. Also, the signs of rotated loadings are undefined. rotatemat, the computational engine of rotate, makes sure that all columns have a positive orientation, i.e., have a positive sum. rotate, after factor and pca, also sorts the columns into decreasing order of explained variance.

❑

Additional types of rotation

rotate supports a few rotation methods that do not fit into the scheme of "simplicity maximization". The first of these methods is known as the target rotation, which seeks to rotate the factor loading matrix to approximate as much as possible a target matrix of the same size as the factor loading matrix.

▷ Example 5

We continue with our same example. If we had expected a factor loading structure in which the first group of four variables would load especially high on the first factor and the second group of four variables on the second factor, we could have set up the following target matrix.

```
. matrix W = ( 1,0 \ 1,0 \ 1,0 \ 1,0 \ 0,1 \ 0,1 \ 0,1 \ 0,1 )

. matrix list W

W[8,2]
    c1  c2
r1   1   0
r2   1   0
r3   1   0
r4   1   0
r5   0   1
r6   0   1
r7   0   1
r8   0   1
```

It is also possible to request an orthogonal or oblique rotation toward the target **W**.

```
. rotate, target(W) horst
```

Factor analysis/correlation

	Number of obs	=	305
Method: iterated principal factors	Retained factors	=	2
Rotation: orthogonal target (Horst on)	Number of params	=	15

Factor	Variance	Difference	Proportion	Cumulative
Factor1	3.30616	0.65307	0.5548	0.5548
Factor2	2.65309	.	0.4452	1.0000

LR test: independent vs. saturated: chi2(28) = 2092.68 Prob>chi2 = 0.0000

Rotated factor loadings (pattern matrix) and unique variances

Variable	Factor1	Factor2	Uniqueness
height	0.8715	0.2802	0.1620
arm_span	0.9197	0.2073	0.1112
fore_arm	0.8869	0.1843	0.1795
lower_leg	0.8631	0.2505	0.1923
weight	0.2304	0.9144	0.1108
bitrod	0.1861	0.7780	0.3600
ch_girth	0.1268	0.7530	0.4169
ch_width	0.2561	0.6530	0.5081

Factor rotation matrix

	Factor1	Factor2
Factor1	0.7817	0.6236
Factor2	-0.6236	0.7817

With this target matrix, the result is not far different from the varimax and other orthogonal rotations.

◁

▷ Example 6

For our last example, we return to the early days of factor analysis, the time before fast computing. Analytical methods for orthogonal rotation, such as varimax, were developed relatively early. Analogous methods for oblique rotations proved more complicated. Hendrickson and White (1964) proposed a computationally simple method to obtain an oblique rotation that is comprised of an oblique Procrustes rotation of the factor loadings toward a signed power of the varimax rotation of the factor loadings. The promax method has a single parameter, the power to which the varimax loadings are raised. Larger promax powers simplify the factor loadings (i.e., generate more zeros and ones) at the cost of more correlation between the common factors. Generally, we recommend that you keep the power in the range (1,4] and not restricted to integers. Specifying promax is equivalent to promax(3).

```
. rotate, promax horst
Factor analysis/correlation                    Number of obs    =      305
      Method: iterated principal factors       Retained factors =        2
      Rotation: oblique promax (Horst on)      Number of params =       15
```

Factor	Variance	Proportion	Rotated factors are correlated
Factor1	3.92727	0.6590	
Factor2	3.31295	0.5559	

```
    LR test: independent vs. saturated:  chi2(28) = 2092.68 Prob>chi2 = 0.0000
Rotated factor loadings (pattern matrix) and unique variances
```

Variable	Factor1	Factor2	Uniqueness
height	0.8797	0.0744	0.1620
arm_span	0.9505	-0.0176	0.1112
fore_arm	0.9205	-0.0340	0.1795
lower_leg	0.8780	0.0443	0.1923
weight	0.0214	0.9332	0.1108
bitrod	0.0074	0.7966	0.3600
ch_girth	-0.0509	0.7851	0.4169
ch_width	0.1152	0.6422	0.5081

Factor rotation matrix

	Factor1	Factor2
Factor1	0.9069	0.7832
Factor2	-0.4214	0.6218

In this simple two-factor example, the promax solution is similar to the oblique oblimin solution.

◁

Saved Results

rotate is implemented as an ado-file. It adds fields (macros, scalars, matrices) named e(*r_name*) to e(), overwriting any such fields that may already be defined. The actual fields defined by rotate will vary between the estimation command (factor or pca) after which rotate was invoked. See [MV] **factor postestimation** and [MV] **pca postestimation** for details.

❑ Technical Note

The remainder of this section contains information of interest to programmers who want to provide rotate support to other estimation commands. Similar to other postestimation commands, such as estat and predict, rotate invokes a handler command. The name of this command is extracted from the field e(rotate_cmd). The estimation command *cmd* should set this field appropriately. For instance, pca sets the macro e(rotate_cmd) to pca_rotate. The command pca_rotate implements rotation after pca and pcamat, using rotatemat as the computational engine. pca_rotate does not display output itself; it relies on pca to do so.

For consistent behavior for end users and programmers alike, we recommend that the estimation command *cmd*, the driver commands, and other postestimation commands adhere to the following guidelines:

Driver command

- The rotate driver command for *cmd* should be named *cmd*_rotate.

- *cmd*_rotate should be an e-class command, i.e., returning in e().

- Make sure that *cmd*_rotate is invoked after the correct estimation command (e.g., if "`e(cmd)`'" != "pca" ...).

- Allow at least the option <u>det</u>ail and any option available to rotatemat.

- Extract from e() the matrix you want to rotate; invoke rotatemat on the matrix; and run this command quietly (i.e., suppress all output) unless the option detail was specified.

- Extract the r() objects returned by rotatemat; see *Methods and Formulas* of [MV] **rotatemat** for details.

- Compute derived results needed for your estimator.

- Store in e() fields (macros, scalars, matrices) named *r_name*, adding to the existing e() fields.

  Store the macros returned by rotatemat under the same named prefixed with r_. In particular, the macro e(r_criterion) should be set to the name of the rotation criterion returned by rotatemat as r(criterion). Other commands can check this field to find out whether rotation results are available.

  We suggest that only the most recent rotation results be stored, overwriting any existing e(r_*) results. The programmer command _rotate_clear clears any existing r_* fields from e().

- Display the rotation results by replaying *cmd*.

Estimation command cmd

- In *cmd*, define e(rotate_cmd) to *cmd*_rotate.

- *cmd* should be able to display the rotated results and should default to do so if rotated results are available. Include an option no<u>ROT</u>ated to display the unrotated results.

- You may use the programmer command _rotate_text to obtain a standard descriptive text for the rotation method.

Other postestimation commands

- Other postestimation commands after *cmd* should operate on the rotated results whenever they are appropriate and available, unless the option noROTated specifies otherwise.

- Mention that you operate on the unrotated results only if rotated results are available, but the user or you as the programmer decided not to use them.

❏

Methods and Formulas

See *Methods and Formulas* of [MV] **rotatemat**.

References

Bentler, P. M. 1977. Factor simplicity index and transformations. *Psychometrika* 42: 277–295.

Comrey, A. L. 1967. Tandem criteria for analytic rotation in factor analysis. *Psychometrika* 32: 277–295.

Crawford, C. B. and G. A. Ferguson. 1970. A general rotation criterion and its use in orthogonal rotation. *Psychometrika* 35: 321–332.

Gorsuch, R. L. 1983. *Factor Analysis.* 2nd ed. Hillsdale, NJ: Lawrence Erlbaum.

Harman, H. H. 1976. *Modern Factor Analysis.* 3rd ed. Chicago: University of Chicago Press.

Hendrickson, A. E. and P. O. White. 1964. Promax: A quick method for rotation to oblique simple structure. *British Journal of Statistical Psychology* 17: 65–70.

Horst, P. 1965. *Factor Analysis of Data Matrices.* New York: Holt, Rinehart, & Winston.

Jennrich, R. I. 1979. Admissible values of γ in direct oblimin rotation. *Psychometrika* 44: 173–177.

———. 2004. Rotation to simple loadings using component loss functions: the orthogonal case. *Psychometrika.* 69: 257–273.

Kaiser, H. F. 1958. The varimax criterion for analytic rotation in factor analysis. *Psychometrika* 23: 187–200.

Thurstone, L. L. 1935. *The Vectors of Mind.* Chicago: University of Chicago Press.

For additional references, see [MV] **rotatemat**.

Also See

Complementary: [MV] **factor**, [MV] **factor postestimation**, [MV] **pca**, [MV] **pca postestimation**

Related: [MV] **procrustes**

Background: [MV] **rotatemat**

Title

rotatemat — Orthogonal and oblique rotation of a Stata matrix

Syntax

rotatemat *matrix_L* [, *options*]

options	description
Main	
orthogonal	restrict to orthogonal rotations; the default, except with promax()
oblique	allow oblique rotations
rotation_methods	rotation criterion
normalize	rotate Horst normalized matrix
horst	synonym for normalize
Reporting	
format(%*fmt*)	display format for matrices; default is format(%9.5f)
blanks(#)	display numbers as blanks when |number| < #; default is blanks(0)
nodisplay	suppress all output except log and trace
noloading	suppress display of rotated loadings
norotation	suppress display of rotation matrix
matname(*str*)	descriptive label of the matrix to be rotated
colnames(*str*)	descriptive name for columns of the matrix to be rotated
Opt options	
optimize_options	control the optimization process; seldom used

(Continued on next page)

445

rotation_methods	description
*varimax	varimax (orthogonal only); the default
vgpf	varimax via the GPF algorithm (orthogonal only)
quartimax	quartimax (orthogonal only)
equamax	equamax (orthogonal only)
parsimax	parsimax (orthogonal only)
entropy	minimum entropy (orthogonal only)
tandem1	Comrey's tandem 1 principle (orthogonal only)
tandem2	Comrey's tandem 2 principle (orthogonal only)
*promax$\left[(\#)\right]$	promax power # (implies oblique); default is promax(3)
oblimin$\left[(\#)\right]$	oblimin with $\gamma = \#$; default is oblimin(0)
cf(#)	Crawford–Ferguson family with $\kappa = \#$, $0 \leq \# \leq 1$
bentler	Bentler's invariant pattern simplicity
oblimax	oblimax
quartimin	quartimin
target(Tg)	rotate toward matrix Tg
partial(Tg W)	rotate toward matrix Tg, weighted by matrix W

* varimax and promax ignore all *optimize_options*.

Description

rotatemat applies a linear transformation **T** to the matrix *matrix_L*, which we will call **A**, so that the result $c(\mathbf{A}(\mathbf{T}')^{-1})$ maximizes some criterion function $c()$ over all matrices **T** in a class of feasible transformations. Two classes are supported: orthogonal (orthonormal) and oblique. Orthonormal rotations comprise all orthonormal matrices **T**, such that $\mathbf{T}'\mathbf{T} = \mathbf{T}\mathbf{T}' = \mathbf{I}$; in this case $\mathbf{A}(\mathbf{T}')^{-1}$ simplifies to **AT**. Oblique rotations are characterized by $\text{diag}(\mathbf{T}'\mathbf{T}) = \mathbf{1}$. A wide variety of criteria $c()$ is available, representing different ways to measure the "simplicity" of a matrix. Most of these criteria can be applied with both orthogonal and oblique rotations.

If you are interested in rotation after factor, factormat, pca, or pcamat, see [MV] **factor postestimation**, [MV] **pca postestimation**, and the general description of rotate as a postestimation facility in [MV] **rotate**.

This entry describes the computation engine for orthogonal and oblique transformations of Stata matrices. This command may be used directly on any Stata matrix.

Options

    ⌐ Main ⌐

orthogonal specifies that an orthogonal rotation be applied. This is the default.

    See *Rotation criteria* below for details on the *rotation_methods* available with orthogonal.

oblique specifies that an oblique rotation be applied. This often yields more interpretable factors with a simpler structure than that obtained with an orthogonal rotation. In many applications (e.g., after factor and pca), the factors before rotation are orthogonal (uncorrelated), while the oblique rotated factors are correlated.

    See *Rotation criteria* below for details on the *rotation_methods* available with oblique.

normalize, and synonym horst, request that the rotation be applied to the Horst (1965) normalization of the matrix **A** so that the rowwise sums of squares equal 1.

format(%*fmt*) specifies the display format for matrices. The default is format(%9.5f).

blanks(#) specifies that small values of the rotated matrix—i.e., those elements of $\mathbf{A}(\mathbf{T}')^{-1}$ that are less than # in absolute value—are displayed as spaces.

nodisplay suppresses all output except the log and trace.

noloading suppresses the display of the rotated loadings.

norotation suppresses the display of the optimal rotation matrix.

matname(*str*) is a rarely used output option; it specifies a descriptive label of the matrix to be rotated.

colnames(*str*) is a rarely used output option; it specifies a descriptive name to refer to the columns of the matrix to be rotated. For instance, colnames(components) specifies that the output label the columns as "components". The default is "factors".

optimize_options control the iterative optimization process. These options are seldom used.

iterate(#) is a rarely used option; it specifies the maximum number of iterations. The default is iterate(1000).

log specifies that an iteration log be displayed.

trace is a rarely used option; it specifies that the rotation be displayed at each iteration.

tolerance(#) is one of three criteria for declaring convergence and is rarely used. The tolerance() convergence criterion is satisfied when the relative change in the rotation matrix **T** from one iteration to the next is less than or equal to #. The default is tolerance(1e-6).

gtolerance(#) is one of three criteria for declaring convergence and is rarely used. The gtolerance() convergence criterion is satisfied when the Frobenius norm of the gradient of the criterion function $c()$ projected on the manifold of orthogonal matrices or of normal matrices is less than or equal to #. The default is gtolerance(1e-6).

ltolerance(#) is one of three criteria for declaring convergence and is rarely used. The ltolerance() convergence criterion is satisfied when the relative change in the minimization criterion $c()$ from one iteration to the next is less than or equal to #. The default is ltolerance(1e-6).

protect(#) requests that # optimizations with random starting values be performed and that the best of the solutions be reported. The output also indicates whether all starting values converged to the same solution. When specified with a large number, such as protect(50), this provides reasonable assurance that the solution found is the global maximum and not just a local maximum. If trace is also specified, the rotation matrix and rotation criterion value of each optimization will be reported.

maxstep(#) is a rarely used option; it specifies the maximum number of step-size halvings. The default is maxstep(20).

init(*matname*) is a rarely used option; it specifies the initial rotation matrix. *matname* should be square and regular (nonsingular) and have the same number of columns as the matrix *matrix_L* to be rotated. It should be orthogonal ($\mathbf{T}'\mathbf{T} = \mathbf{T}\mathbf{T}' = \mathbf{I}$) or normal (diag($\mathbf{T}'\mathbf{T}$) = $\mathbf{1}$), depending on whether orthogonal or oblique rotations are performed. init() can not be combined with random. If neither init() nor random is specified, the identity matrix is used as the initial rotation.

random is a rarely used option; it specifies that a random orthogonal or random normal matrix be used as the initial rotation matrix. random cannot be combined with init(). If neither init() nor random is specified, the identity matrix is used as the initial rotation.

Rotation criteria

In the descriptions below, the matrix to be rotated is denoted as $\mathbf{A}$, p denotes the number of rows of $\mathbf{A}$, and f denotes the number of columns of $\mathbf{A}$ (factors or components). If $\mathbf{A}$ is a loading matrix from factor or pca, p is the number of variables, and f is the number of factors or components.

Criteria suitable only for orthogonal rotations

varimax and vgpf apply the orthogonal varimax rotation (Kaiser 1958). varimax maximizes the variance of the squared loadings within factors (columns of $\mathbf{A}$). It is equivalent to cf($1/p$) and to oblimin(1). varimax, the most popular rotation, is implemented with a dedicated fast algorithm and ignores all *optimize_options*. Specify vgpf to switch to the general GPF algorithm used for the other criteria.

quartimax uses the quartimax criterion (Harman 1976). quartimax maximizes the variance of the squared loadings within the variables (rows of $\mathbf{A}$). For orthogonal rotations, quartimax is equivalent to cf(0) and to oblimax.

equamax specifies the orthogonal equamax rotation. equamax maximizes a weighted sum of the varimax and quartimax criteria, reflecting a concern for simple structure within variables (rows of $\mathbf{A}$) as well as within factors (columns of $\mathbf{A}$). equamax is equivalent to oblimin($p/2$) and cf(#), where $\# = f/(2p)$.

parsimax specifies the orthogonal parsimax rotation. parsimax is equivalent to cf(#), where $\# = (f-1)/(p+f-2)$.

entropy applies the minimum entropy rotation criterion (Jennrich 2004).

tandem1 specifies that the first principle of Comrey's tandem be applied. According to Comrey (1967), this principle should be used to judge which "small" factors be dropped.

tandem2 specifies that the second principle of Comrey's tandem be applied. According to Comrey (1967), tandem2 should be used for "polishing".

Criteria suitable only for oblique rotations

promax$\left[(\#)\right]$ specifies the oblique promax rotation. The optional argument specifies the promax power. Not specifying the argument is equivalent to specifying promax(3). Values less than 4 are recommended, but the choice is yours. Larger promax powers simplify the loadings (generate numbers closer to zero and one) but at the cost of additional correlation between factors. Choosing a value is a matter of trial and error, but most sources find values in excess of 4 undesirable in practice. The power must be greater than 1 but is not restricted to integers.

Promax rotation is an oblique rotation method that was developed before the "analytical methods" (based on criterion optimization) became computationally feasible. Promax rotation is comprised of an oblique Procrustean rotation of the original loadings $\mathbf{A}$ toward the element-wise #-power of the orthogonal varimax rotation of $\mathbf{A}$.

Criteria suitable for orthogonal and oblique rotations

`oblimin`$\left[(\#)\right]$ specifies that the oblimin criterion with $\gamma = \#$ be used. When restricted to orthogonal transformations, the `oblimin()` family is equivalent to the orthomax criterion function. Special cases of `oblimin()` include

γ	special cases
0	quartimax / quartimin
1/2	biquartimax / biquartimin
1	varimax / covarimin
$p/2$	equamax

p = number of rows of $\mathbf{A}$

γ defaults to zero. Jennrich (1979) recommends $\gamma \leq 0$ for oblique rotations. For $\gamma > 0$, it is possible that optimal oblique rotations do not exist; the iterative procedure used to compute the solution will wander off to a degenerate solution.

`cf(#)` specifies that a criterion from the Crawford–Ferguson (1970) family be used with $\kappa = \#$. `cf($\kappa$)` can be seen as $(1 - \kappa)\mathrm{cf}_1(\mathbf{A}) + (\kappa)\mathrm{cf}_2(\mathbf{A})$, where $\mathrm{cf}_1(\mathbf{A})$ is a measure of row parsimony and $\mathrm{cf}_2(\mathbf{A})$ is a measure of column parsimony. $\mathrm{cf}_1(\mathbf{A})$ attains its greatest lower bound when no row of $\mathbf{A}$ has more than one nonzero element, while $\mathrm{cf}_2(\mathbf{A})$ reaches zero if no column of $\mathbf{A}$ has more than one nonzero element.

For orthogonal rotations, the Crawford–Ferguson family is equivalent to the `oblimin()` family. For orthogonal rotations, special cases include the following:

κ	special cases
0	quartimax / quartimin
$1/p$	varimax / covarimin
$f/(2p)$	equamax
$(f - 1)/(p + f - 2)$	parsimax
1	factor parsimony

p = number of rows of $\mathbf{A}$
f = number of columns of $\mathbf{A}$

`bentler` specifies that the "invariant pattern simplicity" criterion (Bentler 1977) be used.

`oblimax` specifies the oblimax criterion, which maximizes the number of high and low loadings. `oblimax` is equivalent to `quartimax` for orthogonal rotations.

`quartimin` specifies that the quartimin criterion be used. For orthogonal rotations, `quartimin` is equivalent to `quartimax`.

`target`(Tg) specifies that $\mathbf{A}$ be rotated as near as possible to the conformable matrix Tg. Nearness is expressed by the Frobenius matrix norm.

`partial`($Tg\ W$) specifies that $\mathbf{A}$ be rotated as near as possible to the conformable matrix Tg. Nearness is expressed by a weighted (by W) Frobenius matrix norm. W should be non-negative and usually is zero–one valued, with ones identifying the target values to be reproduced as closely as possible by the factor loadings, and zeros identify loadings to remain unrestricted.

Remarks

Remarks are presented under the headings

Introduction
Orthogonal rotations
Oblique rotations
Promax rotation

Introduction

For an introduction to rotation, see Harman (1976) and Gorsuch (1983).

All supported rotation criteria are invariant with respect to permutations of the columns and change of signs of the columns. `rotatemat` returns the solution with positive column sums and with columns sorted by the L2 norm; columns are ordered with respect to the L1 norm if the columns have the same L2 norm.

A factor analysis of 24 psychological tests on 145 seventh and eighth grade school children with four retained factors is used for illustration. Factors were extracted with maximum likelihood. The loadings are reported by Harman (1976). We enter the factor loadings as a Stata matrix with 24 rows and 4 columns. For additional information, we add full descriptive labels as comments and short labels as row names.

```
. matrix input L = (
    601    019    388    221 \       Visual perception
    372   -025    252    132 \       Cubes
    413   -117    388    144 \       Paper form board
    487   -100    254    192 \       Flags
    691   -304   -279    035 \       General information
    690   -409   -200   -076 \       Paragraph comprehension
    677   -409   -292    084 \       Sentence completion
    674   -189   -099    122 \       Word classification
    697   -454   -212   -080 \       Word meaning
    476    534   -486    092 \       Addition
    558    332   -142   -090 \       Code
    472    508   -139    256 \       Counting dots
    602    244    028    295 \       Straight-curved capitals
    423    058    015   -415 \       Word recognition
    394    089    097   -362 \       Number recognition
    510    095    347   -249 \       Figure recognition
    466    197   -004   -381 \       Object-number
    515    312    152   -147 \       Number-figure
    443    089    109   -150 \       Figure-word
    614   -118    126   -038 \       Deduction
    589    227    057    123 \       Numerical puzzles
    608   -107    127   -038 \       Problem reasoning
    687   -044    138    098 \       Series completion
    651    177   -212   -017 )       Arithmetic problems
. matrix colnames L = F1 F2 F3 F4

. matrix rownames L = visual      cubes      board
                      flags       general    paragraph
                      sentence    wordclas   wordmean
                      add         code       dots
                      capitals    wordrec    numbrec
                      figrec      obj-num    num-fig
                      fig-word    deduct     numpuzz
                      reason      series     arith

. matrix L = L/1000
```

Thus using `rotatemat`, we can study various rotations of L without access to the full data or the correlation matrix.

Orthogonal rotations

We can rotate the matrix L according to an extensive list of criteria, including orthogonal rotations.

▷ Example 1

The default rotation, orthogonal varimax, is probably the most popular method:

```
. rotatemat L, format(%6.3f)

Rotation of L[24,4]

    Criterion              varimax
    Rotation class         orthogonal
    Horst normalization    off

Rotated factors
```

	F1	F2	F3	F4
visual	0.247	0.151	0.679	0.128
cubes	0.171	0.060	0.425	0.078
board	0.206	-0.049	0.549	0.097
flags	0.295	0.068	0.504	0.050
general	0.765	0.214	0.117	0.067
paragraph	0.802	0.074	0.122	0.160
sentence	0.826	0.148	0.117	-0.008
wordclas	0.612	0.230	0.290	0.061
wordmean	0.840	0.049	0.112	0.152
add	0.166	0.846	-0.076	0.082
code	0.222	0.533	0.134	0.313
dots	0.048	0.705	0.257	0.025
capitals	0.240	0.500	0.450	0.020
wordrec	0.249	0.124	0.032	0.526
numbrec	0.178	0.109	0.106	0.499
figrec	0.158	0.076	0.401	0.510
obj-num	0.197	0.262	0.060	0.539
num-fig	0.096	0.352	0.311	0.422
fig-word	0.204	0.175	0.232	0.336
deduct	0.443	0.115	0.365	0.255
numpuzz	0.233	0.428	0.389	0.169
reason	0.432	0.120	0.363	0.256
series	0.440	0.228	0.472	0.184
arith	0.409	0.509	0.150	0.228

```
Orthogonal rotation
```

	F1	F2	F3	F4
F1	0.677	0.438	0.475	0.352
F2	-0.632	0.737	0.049	0.232
F3	-0.376	-0.458	0.760	0.268
F4	-0.011	0.234	0.441	-0.866

◁

The varimax rotation $\mathbf{T}$ of $\mathbf{A}$ maximizes the (raw) varimax criterion over all orthogonal $\mathbf{T}$, which in the case of $p \times f$ matrices is defined as (Harman 1976)

$$c_{\text{varimax}}(\mathbf{A}) = \frac{1}{p}\sum_{j=1}^{f}\left\{\left(\sum_{i=1}^{p}a_{ij}^4\right) - \frac{1}{p}\left(\sum_{i=1}^{p}a_{ij}^2\right)^2\right\}$$

The criterion $c_{\text{varimax}}(\mathbf{A})$ can be interpreted as the sum over the columns of the variances of the squares of the loadings a_{ij}. A column with large variance will typically consist of many small values and a few large values. Achieving such "simple" columnwise distributions is often helpful for interpretation purposes.

❏ Technical Note

The raw varimax criterion as defined here has been criticized because it weights variables by the size of their loadings, i.e., by their communalities. This is often not desirable. A common rotation strategy is to weight all rows equally by rescaling to the same rowwise sum of squared loadings. This is known as the Horst normalization. You may request this normalized solution with the `horst` option. The default in `rotatemat` and also in `rotate` (see [MV] **rotate**) is not to normalize.

❏

Many other criteria for the rotation of matrices have been proposed and studied in the literature. Most of these criteria can be stated in terms of a "simplicity function". For instance, quartimax rotation (Carroll 1953) seeks to achieve interpretation within rows—in a factor analytic setup, this means that variables should have a high loading on a few factors and a low loading on the other factors. The quartimax criterion is defined as (Harman 1976)

$$c_{\text{quartimax}}(\mathbf{A}) = \left(\frac{1}{pf}\sum_{i=1}^{p}\sum_{j=1}^{f}a_{ij}^4\right) - \left(\frac{1}{pf}\sum_{i=1}^{p}\sum_{j=1}^{f}a_{ij}^2\right)^2$$

▷ Example 2

We display the quartimax solution, use blanks to represent loadings with absolute values smaller than 0.3, and suppress the display of the rotation matrix.

```
. rotatemat L, quartimax format(%6.3f) norotation blanks(0.3)
```

Rotation of L[24,4]

     Criterion   quartimax
     Rotation class   orthogonal
     Horst normalization   off
     Criterion value   -1.032898
     Number of iterations   35

Rotated factors (blanks represent abs()<.3)

	F1	F2	F3	F4
visual	0.374		0.630	
cubes			0.393	
board			0.513	
flags	0.379		0.450	
general	0.791			
paragraph	0.827			
sentence	0.838			
wordclas	0.669			
wordmean	0.860			
add		0.829		
code	0.316	0.521		
dots		0.701		
capitals	0.348	0.482	0.393	
wordrec	0.316			0.492
numbrec				0.469
figrec			0.382	0.470
obj-num				0.503
num-fig		0.357		0.383
fig-word				
deduct	0.528			
numpuzz	0.342	0.414	0.340	
reason	0.517			
series	0.543		0.395	
arith	0.490	0.478		

◁

Some of the criteria supported by rotatemat are defined as one-parameter families. The oblimin(γ) criterion and the Crawford and Ferguson cf(κ) criterion families contain the varimax and quartimax criteria as special cases; i.e., they can be obtained by certain values of γ and κ respectively. Intermediate parameter values provide compromises between varimax's aim of column simplification and quartimax's aim of row simplification. Varimax and quartimax are equivalent to oblimin(1) and oblimin(0) respectively. A compromise, oblimin(0.5), is also known as biquartimax.

▷ Example 3

Since the varimax and quartimax solutions are so close for our matrix L, the biquartimax compromise will also be rather close.

```
. rotatemat L, oblimin(0.5) format(%6.3f) norotation
(output omitted )
```

◁

❑ Technical Note

You may have noticed a difference between the output of rotatemat in the default case or equivalently when we type

        . rotatemat L, varimax

and in other cases. In the default case, no mention is made of the criterion value and the number of iterations. The reason is that rotatemat uses a fast special algorithm for this most common case, while for other rotations it uses a general gradient projection algorithm (GPF) proposed by Jennrich (2001, 2002); see also Bernaards and Jennrich (forthcoming). The general algorithm is used to obtain the varimax rotation if you specify the option vgpf rather than varimax.

                                                                                        ❑

The rotations we have illustrated are orthogonal—the lengths of the rows and the angles between the rows are not affected by the rotations. We may verify—we do not show this in the manual to conserve paper—that after an orthogonal rotation of L

        . matlist L*L'

and

        . matlist r(AT)*r(AT)'

return the same 24 by 24 matrix, while

        . matlist r(T)*r(T)'

and

        . matlist r(T)'*r(T)

both return a 2 by 2 identity matrix. Note that rotatemat returns in r(AT) the rotated matrix and in r(T) the rotation matrix.

Oblique rotations

rotatemat provides a second class of rotations: oblique rotations. These rotations maintain the norms of the rows of the matrix but not their inner products. In geometric terms, interpreting the rows of the matrix to be rotated as vectors, both the orthogonal and the oblique rotations maintain the lengths of the vectors. Under orthogonal transformations, the angles between the vectors are also left unchanged—these transformations are comprised of true re-orientations in space and reflections. Oblique rotations do not conserve angles between vectors. If the vectors are orthogonal before rotations—as will be the case if we are rotating factor or component loading matrices—this will no longer be the case after the rotation. The "freedom" to select angles between the rows allows oblique rotations to generate simpler loading structures than the orthogonal rotations—sometimes much simpler. In a factor analytic setting, the disadvantage is, however, that the rotated factors are correlated.

rotatemat can obtain oblique rotations for most of the criteria that are available for orthogonal rotations; some of the criteria (such as the entropy criterion) are only available for the orthogonal case.

▷ Example 4

We illustrate with the psychological tests matrix L and apply the oblique oblimin criterion.

```
. rotatemat L, oblimin oblique format(%6.3f) blanks(0.3)
```

Rotation of L[24,4]

Criterion	oblimin(0)
Rotation class	oblique
Horst normalization	off
Criterion value	.1957363
Number of iterations	78

Rotated factors (blanks represent abs()<.3)

	F1	F2	F3	F4
visual		0.686		
cubes		0.430		
board		0.564		
flags		0.507		
general	0.771			
paragraph	0.808			
sentence	0.865			
wordclas	0.560			
wordmean	0.857			
add			0.864	
code			0.460	0.305
dots			0.701	
capitals		0.437	0.442	
wordrec				0.571
numbrec				0.543
figrec		0.314		0.540
obj-num				0.584
num-fig				0.438
fig-word				0.341
deduct	0.325			
numpuzz		0.344	0.347	
reason	0.311			
series		0.417		
arith			0.428	

Oblique rotation

	F1	F2	F3	F4
F1	0.823	0.715	0.584	0.699
F2	-0.483	0.019	0.651	0.213
F3	-0.299	0.587	-0.435	0.207
F4	-0.006	0.379	0.213	-0.651

The option `oblique` requested an oblique rotation rather than the default `orthogonal`. You may verify that `r(AT)` equals L * inv(r(T)') within reasonable roundoff with

```
. matlist r(AT) - inv(r(T)')
  (output omitted )
```

The correlation between the rotated dimensions is easily obtained.

```
. matlist r(T)' * r(T)
```

	F1	F2	F3	F4
F1	1			
F2	.4026978	1		
F3	.294928	.2555824	1	
F4	.4146879	.3784689	.3183115	1

◁

Promax rotation

rotatemat also offers promax rotation.

▷ Example 5

We use the matrix L to illustrate promax rotation.

```
. rotatemat L, promax blanks(0.3) format(%6.3f)
Rotation of L[24,4]
    Criterion              promax(3)
    Rotation class         oblique
    Horst normalization    off
Rotated factors (blanks represent abs()<.3)
```

	F1	F2	F3	F4
visual		0.775		
cubes		0.487		
board		0.647		
flags		0.572		
general	0.786			
paragraph	0.825			
sentence	0.888			
wordclas	0.543			
wordmean	0.878			
add			0.921	
code			0.466	
dots			0.728	
capitals		0.468	0.441	
wordrec				0.606
numbrec				0.570
figrec		0.364		0.539
obj-num				0.610
num-fig				0.425
fig-word				0.337
deduct		0.323		
numpuzz		0.369	0.336	
reason		0.322		
series		0.462		
arith			0.436	

Oblique rotation

	F1	F2	F3	F4
F1	0.841	0.829	0.663	0.743
F2	-0.462	0.020	0.614	0.215
F3	-0.282	0.478	-0.386	0.159
F4	-0.012	0.290	0.184	-0.614

The correlation between the rotated dimensions can be obtained as

. matlist r(T)' * r(T)

	F1	F2	F3	F4
F1	1			
F2	.5491588	1		
F3	.3807942	.4302401	1	
F4	.4877064	.5178414	.4505817	1

◁

Saved Results

rotatemat saves in r():

Scalars
    r(f)   criterion value
    r(iter)   number of GPF iterations
    r(rc)   return code
    r(nnconv)   number of nonconvergent trials; protect() only

Macros
    r(ctitle)   descriptive label of rotation method
    r(ctitle12)   version of r(ctitle) at most 12 characters long
    r(criterion)   criterion name (e.g., oblimin)
    r(class)   orthogonal or oblique
    r(normalization) horst or none
    r(carg)   criterion argument
    r(cmd)   rotatemat

Matrices
    r(T)   optimal transformation $\mathbf{T}$
    r(AT)   optimal $\mathbf{AT} = \mathbf{A}(\mathbf{T}')^{-1}$
    r(fmin)   minimums found; protect() only

Methods and Formulas

rotatemat is implemented as an ado-file.

rotatemat minimizes a scalar-valued criterion function $c(\mathbf{AT})$ with respect to the set of orthogonal matrices $\mathbf{T}'\mathbf{T} = \mathbf{I}$, or $c(\mathbf{A}(\mathbf{T}')^{-1})$ with respect to the normal matrix, $\mathrm{diag}(\mathbf{T}'\mathbf{T}) = \mathbf{1}$. Note that for orthonormal $\mathbf{T}$, $\mathbf{T} = (\mathbf{T}')^{-1}$.

The rotation criteria can be conveniently written in terms of scalar valued functions; see Bernaards and Jennrich (forthcoming). Define the inner product $< \mathbf{A}, \mathbf{B} >= \text{trace}(\mathbf{A}'\mathbf{B})$. $|\mathbf{A}| = \sqrt{< \mathbf{A}, \mathbf{A} >}$ is called the Frobenius norm of the matrix $\mathbf{A}$. Let $\mathbf{\Lambda}$ be a $p \times k$ matrix. Denote by $\mathbf{X}^2$ the direct product $\mathbf{X} \cdot \mathbf{X}$. See Harman (1976) for information on many of the rotation criteria and references to the authors originally proposing the criteria. In some cases, we list an alternative reference. Our notation is similar to that of Bernaards and Jennrich (forthcoming).

`rotatemat` uses the iterative "gradient projection algorithm" (Jennrich 2001, 2002) for the optimization of the criterion over the permissible transformations. Different versions are provided for optimal orthogonal and oblique rotations; see Bernaards and Jennrich (forthcoming).

Varimax (orthogonal only)

Varimax is equivalent to oblimin with $\gamma = 1$ or to the Crawford–Ferguson family with $\kappa = 1/p$; see below.

Quartimax (orthogonal only)

$$c(\mathbf{\Lambda}) = \sum_i \sum_r \lambda_{ir}^4 = -\frac{1}{4} < \mathbf{\Lambda}^2, \mathbf{\Lambda}^2 >$$

Equamax (orthogonal only)

Equamax is equivalent to oblimin with $\gamma = p/2$ or to the Crawford–Ferguson family with $\kappa = f/(2p)$; see below.

Parsimax (orthogonal only)

Parsimax is equivalent to the Crawford–Ferguson family with $\kappa = (f-1)/(p+f-2)$; see below.

Entropy (orthogonal only); see Jennrich (2004)

$$c(\mathbf{\Lambda}) = -\frac{1}{2} < \mathbf{\Lambda}^2, \log \mathbf{\Lambda}^2 >$$

Tandem principal 1 (orthogonal only); see Comrey (1967)

$$c(\mathbf{\Lambda}) = - < \mathbf{\Lambda}^2, (\mathbf{\Lambda}\mathbf{\Lambda}')^2 \mathbf{\Lambda}^2 >$$

Tandem principal 2 (orthogonal only); see Comrey (1967)

$$c(\mathbf{\Lambda}) = < \mathbf{\Lambda}^2, \{\mathbf{1}\mathbf{1}' - (\mathbf{\Lambda}\mathbf{\Lambda}')^2\}\mathbf{\Lambda}^2 >$$

Promax (oblique only)

Promax does not fit in the maximizing-of-a-simplicity-criterion framework that is at the core of `rotatemat`. The promax method (Hendrickson and White 1964) was proposed before computing power became widely available. The promax rotation is comprised of three steps:

1. Perform an orthogonal rotation on $\mathbf{A}$; `rotatemat` uses varimax.

2. Raise the elements of the rotated matrix to some power, preserving the signs of the elements. Typically, the power is taken from the range [2,4]. This operation is meant to distinguish more clearly between small and large values.

3. The matrix from step 2 is used as the target for an oblique Procrustean rotation from the original matrix $\mathbf{A}$. The method to compute this rotation in promax is different from the method in the `procrustes` command (see [MV] **procrustes**). The latter produces the real least-squares oblique rotation; promax uses an approximation.

Oblimin; see Jennrich (1979)

$$c(\mathbf{\Lambda}) = \frac{1}{4} < \mathbf{\Lambda}^2, \{\mathbf{I} - (\gamma/p)\mathbf{1}\mathbf{1}'\}\mathbf{\Lambda}^2(\mathbf{1}\mathbf{1}' - \mathbf{I}) >$$

orthomax and oblimin are equivalent when restricted to orthogonal rotations. Special cases of `oblimin()` include the following:

γ	special cases
0	quartimin
1/2	biquartimin
$p/2$	equamax
1	varimax

Crawford and Ferguson (1970) family

$$c(\mathbf{\Lambda}) = \frac{1-\kappa}{4} < \mathbf{\Lambda}^2, \mathbf{\Lambda}^2(\mathbf{1}\mathbf{1}' - \mathbf{I}) > + \frac{\kappa}{4} < \mathbf{\Lambda}^2, (\mathbf{1}\mathbf{1}' - \mathbf{I})\mathbf{\Lambda}^2 >$$

When restricted to orthogonal transformations, `cf()` and `oblimin()` are in fact equivalent. Special cases of `cf()` include the following:

κ	special cases
0	quartimax
$1/p$	varimax
$f/(2p)$	equamax
$(f-1)/(p+f-2)$	parsimax
1	factor parsimony

Bentler's invariant pattern simplicity; see Bentler (1977)

$$c(\mathbf{\Lambda}) = \log[\det\{(\mathbf{\Lambda}^2)'\mathbf{\Lambda}^2\}] - \log(\det[\operatorname{diag}\{(\mathbf{\Lambda}^2)'\mathbf{\Lambda}^2\}])$$

Oblimax

$$c(\mathbf{\Lambda}) = -\log(<\mathbf{\Lambda}^2, \mathbf{\Lambda}^2>) + 2\log(<\mathbf{\Lambda}, \mathbf{\Lambda}>)$$

For orthogonal transformations, oblimax is equivalent to quartimax.

Quartimin

$$c(\mathbf{\Lambda}) = \sum_{r \neq s} \sum_{i} \lambda_{ir}^2 \lambda_{is}^2 = -\frac{1}{4} < \mathbf{\Lambda}^2, \mathbf{\Lambda}^2(\mathbf{11'} - \mathbf{I}) >$$

Target

$$c(\mathbf{\Lambda}) = \frac{1}{2}|\mathbf{\Lambda} - \mathbf{H}|^2$$

for given target matrix $\mathbf{H}$.

Partially specified target

$$c(\mathbf{\Lambda}) = |\mathbf{W} \cdot (\mathbf{\Lambda} - \mathbf{H})|^2$$

for given target matrix $\mathbf{H}$, non-negative weighting matrix $\mathbf{W}$ (usually zero–one valued) and with $\cdot$ denoting the direct product.

References

Bentler, P. M. 1977. Factor simplicity index and transformations. *Psychometrika* 42: 277–295.

Bernaards, C. A. and R. I. Jennrich. Forthcoming. Gradient projection algorithms for arbitrary rotation criteria in factor analysis. *Educational and Psychological Measurement*.

Carroll, J. B. 1953. An analytical solution for approximating simple structure in factor analysis. *Psychometrika* 18: 23–38.

Comrey, A. L. 1967. Tandem criteria for analytic rotation in factor analysis. *Psychometrika* 32: 277–295.

Crawford, C. B. and G. A. Ferguson. 1970. A general rotation criterion and its use in orthogonal rotation.. *Psychometrika* 35: 321–332.

Gorsuch, R. L. 1983. *Factor Analysis*. 2nd ed. Hillsdale, New Jersey: Lawrence Erlbaum.

Harman, H. H. 1976. *Modern Factor Analysis*. 3rd ed. Chicago: University of Chicago Press.

Hendrickson, A. E. and P. O. White. 1964. Promax: A quick method for rotation to oblique simple structure. *British Journal of Statistical Psychology* 17: 65–70.

Horst, P. 1965. *Factor Analysis of Data Matrices*. New York: Holt, Rinehart, & Winston.

Jennrich, R. I. 1979. Admissible values of γ in direct oblimin rotation. *Psychometrika* 44: 173–177.

——. 2001. A simple general procedure for orthogonal rotation. *Psychometrika* 66: 289–306.

——. 2002. A simple general method for oblique rotation. *Psychometrika* 67: 7–19.

——. 2004. Rotation to simple loadings using component loss functions: the orthogonal case. *Psychometrika* 69: 257–273.

Kaiser, H. F. 1958. The varimax criterion for analytic rotation in factor analysis. *Psychometrika* 23: 187–200.

Also See

Complementary:	[MV] **rotate**
Related:	[MV] **procrustes**

Title

scoreplot — Score and loading plots after factor and pca

Syntax

Plot score variables

scoreplot [, *options*]

Plot the loadings (factors or components)

loadingplot [, *options*]

options	description
Main	
factors(#)	number of factors/scores to be plotted; default is factors(2)
components(#)	synonym for factors()
norotated	use unrotated factors or scores, even if rotated results exist
marker_options	change look of markers (color, size, etc.)
marker_label_options	change look or position of marker labels
*maxlength(#)	abbreviate variable names to # characters; default is maxlength(12)
matrix	graph as a scatterplot matrix when multiple plots are produced
combined	graph as a combined graph when multiple plots are produced
graph_matrix_options	affect the rendition of the matrix graph
combine_options	affect the rendition of the combined graph
†scoreopt(*predict_opts*)	options for predict generating score variables
Y-Axis, X-Axis, Title, Caption, Overall	
twoway_options	any options other than by() documented in [G] *twoway_options*

*maxlength(#) is only allowed with loadingplot.

†scoreopt(*predict_opts*) is only allowed with scoreplot.

Description

scoreplot produces scatterplots of the score variables after factor, factormat, pca, or pcamat.

loadingplot produces scatterplots of the loadings (factors or components) after factor, factormat, pca, or pcamat.

Options

⌐ Main ⌐

factors(#) produces plots for all combinations of score variables up to #. # should not exceed the number of retained factors (components) and defaults to 2. components() is a synonym. No plot is produced with factors(1).

462

norotated specifies that the unrotated loadings (loadingplot) or the unrotated scores (scoreplot) be used. The default is to use the rotated loadings or scores from the last rotation, if any; see [MV] **rotate**.

marker_options affect the rendition of markers drawn at the plotted points, including their shape, size, color, and outline; see [G] *marker_options*.

marker_label_options specify if and how the markers are to be labeled; see [G] *marker_label_options*.

maxlength(#), an option used with loadingplot, causes the variable names (used as point markers) to be abbreviated to # characters. The abbrev() function performs the abbreviation, and if # is less than 5, it is treated as 5; see [D] **functions**.

matrix specifies that plots be produced using graph matrix. This is the default when 3 or more factors are specified. This option may not be used with combined.

combined specifies that plots be produced using graph combine. This option may not be used with matrix.

graph_matrix_options affect the rendition of the matrix plot; see [G] **graph matrix**.

combine_options affect the rendition of the combined plot; see [G] **graph combine**. *combine_options* may not be specified unless factors() is greater than 2.

scoreopt(*predict_opts*), an option used with scoreplot, specifies options for predict to generate the score variables. For example, after factor, scoreopt(bartlett) specifies that Bartlett scoring be applied.

Y-Axis, X-Axis, Title, Caption, Overall

twoway_options are any of the options documented in [G] *twoway_options*, excluding by(). These include options for titling the graph (see [G] *title_options*) and options for saving the graph to disk (see [G] *saving_option*).

Remarks

One of the main results from a principal component analysis or a factor analysis is a set of eigenvectors that are called components or factors. These are saved in what is called a loading matrix. pca, pcamat, factor, and factormat save the loading matrix in e(L). If there were p variables involved in the PCA or factor analysis, and f components or factors were retained, there will be p rows and f columns in the resulting loading matrix.

The columns of the loading matrix are in order of importance. For instance, with PCA, the first column of the loading matrix is the component that accounts for the most variance, the second column accounts for the next most variance, and so on.

In a loading plot, the values from one column of the loading matrix are plotted against the values from another column of the loading matrix. Of most interest is the plot of the first and second columns (the first and second components or factors), and this is what loadingplot produces by default. The rows of the loading matrix provide the points to be graphed. Variable names are automatically used as the marker labels for these points.

▷ Example 1

We use the Renaissance painters' data introduced in example 2 of [MV] **biplot**. There are four attribute variables recorded for ten painters. We examine the first two principal component loadings for this dataset.

```
. use http://www.stata-press.com/data/r9/renpainters
(Scores by Roger de Piles for Renaissance Painters)

. pca composition drawing colour expression
  (output omitted )

. loadingplot
```

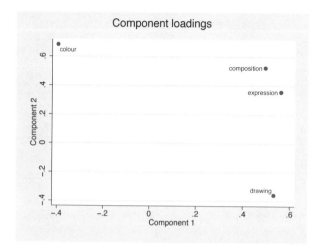

From the first component, we see that color (or colour, if you prefer) is separated from the other three attributes (variables): composition, expression, and drawing. From the second component, the difference in drawing stands out.

◁

Score plots approach the view of the loading matrix from the perspective of the observations. predict after pca and factor produces scores; see [MV] **pca postestimation** and [MV] **factor postestimation**. A score for an observation from a particular column of the loading matrix is obtained as the linear combination of that observation's data using the coefficients found in the loading. In other words, from the raw variables, the linear combinations described in the columns of the loading matrix are applied to generate new component or factor score variables. A score plot graphs one score variable against another. By default, scoreplot graphs the scores generated from the first and second columns of the loading matrix (the first two components or factors).

▷ Example 2

We continue with the PCA of the Renaissance painters.

```
. scoreplot
```

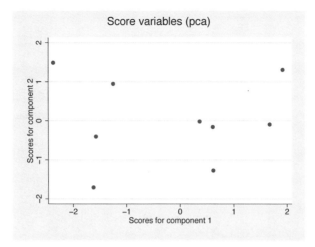

Unlike `loadingplot`, which can use the variable names as labels, `scoreplot` does not automatically know how to label the points. The graph above is not very helpful. The *marker_label_option* `mlabel()` takes care of this.

```
. scoreplot, mlabel(painter) aspect(1) xlabel(-2(1)3) ylabel(-2(1)3)
          title(Renaissance painters)
```

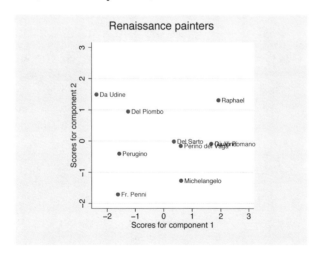

We added a few other options to improve the graph. We extended the axes to include the value 3 so that there would be room for the marker labels, imposed an aspect ratio of 1, and added a title.

The score plot gives us a feeling for the similarities and differences between the painters. Da Udine is in an opposite corner from Michelangelo. The other two corners have Fr. Penni and Raphael.

You could refer back to the loading plot and compare it with this score plot to see that the corner with Da Udine is most associated with the `colour` variable, while the corner with Michelangelo is best associated with `drawing`. Raphael is in the corner where the variables `composition` and `expression` are predominant.

If you like to make these kinds of associations between the variables and the observations, you will enjoy using the biplot command. It provides a joint view of the variables and observations; see [MV] **biplot**.

◁

After a rotation, the rotated factor or component loading matrix is saved in e(r_L). By default, both loadingplot and scoreplot work with the rotated loadings if they are available. The norotated option allows you to obtain the graphs for the unrotated loadings and scores.

You can also request a matrix or combined graph for all combinations of the first several components or factors with the components() option (or the alias factors()).

▷ Example 3

Even though the results from our initial look at the principal components of the Renaissance painters seem clear enough, we continue our demonstration by showing the score plots for the first three components after a rotation. See [MV] **rotate** for information on rotation in general, and see [MV] **pca postestimation** for specific guidance and warnings concerning rotation after a PCA.

```
. rotate
(output omitted )

. scoreplot, mlabel(painter) components(3) combined
              aspect(.8) xlabel(-2(1)3) ylabel(-2(1)2)
```

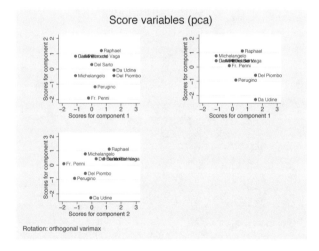

By default, the rotation information was included as a note in the graph. We specified components(3) to obtain all paired plots between the first three components. The combined option selected the combined view instead of the default matrix view of the graphs. The aspect(), xlabel(), and ylabel() options provide reasonable use of the graphing region while providing the same scale in the horizontal and vertical directions.

◁

As the number of factors() or components() increases, the graphing area for each plot gets smaller. While the default matrix view (option matrix) may be the most natural, the combined view (option combined) displays half as many graphs. However, the combined view uses more space for the labeling of axes than the matrix view. Regardless of the choice, with numerous requested factors or components, the graphs become too small to be of any use. In loadingplot, the maxlength()

option will trim the variable name marker labels that are automatically included. This may help reduce overlap when multiple small graphs are shown. You can go further and remove these marker labels by using the `mlabel("")` option.

Other examples of `loadingplot` and `scoreplot` are found in [MV] **pca postestimation** and [MV] **factor postestimation**.

Methods and Formulas

`scoreplot` and `loadingplot` are implemented as ado-files.

Also See

Complementary: [MV] **factor**, [MV] **factor postestimation**,

[MV] **pca**, [MV] **pca postestimation**, [MV] **screeplot**

Title

> **screeplot** — Scree plot of eigenvalues

Syntax

> screeplot [*eigvals*] [, *options*]

scree is a synonym for screeplot.

options	description
Main	
<u>n</u>eigen(*#*)	graph only largest *#* eigenvalues; default is to plot all eigenvalues
<u>mean</u>	graph horizontal line at the mean of the eigenvalues
<u>meanl</u>opts(*line_options*)	affect rendition of the mean line
ci	same as ci(asymptotic) (after pca only)
ci(*ci_options*)	graph confidence intervals (after pca only)
Plot	
cline_options	affect rendition of the lines connecting points
Add plot	
addplot(*plot*)	add other plots to the generated graph
Y-Axis, X-Axis, Title, Caption, Legend, Overall	
twoway_options	any options other than by() documented in [G] ***twoway_options***

ci_options	description
Main	
<u>asy</u>mptotic	compute asymptotic confidence intervals; the default
<u>heter</u>oskedastic	compute heteroskedastic bootstrap confidence intervals
<u>homo</u>skedastic	compute homoskedastic bootstrap confidence intervals
<u>tab</u>le	produce a table of confidence intervals
<u>l</u>evel(*#*)	set confidence level; default is level(95)
<u>r</u>eps(*#*)	number of bootstrap simulations; default is reps(200)
seed(*str*)	random-number seed used for the bootstrap simulations
CI plot	
area_options	affect the rendition of the confidence bands

Description

screeplot produces a scree plot of the eigenvalues of a covariance or correlation matrix.

screeplot automatically obtains the eigenvalues after estimation commands that have eigen as one of their e(properties) and that store the eigenvalues in the matrix e(Ev).

screeplot lets you obtain a scree plot in other cases by directly specifying *eigvals*, a vector containing the eigenvalues.

Options

___Main___

neigen(*#*) specifies the number of eigenvalues to plot. The default is to plot all eigenvalues.

mean displays a horizontal line at the mean of the eigenvalues.

meanlopts(*line_options*) affects the rendition of the mean reference line added using the mean option; see [G] **graph twoway line**.

ci, synonym for ci(asymptotic), displays confidence intervals for the eigenvalues using the asymptotic method.

ci(*ci_options*) displays confidence intervals for the eigenvalues. The following methods for estimating confidence intervals are available:

ci(asymptotic) specifies the asymptotic distribution of the eigenvalues of a central Wishart distribution, the distribution of the covariance matrix of a sample from a multivariate normal distribution. We remind the user that the asymptotic theory applied to correlation matrices is not fully correct, likely giving confidence intervals that are somewhat too narrow.

ci(heteroskedastic) specifies a parametric bootstrap using the percentile method and assuming that the eigenvalues are from a matrix that is multivariate normal with the same eigenvalues as observed.

ci(homoskedastic) specifies a parametric bootstrap using the percentile method and assuming that the eigenvalues are from a matrix that is multivariate normal with all eigenvalues equal to the mean of the observed eigenvalues. For a PCA of a correlation matrix, this mean is 1.

Note that the confidence intervals are not adjusted for "simultaneous inference".

Additional ci(*ci_options*) include

ci(table) produces a table with the confidence intervals.

ci(level(*#*)) specifies the confidence level, as a percentage, for confidence intervals. The default is level(95) or as set by set level; see [U] **20.6 Specifying the width of confidence intervals**.

ci(reps(*#*)) specifies the number of simulations to be performed for estimating the confidence intervals. This option is only valid when heteroskedastic or homoskedastic is specified. The default is reps(200).

ci(seed(*str*)) sets the random-number seed used for the parametric bootstrap. Setting the seed makes sure that results are reproducible. See [D] **generate**. This option is only valid when heteroskedastic or homoskedastic is specified.

___CI plot___

ci(*area_options*) affects the rendition of the confidence bands; see [G] ***area_options***.

___Plot___

cline_options affect the rendition of the lines connecting the plotted points; see [G] ***cline_options***.

⌐ Add plot ⌐

addplot(*plot*) provides a way to add other plots to the generated graph. See [G] ***addplot_option***.

⌐ Y-Axis, X-Axis, Title, Caption, Legend, Overall ⌐

twoway_options are any of the options documented in [G] ***twoway_options***, excluding by(). These include options for titling the graph (see [G] ***title_options***) and options for saving the graph to disk (see [G] ***saving_option***).

Remarks

Cattell (1966) introduced scree plots, which are visual tools used to help determine the number of important components or factors in multivariate settings, such as principal component analysis and factor analysis; see [MV] **pca** and [MV] **factor**. The scree plot is examined for a natural break between the large eigenvalues and the remaining small eigenvalues. The word "scree" is used in reference to the appearance of the large eigenvalues as the hill and the small eigenvalues as the debris of loose rocks at the bottom of the hill. Examples of scree plots can be found in most books that discuss principal component analysis or factor analysis, including Rabe-Hesketh and Everitt (2000), Hamilton (1992, 249–288), Rencher (2002), and Hamilton (2004, chapter 12).

▷ Example 1

Multivariate commands, such as pca and factor (see [MV] **pca** and [MV] **factor**), produce eigenvalues and eigenvectors. The screeplot command graphs the eigenvalues, so you can decide how many components or factors to retain.

We demonstrate scree plots after a principal component analysis. See example 1 of [D] **impute** for a description of the data.

```
. use http://www.stata-press.com/data/r9/emd
(1980 City Data)
. pca ln_eat - hhsize
  (output omitted )
. screeplot
```

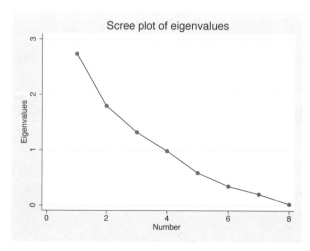

This scree plot does not suggest a natural break between high and low eigenvalues.

We render this same scree plot with the addition of confidence bands using the `ci()` option. The `asymptotic` suboption selects confidence intervals that are based on the assumption of asymptotic normality. Because the asymptotic theory applied to correlation matrices is not fully correct, we also use the `level()` suboption to adjust the confidence interval. The `table` suboption displays a table of the eigenvalues and lower and upper confidence-interval values.

```
. scree, ci(asympt level(90) table)
(caution is advised in interpreting an asymptotic theory-based confidence
 interval of eigenvalues of a correlation matrix)
```

	eigval	low	high
Comp1	2.733531	2.497578	2.991777
Comp2	1.795623	1.640628	1.965261
Comp3	1.318192	1.204408	1.442725
Comp4	.9829996	.8981487	1.075867
Comp5	.5907632	.5397695	.6465744
Comp6	.3486276	.3185346	.3815635
Comp7	.2052582	.1875407	.2246496
Comp8	.0250056	.0228472	.027368

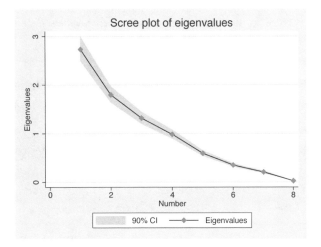

`screeplot` warned us about the use of asymptotic confidence intervals with eigenvalues based on a correlation matrix. `screeplot` knew that the eigenvalues were based on a correlation matrix instead of a covariance matrix by examining the information available in the `e()` results from the `pca` that we ran.

Instead of displaying an asymptotic confidence band, we now display a parametric bootstrap confidence interval. We select the `heteroskedastic` suboption of `ci()` that allows for unequal eigenvalues. The `reps()` and `seed()` suboptions control the execution of the bootstrap. The `mean` option provides a horizontal line at the mean of the eigenvalues, in this case at 1 since we are dealing with a principal component analysis performed on a correlation matrix.

```
. scree, ci(hetero reps(1000) seed(18228)) mean
```

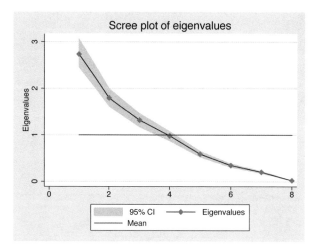

Our final scree plot switches to computing the bootstrap confidence intervals based on the assumption that the eigenvalues are equal to the mean of the observed eigenvalues (using the `homoskedastic` suboption of `ci()`). We again set the seed using the `seed()` suboption.

```
. scree, ci(homo seed(56227))
```

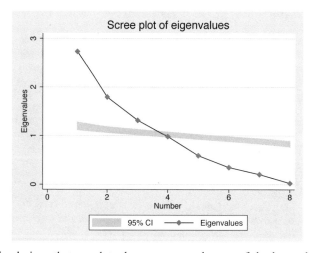

From the graph, it is obvious that our data do not support the use of the homoskedastic assumption.

◁

Methods and Formulas

`screeplot` is implemented as an ado-file.

References

Cattell, R. B. 1966. The scree test for the number of factors. *Multivariate Behavioral Research* 1: 245–276.

Hamilton, L. C. 1992. *Regression with Graphics.* Pacific Grove, CA: Brooks/Cole.

———. 2004. *Statistics with Stata.* Belmont, CA: Brooks/Cole.

Rabe-Hesketh, S. and B. S. Everitt. 2000. *A Handbook of Statistical Analysis using Stata.* 2nd ed. Boca Raton, FL: Chapman & Hall/CRC.

Rencher, A. C. 2002. *Methods of Multivariate Analysis.* 2nd ed. New York: Wiley.

Also See

Complementary:	[MV] **factor**, [MV] **pca**, [MV] **mds**
Background:	*Stata Graphics Reference Manual*

Subject and author index

This is the subject and author index for the *Stata Multivariate Statistics Reference Manual*. Readers interested in topics other than cluster analysis should see the combined subject index in the *Stata Quick Reference and Index*, which indexes the *Getting Started with Stata for Macintosh Manual*, the *Getting Started with Stata for Unix Manual*, the *Getting Started with Stata for Windows Manual*, the *Stata User's Guide*, the *Stata Base Reference Manual*, the *Stata Data Management Reference Manual*, the *Stata Graphics Reference Manual*, the *Stata Programming Reference Manual*, the *Stata Longitudinal/Panel Data Reference Manual*, the *Stata Survey Data Reference Manual*, the *Stata Survival Analysis & Epidemiological Tables Reference Manual*, the *Stata Time-Series Reference Manual*, and this manual. Readers interested in Mata topics should see the index at the end of the *Mata Reference Manual*.

Semicolons set off the most important entries from the rest. Sometimes no entry will be set off with semicolons; this means all entries are equally important.

A

absolute value dissimilarity measure,
 [MV] *measure_option*
Akaike, H., [MV] **factor postestimation**
Aldenderfer, M. S., [MV] **cluster**
Allison, M. J., [MV] **manova**
Anderberg coefficient similarity measure,
 [MV] *measure_option*
Anderberg, M. R., [MV] **cluster**,
 [MV] *measure_option*
Anderson, E., [MV] **clustermat**
Anderson, T. W., [MV] **manova**, [MV] **pca**
Andrews, D. F., [MV] **manova**
angular similarity measure, [MV] *measure_option*
anti-image
 correlation matrix, [MV] **factor postestimation**,
  [MV] **pca postestimation**
 covariance matrix, [MV] **factor postestimation**,
  [MV] **pca postestimation**
approximating Euclidean distances, [MV] **mds postestimation**
average-linkage clustering, [MV] **cluster**, [MV] **cluster averagelinkage**, [MV] **clustermat**
averagelinkage, cluster subcommand,
 [MV] **cluster averagelinkage**

B

Bartlett, M. S., [MV] **factor**, [MV] **factor postestimation**
Bartlett scoring, [MV] **factor postestimation**
Basilevsky, A. T., [MV] **pca**
Beerstecher, E., [MV] **manova**

Bentler rotation, [MV] **rotate**, [MV] **rotatemat**
Bentler, P. M., [MV] **rotate**, [MV] **rotatemat**
Bernaards, C. A., [MV] **rotatemat**
Bibby, J. M., [MV] **factor**, [MV] **manova**, [MV] **mds**,
 [MV] **mds postestimation**, [MV] **mdslong**,
 [MV] **mdsmat**, [MV] **pca**, [MV] **procrustes**
biplot command, [MV] **biplot**
biplots, [MV] **biplot**, [MV] **ca postestimation**
biquartimax rotation, [MV] **rotate**, [MV] **rotatemat**
biquartimin rotation, [MV] **rotate**, [MV] **rotatemat**
Blashfield, R. K., [MV] **cluster**
Blasius, J., [MV] **ca**
Bollen, K. A., [MV] **factor postestimation**
Borg, I., [MV] **mds**, [MV] **mdslong**, [MV] **mdsmat**
Box, G. E. P., [MV] **manova**
Bray, R. J., [MV] **clustermat**
Brown, J. D., [MV] **manova**
Browne, M. W., [MV] **procrustes**

C

ca command, [MV] **ca**
cabiplot command, [MV] **ca postestimation**
Cailliez, F., [MV] **mdsmat**
Caliński and Harabasz index stopping rules,
 [MV] **cluster stop**
Caliński, T., [MV] **cluster**, [MV] **cluster stop**
camat command, [MV] **ca**
Canberra dissimilarity measure, [MV] *measure_option*
canon command, [MV] **canon**
canonical correlations, [MV] **canon**
caprojection command, [MV] **ca postestimation**
Carroll, J. B., [MV] **rotatemat**
categorical data, [MV] **ca**, [MV] **manova**
Cattell, R. B., [MV] **factor postestimation**,
 [MV] **pca postestimation**, [MV] **procrustes**,
 [MV] **screeplot**
centroid-linkage clustering, [MV] **cluster**, [MV] **cluster centroidlinkage**, [MV] **clustermat**
centroidlinkage, cluster subcommand,
 [MV] **cluster centroidlinkage**
Clarke, M. R. B., [MV] **factor**
cluster analysis, [MV] **cluster**, [MV] **cluster averagelinkage**, [MV] **cluster centroidlinkage**,
 [MV] **cluster completelinkage**, [MV] **cluster dendrogram**, [MV] **cluster generate**,
 [MV] **cluster kmeans**, [MV] **cluster kmedians**, [MV] **cluster medianlinkage**,
 [MV] **cluster singlelinkage**, [MV] **cluster stop**, [MV] **cluster utility**, [MV] **cluster wardslinkage**, [MV] **cluster waveragelinkage**
dendrograms, [MV] **cluster dendrogram**
dir, [MV] **cluster utility**
drop, [MV] **cluster utility**
hierarchical, [MV] **cluster**, [MV] **cluster averagelinkage**, [MV] **cluster centroidlinkage**,
 [MV] **cluster completelinkage**, [MV] **cluster medianlinkage**, [MV] **cluster singlelinkage**,